Thermal Spray Coatings

Thermal Spray Coatings

Edited by
Lalit Thakur
Hitesh Vasudev

CRC Press
Taylor & Francis Group
Boca Raton London New York

CRC Press is an imprint of the
Taylor & Francis Group, an **informa** business

First edition published 2022
by CRC Press
6000 Broken Sound Parkway NW, Suite 300, Boca Raton, FL 33487-2742

and by CRC Press
2 Park Square, Milton Park, Abingdon, Oxon, OX14 4RN

CRC Press is an imprint of Taylor & Francis Group, LLC

Library of Congress Cataloging-in-Publication Data

ISBN: 978-1-032-08148-9 (hbk)
ISBN: 978-1-032-08153-3 (pbk)
ISBN: 978-1-003-21318-5 (ebk)

DOI: 10.1201/9781003213185

Typeset in Times
by SPi Technologies India Pvt Ltd (Straive)

Contents

Preface

Thermal spraying has emerged as a unique family of techniques to deposit the thick protective coatings of different materials (metals, alloys, polymers, ceramics, and cermets) over various engineering equipment with economy and ease. First and foremost, the current book covers the fundamental science and engineering aspects of thermal spray coating technology, historical developments, different thermal spraying techniques, general applications, advantages, and challenges.

Furthermore, technological progress has been addressed on developing new spray techniques, such as the use of cold gas spray method as cold spray additive manufacturing, which enables the development of 3D engineering components. Recent advancements in high-temperature corrosion- and erosion-resistant coatings, and thermal barrier coatings for power plants, the automotive sector, and jet engines have been reported in this book.

The present book provides a state-of-the-art in thermal spray coatings by documenting the recent research trends that include the development of coatings of different materials for friction and wear (hot forming dies), corrosion (hot corrosion and oxidation in boiler steels), thermal shielding, bio-medical (implants), hydrophobicity, etc. A newly introduced post-treatment method of microwave processing of thermal spray coated parts have been discussed. Finally, a case study for the failure analysis of boiler tubes due to hot corrosion and erosion in power plants has been considered, and recommendations have been made to the industry by suggesting relevant materials in the form of coatings.

This book covers the studies and research works of reputed scientists and engineers who have developed thermal spray coatings for various engineering applications. Hence, the book will serve as a valuable resource for fundamentals and the latest advancement of thermal spray coatings and consolidated references for the aspirants and professionals of the thermal spray community. Due care has been taken to explain and present the chapters with the help of self-explanatory schematic diagrams, microstructural images, data-enriched tables, etc. We express sincere gratitude toward the contributing authors of different chapters of this book and the authors of different research articles, books, internet sites, and scientific reports referred to in this book.

Dr. Lalit Thakur

Dr. Hitesh Vasudev

Editors

Dr. Lalit Thakur is currently working as an assistant professor in the Department of Mechanical Engineering at the National Institute of Technology Kurukshetra, India. He has obtained his M.Tech. (Welding Engineering) and Ph.D. degrees from the Indian Institute of Technology (IIT) Roorkee, India. For the last 12 years, he has been continuously exploring new possibilities in Welding Engineering and Thermal Spray Technology. His current research areas are thermal spray coatings and TIG weld claddings for wear, corrosion, bio-medical applications, ferromagnetic lubricants, advanced composite development by powder metallurgy, stir casting, and friction stir processing. He has authored more than 50 research publications in various international journals, books, and conferences of repute. He has published two patents in the field of thermal spray coating and friction stir processing. The research outcomes have been published in reputed journals (SCI/Scopus) such as *Surface and Coatings Technology*, *Surface Engineering*, *Materials Research Express*, *Tribology Transactions*, *Wear*, *JMEP*, *JMST*, *Applied Surface Science*, etc. Moreover, he is a dedicated reviewer of reputed journals such as *JOM*, *Surface and Coatings Technology*, *JTST*, *Ceramics International*, *JMEP*, *Progress in Natural Science*, etc. He is also a member of the American Society of Metals (ASM), Thermal Spray Society (TSS), Institution of Engineers India (IEI), and life member of Indian Structural Integrity Society (INSIS). He has guided many Master's and Ph.D. scholars in thermal spraying, advanced manufacturing, and welding technology.

Dr. Hitesh Vasudev is currently working as Associate Professor at Lovely Professional University, Phagwara, India. He has received Ph.D. in Mechanical Engineering from Guru Nanak Dev Engineering College, Ludhiana, India. His area of research is thermal spray coatings, especially for the development of new materials used for high-temperature erosion and oxidation resistance and microwave processing of materials. He has contributed extensively in thermal spray coatings in repute journals, including *Surface coatings and Technology*, *Materials Today Communications*, *Engineering Failure Analysis*, *Journal of Cleaner Production*, *Surface Topography: Metrology and Properties*, and *Journal of Failure Prevention and Control*, in various publications such as Elsevier, Taylor & Francis, Springer Nature, IGI Global, and InTech Open. Moreover, he is a dedicated reviewer of reputed journals such as *Surface Coatings and Technology*, *Ceramics International*, *Journal of Material Engineering Performance*, *Engineering Failure Analysis*,

Surface Topography: Metrology and Properties, *Material Research Express*, *Engineering Research Express*, and IGI Global journals. He has authored more than 30 International publications in various international journals and conferences. He has published 15 book chapters in various books related to surface engineering and manufacturing processes. He has also published a unique patent in the field of thermal spraying. He has teaching experience of more than eight years. He received a "Research Excellence" award in 2019 from Lovely Professional University, Phagwara, India. He has organized a National Conference and has been a part of many international conferences.

Contributors

N. Arivazhagan
School of Mechanical Engineering
Vellore Institute of Technology
Vellore, India

Amit Bansal
Mechanical Engineering Department
IKGPTU
Jalandhar, India

A. Behera
Department of Metallurgical and
    Materials Engineering
National Institute of Technology
    Rourkela
Rourkela, India

Rob Brittain
School of Mechanical Engineering
University of Leeds
Leeds, United Kingdom

Kuiying Chen
Structures, Materials and Performance
    Laboratory
Aerospace Research Centre, National
    Research Council Canada
Ottawa, Canada

Pankaj Chhabra
Baba Banda Singh Bahadur Engineering
    College
Fatehgarh Sahib
Punjab, India

Sihao Deng
Université de Bourgogne Franche-
    Comté – UTBM
Laboratoire Interdisciplinaire Carnot de
    Bourgogne
Belfort, France

Ali Sabea Hammood
Faculty of Engineering
Department of Materials Engineering
University of Kufa
Najaf, Iraq

Akhil Jerri
Department of Mechanical Engineering
National Institute of Technology
    Karnataka
Surathkal, India

Sharnappa Joladarashi
Department of Mechanical
    Engineering
National Institute of Technology
    Karnataka
Surathkal, India

Manpreet Kaur
Baba Banda Singh Bahadur Engineering
    College
Fatehgarh Sahib
Punjab, India

Anup Kumar Keshri
Metallurgical and Materials Engineering
Indian Institute of Technology Patna
Bihar, India

Hanlin Liao
Laboratoire Interdisciplinaire Carnot de
    Bourgogne
Université de Bourgogne Franche-
    Comté – UTBM
Belfort, France

M. Manikandan
School of Mechanical Engineering
Vellore Institute of Technology
Vellore, India

Mahantayya Mathapati
Department of Mechanical
  Engineering
KLE College of Engineering &
Technology
Karnataka, India

Amrinder Mehta
Department of Mechanical Engineering
DA V University Sarmastpur
Jalandhar, India

S. S. Mohapatra
Department of Chemical Engineering
National Institute of Technology
  Rourkela
Rourkela, India

Prakash C. Patnaik
Structures, Materials and Performance
  Laboratory, Aerospace Research
  Centre,
National Research Council Canada
Ottawa, Canada

C. Durga Prasad
Department of Mechanical Engineering
RV Institute of Technology and
  Management
Karnataka, India

Gaurav Prashar
School of Mechanical Engineering
Lovely Professional University
Punjab, India

M. R. Ramesh
Department of Mechanical Engineering
National Institute of Technology
  Karnataka
Surathkal, India

M. Nageswara Rao
School of Mechanical Engineering
Vellore Institute of Technology
Vellore, India

Rija Nirina Raoelison
Laboratoire Interdisciplinaire Carnot de
  Bourgogne
Université de Bourgogne Franche-
  Comté – UTBM
Belfort, France

M. Sathishkumar
Department of Mechanical Engineering
Amrita School of Engineering
Amrita Vishwa Vidyapeetham
Chennai, India

Swati Sharma
Metallurgical and Materials Engineering
Malaviya National Institute of
  Technology Jaipur
Rajasthan, India

Gagandeep Singh
Baba Banda Singh Bahadur Engineering
  College
Fatehgarh Sahib
Punjab, India

Jashanpreet Singh
Mechanical Engineering Department
Thapar Institute of Engineering and
  Technology
Punjab, India

Sharanjith Singh
Department of Mechanical
  Engineering
DAV University Sarmastpur
Jalandhar, India

V. Sreenivasulu
ATS Techno Pvt. Ltd.
Ahmedabad, India

Biswajit Swain
Department of Metallurgical and
  Materials Engineering
National Institute of Technology
Rourkela, India

Lalit Thakur
Department of Mechanical Engineering
National Institute of Technology
Kurukshetra, India

Hitesh Vasudev
School of Mechanical Engineering
Lovely Professional University
Punjab, India

M. Vignesh
Department of Mechanical Engineering
Saveetha School of Engineering,
   SIMATS
Chennai, India

Hongjian Wu
Laboratoire Interdisciplinaire Carnot de
   Bourgogne
Université de Bourgogne Franche-
   Comté – UTBM
Belfort, France

Yicha Zhang
Laboratoire Interdisciplinaire Carnot de
   Bourgogne
Université de Bourgogne Franche-
   Comté – UTBM
Belfort, France

1 Thermal Spraying Fundamentals

Process Applications, Challenges, and Future Market

Gaurav Prashar, Hitesh Vasudev, and Lalit Thakur

CONTENTS

DOI: 10.1201/9781003213185-1

1.1   INTRODUCTION

Thermal spray (TS) technology in the early developmental stage was used mainly for repairing works and protecting the components from corrosion and wear [1]. TS is a general name for a family of coating techniques used to deposit metallic, composite, cermet coatings, and non-metallic coatings. However, the basic fundamental concept or science behind all thermal spraying processes was not understood clearly up to the early 1980s. The process performance characteristics were entirely dependent on the skill, expertise, and experience of the operators. The outcomes were unsatisfactory process reliability and low quality of coatings. After that, the large-scale acceptance of thermal spraying for the industrial sector (mainly in aeronautic and nuclear) had started in the early 1990s. According to empirical research conducted by the Institute for Science Transfer in the year 2000, TS was ranked fourth among the top-quality surface engineering techniques [2]. Table 1.1 lists the principal coating methods, typical coating thicknesses reachable, common coating materials, and typical engineering applications. A few methods are not feasible for the deposition of certain coating materials.

The components used in different applications have to operate under varying load conditions and come across a wide range of problems. Surface modification by these TS techniques allows the deposition of almost every class of material such as metals, alloys, ceramics, and composites with better thickness and possible coating characteristics to protect these components operating in hostile conditions. Figure 1.1 presents the coating process comparison. These modification techniques can be used for applications ranging from resistance against corrosion, erosion, oxidation to electromagnetic shielding, biomedical implants, magnetic coatings, etc. In this chapter, a brief history of TS processes, characteristic features, area of application, and a range of materials that can be sprayed has been discussed, along with the challenges and future perspective of the TS coating market.

TABLE 1.1
Comparison of Different Coating Processes [3]

Coating Methods	Typical Coating Thickness	Coating Materials	Characteristics Features	Applications
Physical vapor deposition	1–5 µm	Ti(C,N)	Wear resistance	Machine tools
Chemical vapor deposition	1–50 µm	Silicon carbide	Wear resistance	Fiber coatings
	1–10 µm	Polymers	Corrosion resistance, aesthetics	Automobile
Baked polymers				
	0.04–3 mm	Ceramics and metallic alloys	Wear resistance, corrosion resistance	Bearings
TS				
Hard chromium plate	10–100 µm	Chrome	Wear resistance	Rolls
Weld overlay	0.5–5 mm	Steel, stellite	Wear resistance, corrosion resistance	Valves
Galvanization	1–5 µm	Zinc		Steel sheet
	10–100 µm	Ni-Cr-B-Si alloys	Very hard and dense surface	Shafts
Braze overlay				
Laser cladding	0.5–4mm	Alloy-625	Corrosion and wear resistance	Coatings of disc harrows, saw blades, hydraulic cylinders, and heat exchangers

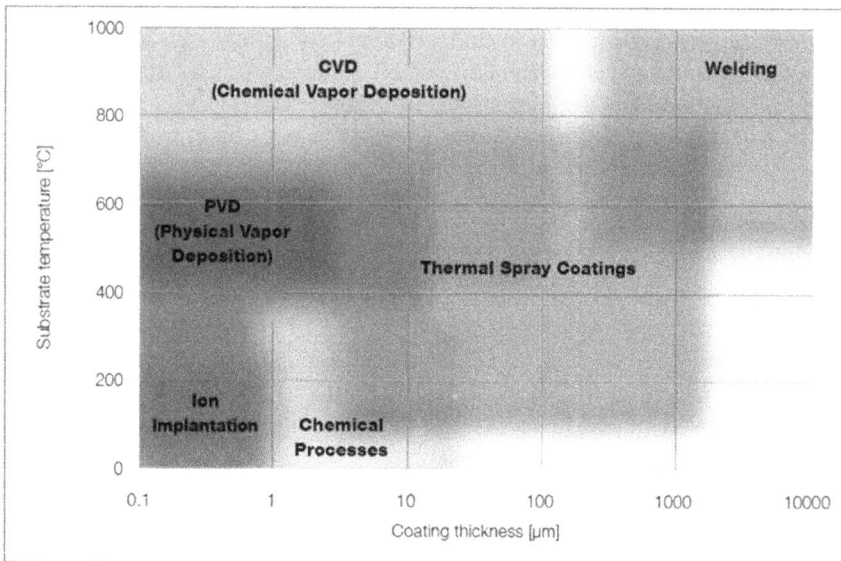

FIGURE 1.1   Coating process comparison [4].

1.2 HISTORICAL DEVELOPMENTS IN THE FIELD OF TS PROCESSING

Historical developments and advancements in the area of TS technology up to 2019 are presented in Figure 1.2 [5]. The TS methods can be categorized into different classes based on the characteristics and specifications. The earliest records of thermal spraying were found in Swiss scientist Max Ulrich Schoop's patents dated 1882–1889. These patents provide information regarding the technique of feeding wires (Pb and Sn) into a re-designed oxyacetylene welding torch. Later on, the torches were modified to accept the feedstock materials in powder form. The feedstock powder particles were caught up in a hot expanding jet stream, where particles get sufficient heat while being propelled toward the substrate's surface. On impacting the surface, these molten and semi-molten particles form splats and solidify rapidly. The outcome was the coating that developed incrementally from these impacting droplets. In 1908, electric arc spraying was developed by Schoop, which was capable of spraying some more metals like steel, stainless steel (SS) and Zn, through some improvements such as better process and equipment control. Schoop developed the TS methods during his whole life and patented some of them.

However, all these early processes were not capable of combating corrosion issues when rapid industrialization development occurred after World War-II, as most of them were designed with the bottom-up approach. During this period, believers of TS technology realized the importance of Surface Engineering in the industries, and they firmly pushed the approach to the upstream by developing TS technology. Countries like the USA, Germany, and Russia were the main contributors to this technology for such applications over the years. The main principle approaches behind the TS during 1920–1950 were powder flame and wire spraying. Primary applications were corrosion control and reclamation of parts in the paper industry. In 1937, first patent on the commercial use of TS coatings for high-temperature corrosion protection for the boiler was introduced by Quinlan and Grobel [6].

Consequently, for applications related to ceramic oxides, the period 1950s noticed the development of DC atmospheric plasma spraying (APS), which uses Ar as the primary gas for the plasma stream generation. With further improvements in TS technology, it becomes a candidate for protecting parts/components and enhancing performance enhancement in the aircraft engine industry between 1970 and 1980. In the early 1980s, by employing the rocket engine technology, two scientists Browning and Witfield, developed high-velocity oxy-fuel (HVOF) technology for metal powder spraying [7]. The powder particles in molten or semi-molten states are deposited onto the substrate at supersonic velocities of approximately $1000ms^{-1}$. This process has applications in various fields where the engineering components are exposed to corrosion and wear-related conditions. Furthermore, incorporating the concept of "Hypervelocity" Impact Browning was the first to develop high-velocity air-fuel (HVAF) [8]. The HVAF technique presents flexibility in fuels used like propane, hydrogen, natural gas, or propylene. As the system uses a significant air volume for combustion, the jet temperatures are restricted between 1900°C and 1950°C. Coatings developed by HVAF exhibited reduced oxidation in comparison with HVOF-processed counterparts. Various TS processes are different based on in-flight particle velocity and temperature, as shown in Figure 1.3.

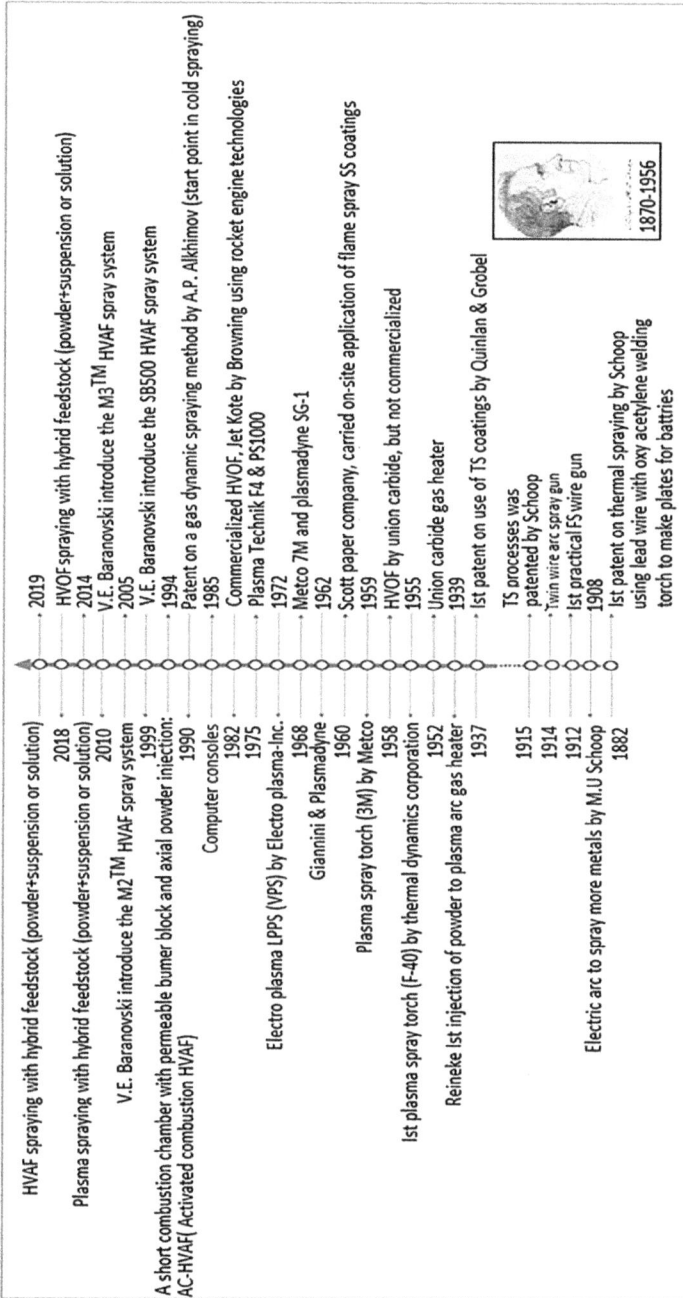

Timeline entries:

- 2019 — HVAF spraying with hybrid feedstock (powder+suspension or solution)
- 2018 — HVAF spraying with hybrid feedstock (powder+suspension or solution)
- 2014 — HVOF spraying with hybrid feedstock (powder+suspension or solution)
- 2010 — Plasma spraying with hybrid feedstock (powder+suspension or solution)
- 2005 — V.E. Baranovski introduce the M3™ HVAF spray system
- 1999 — V.E. Baranovski introduce the M2™ HVAF spray system
- 1994 — V.E. Baranovski introduce the SB500 HVAF spray system
- 1990 — A short combustion chamber with permeable burner block and axial powder injection: AC-HVAF (Activated combustion HVAF)
- 1985 — Patent on a gas dynamic spraying method by A.P. Alkhimov (start point in cold spraying)
- 1982 — Computer consoles
- 1975 — Electro plasma LPPS (VPS) by Electro plasma-Inc.
- 1972 — Commercialized HVOF, Jet Kote by Browning using rocket engine technologies; Plasma Technik F4 & PS1000
- 1968 — Giannini & Plasmadyne
- 1962 — Metco 7M and plasmadyne SG-1
- 1960 — Plasma spray torch (3M) by Metco
- 1959 — Scott paper company, carried on-site application of flame spray SS coatings
- 1958 — 1st plasma spray torch (F-40) by thermal dynamics corporation
- 1955 — HVOF by union carbide, but not commercialized
- 1952 — Reineke 1st injection of powder to plasma arc gas heater
- 1939 — Union carbide gas heater
- 1937 — 1st patent on use of TS coatings by Quinlan & Grobel
- 1915 — TS processes was patented by Schoop
- 1914 — Twin wire arc spray gun
- 1912 — Electric arc to spray more metals by M.U Schoop; 1st practical FS wire gun
- 1908 — 1st patent on thermal spraying by Schoop using lead wire with oxy acetylene welding torch to make plates for batteries
- 1882

1870-1956

FIGURE 1.2 Historic advancements in the field of TS coatings [5].

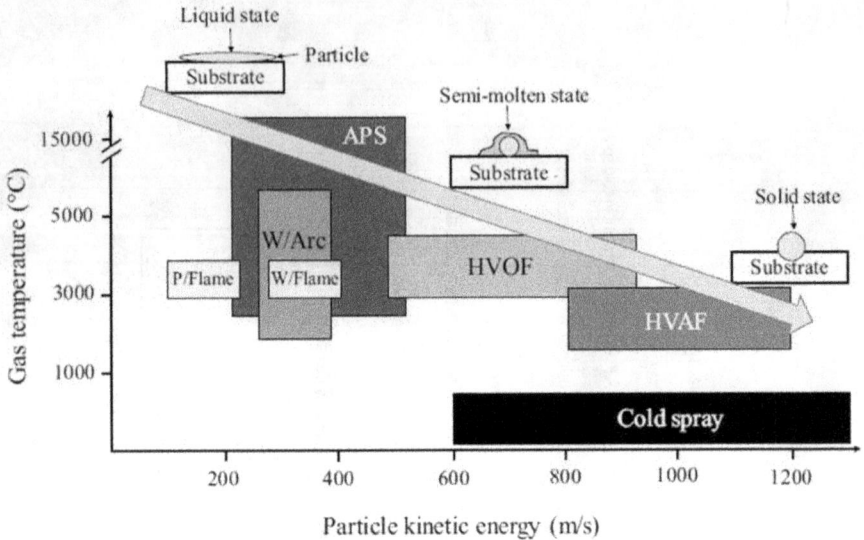

FIGURE 1.3 TS processes categorized in terms of temperature and in-flight particle velocity of the jet stream [5].

1.2.1 THERMAL SPRAYING FUNDAMENTALS

Thermal spraying allows for the deposition of a wide range of coating materials by various spray processes. The standard DIN EN 657 describes the basic concept and the meaning of the word *thermal spraying* [9]. *Thermal Spraying* is a process in which spray material in the form of powder, wire, or rod is fed to the heat source. The spraying material may be fed inside or outside the spraying gun, and due to high flame temperatures, the feedstock material is melted thoroughly or superficially heated until it becomes soft. Then a stream of gas propels or accelerates the melted feedstock particles toward the substrate's surface, whereon impact with the surface, the particles deformed plastically in the shape of splats to form a coating. The basic principle of thermal spraying is shown in Figure 1.4. During the process of coating, the substrate surface is not melted. However, it is subjected to medium thermal stresses. The splat's final shape developed by molten particle impact on solid substrate depends on fluid and heat transfer flow. During impact, a splat can fragment or remain intact, solidifying in the form of a disc. Splat breakup is caused by two different mechanisms and may occur when solidification is gradual or rapid. Splats disintegrate if they don't freeze during impact, forming splats with a small central area enclosed by an annular ring. When splats solidify quickly, a solid sheet forms along the edges, obstructing liquid flow and causing splashing. The splats produced have fingers radiating out from a broader central portion. We get round disc splats with no signs of splashing in between these two extremes [10]. Figure 1.5 shows splats photographs illustrating three distinct modes of splat impact.

According to the heat sources, primary thermal spraying methods mentioned in the standard DIN EN 657 are HVOF, cold spray (CS), arc, and plasma spraying. The classification of thermal spraying is shown in Figure 1.6.

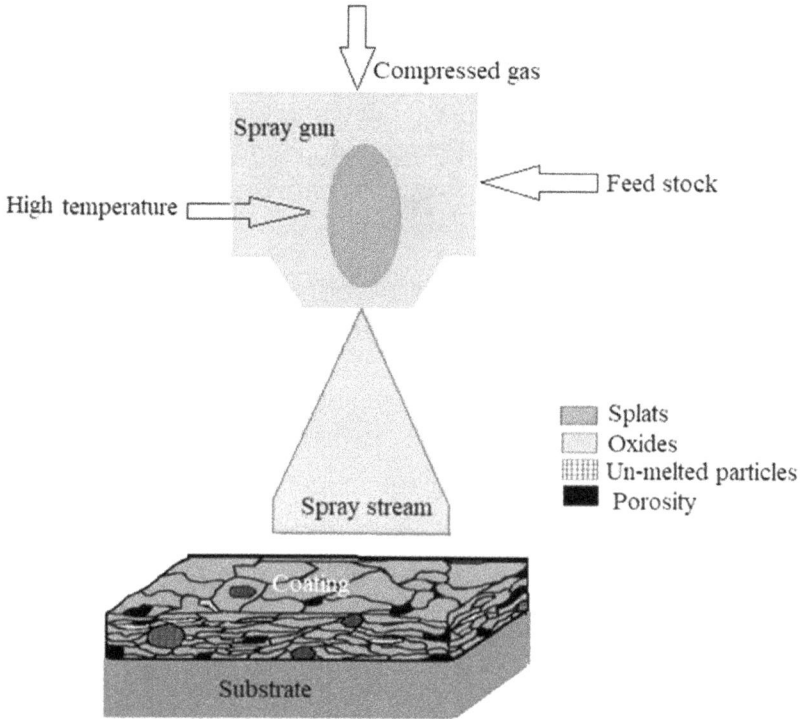

FIGURE 1.4 Schematic of the TS coating process.

1.2.2 COATING STRUCTURE

The coating structure depends on the spray conditions (process and parameters), especially the coating material selected for spraying. The most common feature noticed in any TS coating is its lamellar grain structure. This type of structure is formed by the flattening of molten and semi-molten in-fight particles upon striking the substrate's cold surface, followed by rapid solidification. The flattened particles are known as splats which have different morphologies.

The general TS coating formation structure is shown in Figure 1.7. Depending upon the spray process and material selected, TS coatings have microcracks, porosity, and heterogeneous and anisotropic structures. The heterogeneous structure is formed due to the different conditions of the in-flight particles as it is practically impossible to achieve the same temperature and velocity conditions for all the particles because of different shapes and sizes. Coatings that protect components from wear and corrosion must have low porosity since less porosity also improves adhesion strength. Hence, it is crucial to control and predict the porosity during the deposition of the coating. Various distinct mechanisms have been proposed for generating porosity; splat curling due to thermal stresses, gas entrapment under impacting particles, and incompletely filled voids when molten particles land on a rough surface generates porosity. Pores created by gas entrapment in TS coatings are usually microscopic and located at the interface between the splats.

	Splat Impact Mode	Pictures during impact	Splat Images
(a)	**Fragmentation** (Slow solidification) Θ~0.01	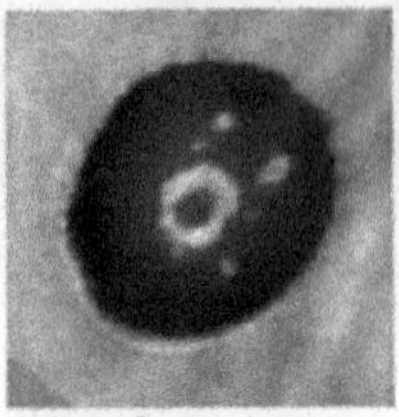 Mo on glass, room temp	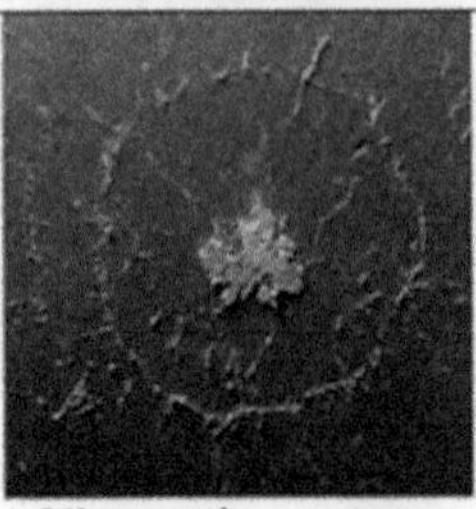 Ni on steel, room temp
(b)	**Disk Splats** (Intermediate rate of solidification) Θ~0.1	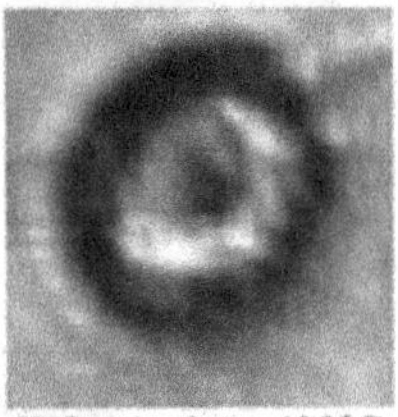 ZrO₂ on glass, 400°C	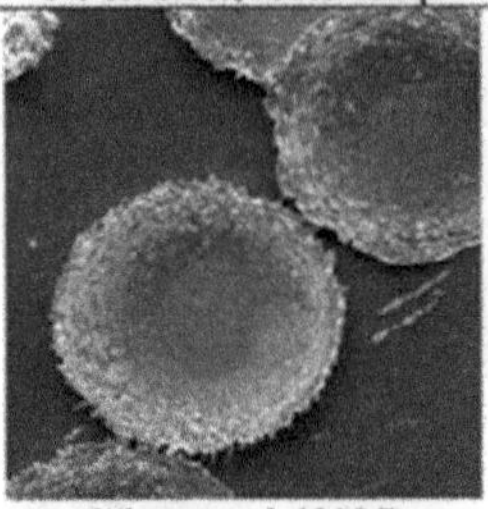 Ni on steel 400°C
(c)	**Freezing induced break-up** (Rapid solidification) Θ~0.4	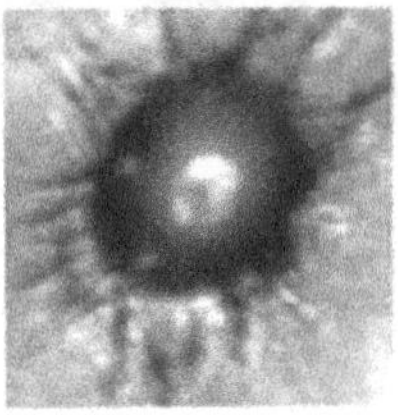 Mo on glass, 400°C	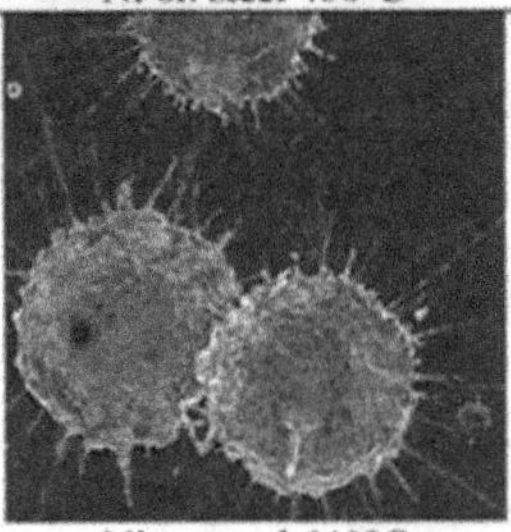 Ni on steel 640°C

FIGURE 1.5 Photographs for splats obtained during and after impact. (a) Fragmenting during impact, (b) producing disk splats, and (c) freezing-induced breakup [10].

On the other hand, splat curling at edges is most common and maybe a significant porosity source. Figure 1.8a shows the SEM micrograph of YSZ splat deposited on steel substrate by the plasma spray process. A section (Figure 1.8b) across a splat along line A–A′ represents a gap between the splat and the substrate, where the splat edge had curled up due to residual stresses set up in the splats. The splat's bottom surface is connected to the substrate and cannot shrink, while the top surface can contract. The stress-relieving is achieved either by splat curling or splat cracking. In comparison to ceramic splats, metallic splats are more ductile and hence resistant to cracking. Therefore, in metallic coatings, curling-up is a much more significant cause of porosity than in ceramic coatings [11].

The structure and final composition of the coatings are affected by the below-mentioned reactions that occur due to the heat input supplied to the spray particles and the residence time these in-flight particles remain in the atmosphere [12]:

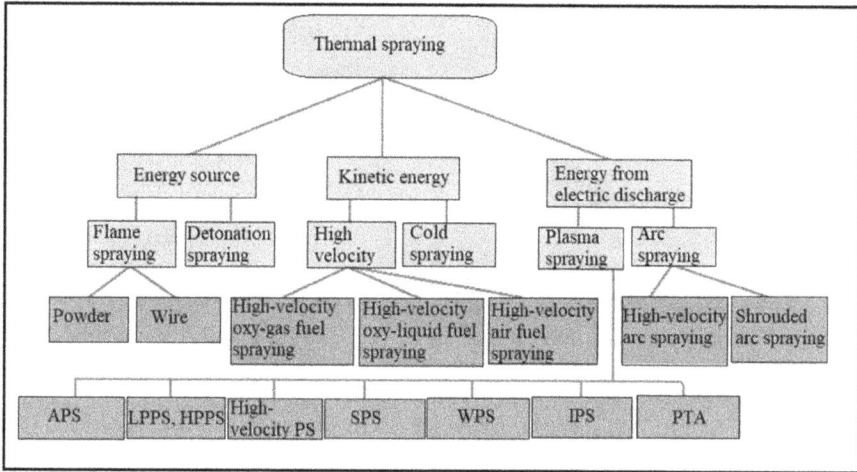

FIGURE 1.6 Classification of the thermal spraying methods [5].

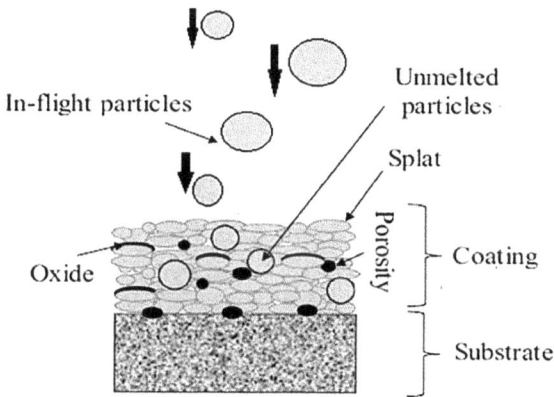

FIGURE 1.7 TS coating formation structure.

- Metal compounds reaction (e.g., hard material decomposes in the presence of oxygen).
- Development of non-volatile metal compounds such as oxides, hydrides, and nitrides, mainly from the highly reactive metals in the presence of oxygen, nitrogen, and hydrogen, respectively.

1.2.3 COATING ADHESION AND COHESION

The mechanical behavior of thermally sprayed coatings strongly depends on its adhesion with the substrate and cohesion among the deposited splats. Adhesion of the coating is the outcome of three distinct mechanisms: diffusion, chemical, and

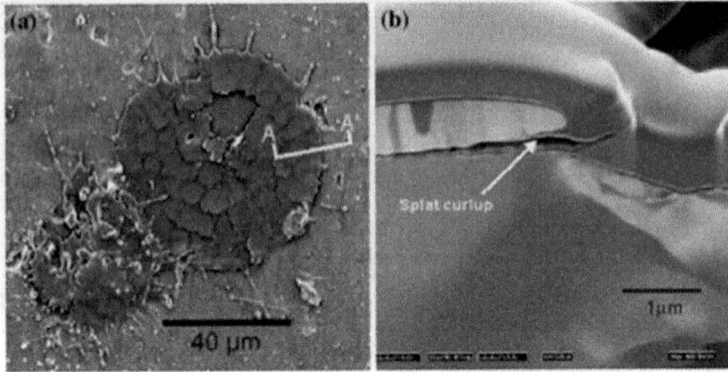

FIGURE 1.8 (a) SEM micrographs of YSZ single splats and (b) cross section through splat at line A–A' showing splat/substrate interface [11].

mechanical [13]. Diffusion adhesion in metals and alloys happens only when the coating application temperature is too high and no oxide layer is present at the surface. For diffusion to occur, the oxide layer present on the surface of the substrate will be removed, and the substrate's temperature must be kept high. During the CS process, the initial particles impacting on substrate surface do not adhere at once; rather, there exists an induction phase during which particles deform and clean the surface and removes partially the oxide layer present on the surface. For this, particle velocity must be greater than critical velocity [14]. The chemical adhesion develops from the chemical reactions occurring between the coating and the workpiece. During chemical adhesion, the impacting particles melt the substrate's surface due to which chemical compound of both liquids exists. For instance, a chemical compound $MoFe_2$ forms when Mo particles strike on steel substrate due to higher melting temperature and effusivity of Mo droplet. A chemical reaction between impacting cast iron particle and Al alloy substrate also occurs, and a thin layer of $FeAl_2O_4$ was noticed between the splat and the substrate [15].

Moreover, if the substrate's melting does not occur upon impact, higher temperature induced in the oxide layer of substrate promotes adhesion. For instance, at the interface between the 304L SS substrate and YSZ splat, a 20–30-nm-thick oxide layer was noticed to extend uniformly over the cross-sectional length of the splat [16]. The entire splat is in good contact with the substrate, as shown in Figure 1.9.

Finally, the mechanical adhesion occurs when the substrate's surface is rough, for instance, by grit blasting. During cooling, splats on the substrate surface shrink, and due to frictional forces that develop, splats adhere to the substrate's surface, as shown in Figure 1.10 [17]. The amplitude of surface roughness, which is defined by the distance among the highest peak and the deepest undercut and the spacing among peaks and valleys, plays a role in adhesion. An excellent correlation exists between the adhesion strength of coatings (NiAl) on a Ti-6Al-4V substrate and root mean square roughness value of substrate, which considers both the peak spacing and amplitude [18]. Furthermore, the adsorbed contaminants on the surface of the

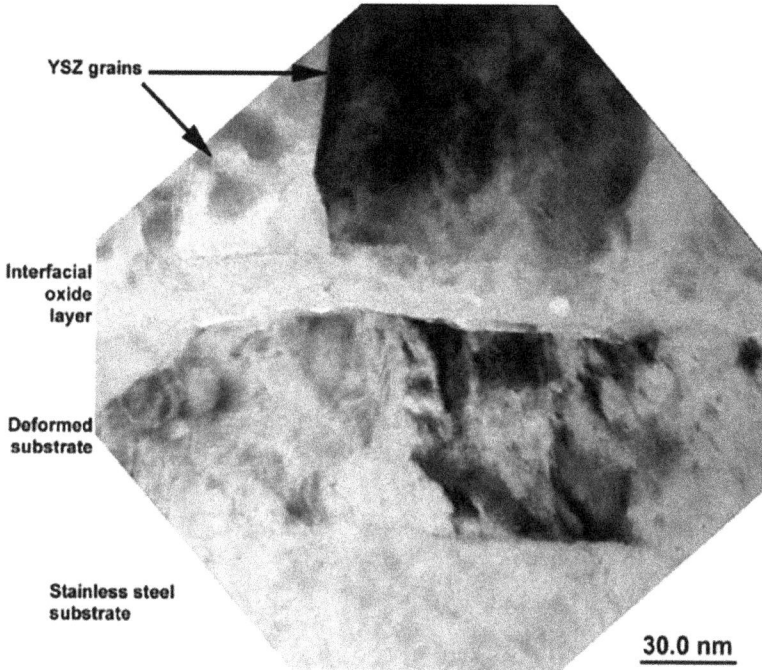

FIGURE 1.9 Bright-field TEM image of YSZ splat/substrate interface [16].

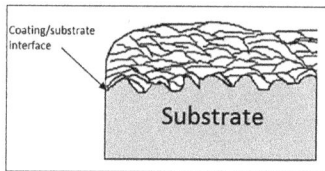

FIGURE 1.10 Mechanical interlocking diagrammatical setup.

substrate also influence adhesion strength. Thus, the coating's adhesion and cohesion are linked strongly to the excellent contact between the splats and substrate.

1.3 TS PROCESSES

This section focuses strongly on the principle, feedstock material, and coating properties of various TS processes.

1.3.1 FLAME SPRAYING

Flame spraying, according to chronology, is the first spraying technique invented by Swiss engineer MU Schoop in 1912, used initially for spraying of metals having low melting points, such as tin and lead. Later on, it was extended to the spraying of refractory metals and ceramics as well. The flame spray process deposits quality coatings by utilizing a heat source to melt the feedstock materials, usually available in powder, wire, and rod forms. The flame spraying is performed using a heat source developed from the chemical reaction between the hydrocarbon gaseous fuel and oxygen. This heat source is in the form of a hot gas stream made up of combustion products that come out of the spray torch at high velocity [19]. After that, spray material in powder form is supplied into the hot flame using the compressed air that gets melted and propelled toward the substrate's surface. If a wire is used as the spraying material, it is melted and atomized before it is propelled toward the substrate surface. Schematic for both the techniques are shown in Figures 1.11 and 1.12 (Table 1.2).

1.3.2   HVOF & HVAF Spraying

Both these spraying processes are combustion-driven with the pressure of combustion less than 1MPa using gaseous fuel. However, it can be more than 1MPa when a

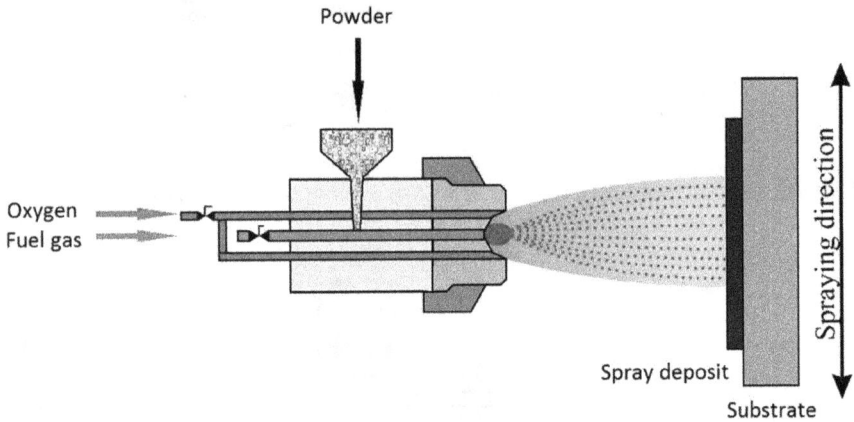

FIGURE 1.11   Schematic of powder flame spraying technique [20].

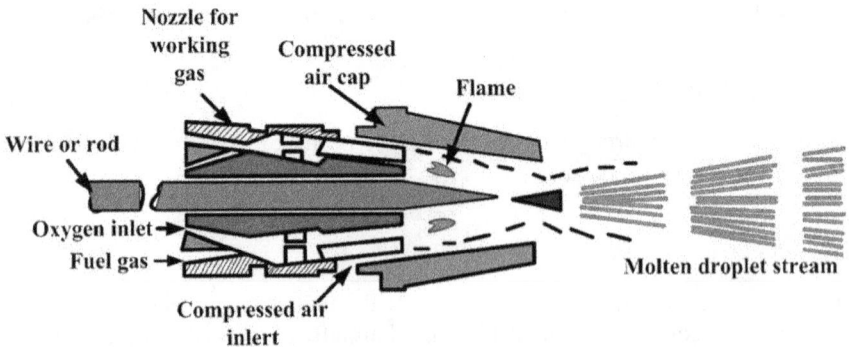

FIGURE 1.12   Schematic of wire or rod spraying [21].

TABLE 1.2
Comparison of Flame and Wire Spraying

	Flame Spraying	Wire or Rod Spraying
Feedstock material	Metals, polymers, and self-fluxing alloys	Wires of metals like molybdenum, zinc, aluminum, and alloys of brass, stainless steels, self-fluxing alloys, etc. Rods are usually of ceramics like aluminum oxide, titanium oxide, chromium oxide, etc.
Porosity	Up to 10% for as-sprayed coatings, and it is pore-free for self-fluxing	Similar as obtained with powder but more for ceramics
Process deposition efficiency	Approximately 50%	Around 70%
Noise level	90–125dBA [19]	For wire spraying: 118–122 dBA For rod spraying: 125dBA [19]
Adhesion strength	In case of metals and alloys: 30MPa For ceramics: 15MPa	Slightly better than powder
Oxide content	6–12%	Less than with powder, around 4–8%
Flow rate (material)	2–10 kgh^{-1} [19]	5–25 kgh^{-1} [19]
Process applications	Abrasion and adhesive wear [19]	Abrasion and adhesive wear [19]

liquid fuel (kerosene) is used. HVOF & HVAF spraying processes produce gas streams of very high velocity, which are achieved using the de Laval nozzle. Using kerosene as liquid fuel, high velocity, and low flame temperatures can be achieved in the HVOF spray system [19–22]. Consequently, unwanted reactions such as oxidation and phase dissolution can be avoided in composite coatings. Thus, low porosity and better mechanical properties such as microhardness and fracture toughness are achievable with the liquid-fueled HVOF spray process.

It has been observed that the medium temperature range of the liquid fuel HVOF system can still influence the properties of some materials and composites that are heat sensitive. In such conditions, the HVAF spraying is an effective solution in which the compressed air replaces pure oxygen. The flame temperature is further reduced because excess nitrogen (up to 2000 slm) is present in the compressed air [23]. The schematic of HVOF and HVAF is exhibited in Figures 1.13 and 1.14 (Table 1.3).

1.3.3 PLASMA SPRAYING METHODS

The gun for plasma spraying consists of a copper anode and a tungsten cathode. An electric arc is discharged across the two electrodes that heat a working gas, typically argon or a mixture of argon+helium, argon+hydrogen, argon+nitrogen. The heated gaseous mixture becomes a high-temperature plasma of about 15,000K (ionized gas) and exits the nozzle with velocities approximately 800 ms^{-1} or above [19, 24, 25]. A feedstock material for coating in the powder form is discharged into the high-temperature plasma jet, mainly in a radial direction, as shown in Figure 1.15.

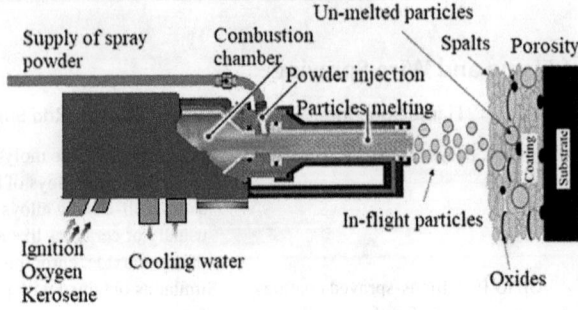

FIGURE 1.13 Cross section of typical HVOF system.

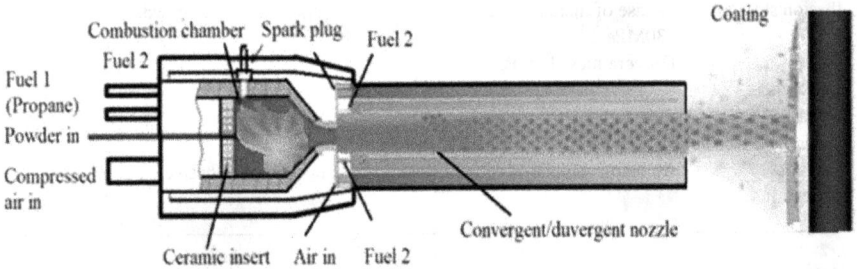

FIGURE 1.14 Schematic diagram of HVAF process.

TABLE 1.3
Comparison of HVOF & HVAF Spraying

	HVOF Spraying	HVAF Spraying
Materials used	Metals, alloys, cermets, composite, and self-fluxing alloys in the powder and wire forms	Metals, alloys, but most successful with the spraying of cermets
Porosity	Less than1–2%	Negligible vol.% (0–0.2%)
Process deposition efficiency	Approximately 60%	Around 70–80%
Noise level	125–135 dBA	133dBA or higher
Adhesion strength	>70MPa [19]	70MPa [19]
Oxide content	1–5% [19]	1.5–2 times the oxide content of feedstock [19]
Flow rate (material)	0.9–3 kg h⁻¹	7.2kg h⁻¹(gaseous fuel gun) 12 kg h⁻¹(liquid fuel gun)
Process applications	Wear and corrosion	Wear and corrosion
Particle temperature	3000°C [19]	2000–3000°C [19]
Particle velocity	600–800 m s⁻¹	600–1200 m s⁻¹
Coating thickness	0.1–2 mm	0.1–12 mm

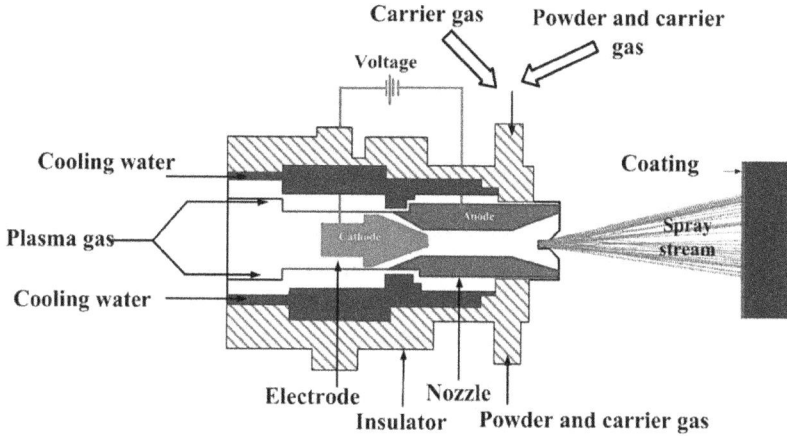

FIGURE 1.15 Schematic of coating deposited by plasma spraying technique [21].

However, the plasma spraying techniques can be categorized depending upon the environment where it is performed. These include:

- *Atmospheric plasma spraying (APS)*
  The APS technique uses a direct current (DC) arc source to produce heat through a non-carburizing or non-oxidizing gas flow. The power output for commonly used plasma spray guns ranges from 20 to 250 kW, and various configurations of spray guns are available according to various spraying requirements [19, 25]. Spray rates may vary depending upon the feedstock material, type of gun, powder injection system, plasma gas, etc. However, APS coatings present good resistance against abrasive, adhesive, and sliding wear. APS also produces thermally resistant or conductive surfaces and can be used commonly for the salvage and restoration of worn-out surfaces.
- *Controlled or Inert Atmosphere Plasma Spraying*
  When the oxidation of in-flight particles is the issue, this technique can be used, and the process is carried out in an inert atmosphere (generally in the presence of argon gas). Antechambers (connecting zones) are required so that air will not enter the main chamber. The atmospheric chamber must be cooled with water if the spray chamber volume is less than 10m^3. The higher cost of inert-APS is the main factor compared to APS, as the equipment required to recycle argon gas is expensive [19]. This method can deposit the materials with a high melting point of approximately 4000°C, such as TaC.
- *Vacuum or Low-Pressure Plasma Spraying (VPS/LPPS)*
  As compared to the inert atmospheric spray process, this method is also carried out in a controlled atmosphere but at a pressure (in the range 10–50 kPa), which is slightly below the atmospheric pressure. The equipment used is more costly than that of inert atmospheric spraying. Transferred arc etching is used

to remove the oxide layer from the substrate surface before the coating deposition, and the coating process is conducted under the reversed polarity. During the deposition operation, the component is heated to a significantly high temperature to achieve a strong diffusion bonding between the component and the coating material. This method is mainly used for the deposition of coatings over the turbine blades. Several advantages of this technique are enhanced bonding, high coating density, better control over the coating thickness (even with a surface of an irregular shape), and excellent deposition efficiency [26]. Moreover, during low pressures, the diameter and length of the plasma become significant, and by the use of nozzles (convergent/divergent), a high-speed plasma jet can be generated. The absence of oxygen and the ability to operate at high temperatures results in dense and more adherent coatings with lower oxide content [24].

- *Induction Plasma Spraying*
  The induction plasma spraying process is conducted in a controlled atmosphere but at atmospheric pressure or under mild vacuum conditions. The process is very similar to VPS, but the spraying torch is fixed, and the component is either rotated or moved in a linear direction [19]. It permits the bigger particles' spraying (about 200–250 µm) than those sprayed using DCAPS torches. The plasma power ratings for installations are in the range of 50–400 kW, with powder feed rates of 2–10 kgh^{-1}, respectively. It is used mainly for metals and ceramics spraying. One of its major applications is powder spheroidization to develop powders with good flowability that are needed for many applications not linked to the spray processes. Plasma spheroidization improves the quality of metal powders for additive manufacturing. Plasma technology develops metal powders that are more spherical and deliver better properties for additive manufacturing.

- *Plasma Transferred Arcs (PTA) Process*
  It develops from the combination of thermal spraying and welding process that needs electrically conductive components acting mainly as the anode. A secondary current is applied through the workpiece and plasma that regulates the melting of surface and penetration depth. Also, it requires less electrical power in comparison with the non-transferred type arc processes. Material for the feedstock purpose may be the wires or powders with particle sizes in the range up to 100 µm. Cermets, metals, and alloys can be sprayed conveniently using PTA method. The different guns may be characterized concerning the maximum current, which varies in the range of 200–600 A. Thicker coatings can be obtained compared to other spray processes (10 mm or even more), and a solid metallurgical bond with the component is achieved. The orientation of the component during spraying must be horizontal as possible. Minimum porosity is reported, and the process deposition efficiency is more than 90%. The wear resistance and high-temperature corrosion resistance are excellent for coatings deposited by PTA process. It is used to deposit the coatings over heavy components working in oil fields and the mining industry [19, 27] (Table 1.4).

TABLE 1.4

Comparison of Various Plasma Spraying Techniques

Conditions	Air Plasma	Inert Atmosphere Plasma	Vacuum Plasma	Induction Plasma	Plasma Transferred Arcs
Materials used	Used for spraying of oxides	Used for materials that have a high melting temperature (4000°C like TaC)	Ceramic or metal powder	Metals and ceramics	Metals, alloys, and cermets
Porosity	5–10%	Up to 10%	1–2%	Up to 2%	Negligible
Noise level	110–125 dBA [19]	As of surrounding fans and pumps [19]	Similar to inert atmosphere plasma [19]	—	—
Adhesion strength	>40–50 MPa	Higher than Air plasma	>68 MPa	>68 MPa	Higher bond strength
Oxide content	1–5 wt%	Low oxidation	Low oxidation	Low oxide content as compared to APS	Low
Flow rate (material)	3–6 kgh^{-1} (for power levels 40–50kW) [19] 15–20 kgh^{-1} (for power levels 250 kW)	5–10 kgh^{-1} (for power levels 50–400 kW) [19]	10 kgh^{-1} (for power 50–100 Kw) [19]	2–10 kgh^{-1} (for power levels 50–400 kW) [19]	18 kgh^{-1} [19]
Process applications	Used for restoration works and in wear-resistant applications [19]	Oxidation resistance applications [19]	Used for turbine blades [19]	Powder spheroidization to develop powders with excellent flowability for additive manufacturing purposes [19]	Coating of large parts, suitable for wear-resistant and elevated temperature corrosion-resistant applications [19]

1.4 COLD SPRAYING

Cold gas spray (CGS) is an emerging technology in which a coating layer is formed by the intensive deformation of feedstock powder particles upon its impact on the surface of a substrate. CGS employs a highly pressurized hot gas (N_2 or He) to carry and accelerate powder particles using a convergent–divergent nozzle to velocities in the range of

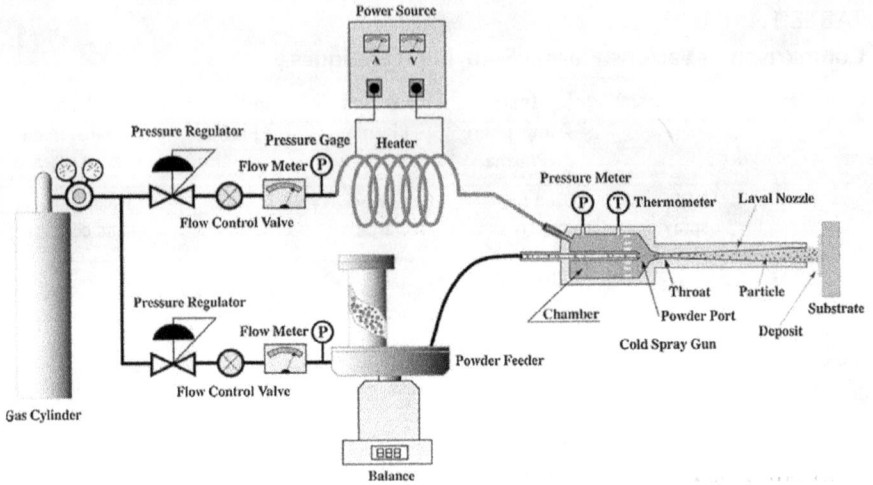

FIGURE 1.16   Schematic representation of CS [28].

300–1200 ms^{-1}, as shown in Figure 1.16. This hot carrier gas also heats the powder particles below their melting point, which assists the deformation of powder particles upon impact. Due to the high kinetic energy transferred to the powdered feedstock, the particles severely deform the substrate's impact, leading to high bonding and densification. Particles bond to the substrate surface when they reach a minimum velocity, known as the critical velocity, and above which specific thermo-mechanical conditions are achieved, allowing the feedstock particles to adhere to the substrate firmly. Critical velocities have been estimated for several materials and defined as a function of material properties and particle sizes. The range of critical velocity reported for different materials deposited by CS to date lies between 150 and 900 ms^{-1} [29, 30]. Expansion of the carrier gas at the nozzle's diverging zone reduces its temperature, which keeps the feedstock particles in the solid-state before impacting the substrate; thus, oxidation is minimized phase transformations, or undesired microstructural changes are avoided. CGS has attracted worldwide interest due to its high deposition efficiency and feasibility to produce highly compact deposits with more thickness.

1.5   DETONATION GUN (D-GUN) SPRAYING

A mixture of fuel and O_2 is ignited in a long barrel using a spark plug in this process. The detonation-pressure wave arises from the explosion of the mixture, which heats up, and propels the powder particles with a high velocity toward the surface of the substrate (Figure 1.17). It is a must that proper cooling arrangements should be provided to both the coating and the substrate during the operation. The process is continuously pulsed in a frequency range of 6–100 Hz. The resulting coating deposit is well bonded, highly dense, and extremely hard. The high noise levels of 145+ dBA restrict its usage in an open environment and make it mandatory to perform inside the acoustical enclosures. Coatings porosity is <2%, and its oxide content is in the range of 0.1 and 0.5 wt%. The powder flow rates are limited to

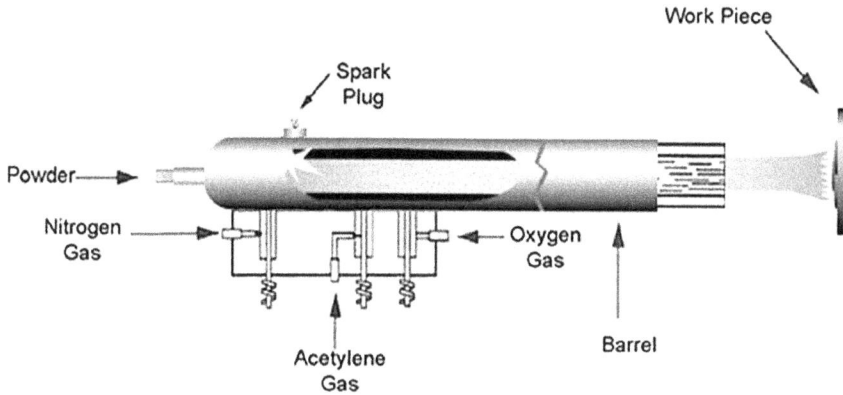

FIGURE 1.17 Schematic of D-gun spraying.

TABLE 1.5
Comparison of CGS & D-Gun Spraying

Conditions	CS	D-Gun
Materials used	Metals, alloys, and cermets	Metals, alloys, and cermets
Porosity	Negligible	<2%
Process deposition efficiency	40–90% [19]	Around 90% [19]
Noise level	About 110 dBA [19]	145 + dBA [19]
Adhesion strength	40MPa	>70 MPa
Oxide content	Negligible	1–5%
Flow rate (material)	0.5–3kgh^{-1} for low-pressure CS 4–12kg h^{-1} for high-pressure CS	1–2 kgh^{-1}
Process applications	Metal restoration, engine blocks, piston heads, copper metallization, and localized corrosion resistance [19]	Abrasion, adhesion, and corrosion resistance [19]

1–2 kgh^{-1}, and the process deposition efficiency is approximately 90%. Sprayed materials are mainly in the powder form of metals, alloys, and cermets. This process can also spray some oxides, but the feedstock particle size is limited to <20 μm. The major application areas include the coatings for abrasion, adhesion, and corrosion resistance [19, 31] (Table 1.5).

Table 1.6 summarizes the prime groups of different coating materials and their specific properties when thermally sprayed by various standard techniques.

1.6 SUSPENSION AND SOLUTION PRECURSOR TS

The need for fine-structured coatings has resulted in the development of suspension plasma and HVOF-based TS processes. Since the feedstock powder must be larger than 10 microns in size, the traditional APS and HVOF methods cannot manufacture

TABLE 1.6

Comparison of Mechanical Properties of Different Coating Processes [32]

Property	Material	TS Processes				
		Flame	HVOF	Arc	Plasma	CGS
Temperature of Gas (°C)		3000	2500–3000	4000	12,000–16,000	730
Spray rate (kg h⁻¹)		2–6	1–9	10–25	2–10	12
Velocity of in-flight particles (m s⁻¹)		Approx. 50	Up to 800	Up to 150	Approx. 450	1200
Bond strength (MPa)	Ferrous alloys	14–21	48–62	28–41	21–34	62–86
	Non-ferrous alloys	7–34	48–62	14–48	14–48	58
	Self-fluxing alloys	83+fused	70–80	15–50	—	—
	Ceramics	14–34	—	—	21–41	
	Carbides	34–48	83+	—	55–69	
Thickness of deposited coating (mm)	Ferrous alloys	0.05–2.0	0.05–2.5	0.1–2.5	0.4–2.5	3
	Non-ferrous alloys	0.05–5.0	0.05–2.5	0.1–5.0	0.05–5.0	Up to 120 micron
	Self-fluxing alloys	0.15–2.5	0.05–2.5	—	—	—
	Ceramics	0.25–2.0	—	—	0.1–2.0	—
	Carbides	0.15–0.8	0.05–5.0	—	0.15–0.8	—
Hardness (HRC)	Ferrous alloys	35	45	40	40	245–247 Hv
	Non-ferrous alloys	20	55	35	50	100Hv
	Self-fluxing alloys	30–60	30–60	—	30–60	—
	Ceramics	40–65	—	—	45–65	—
	Carbides	45–55	55–72	—	50–65	—
Porosity (%)	Ferrous alloys	3–10	<2	3–10	2–5	0.5
	Non-ferrous alloys	3–10	<2	3–10	2–5	<1
	Self-fluxing alloys	<2 fused	<2	—	—	—
	Ceramics	5–15	—	—	1–2	—
	Carbides	5–15	<1	—	2–3	—

fine-structured coatings. As melted by plasma or heat, these particles form "splats" upon impact that are far larger than µm scale. Moreover, the degradation of fine particles in conventional processes is more, which results in inferior coatings. Using fine-particle suspensions with a liquid carrier feed system instead of a gas carrier feed system is one way to get around this constraint. The APS and HVOF torches are both flexible and easy to use. By replacing existing gas-based powder feeders with liquid-based suspension feeders, they can quickly be converted to accommodate

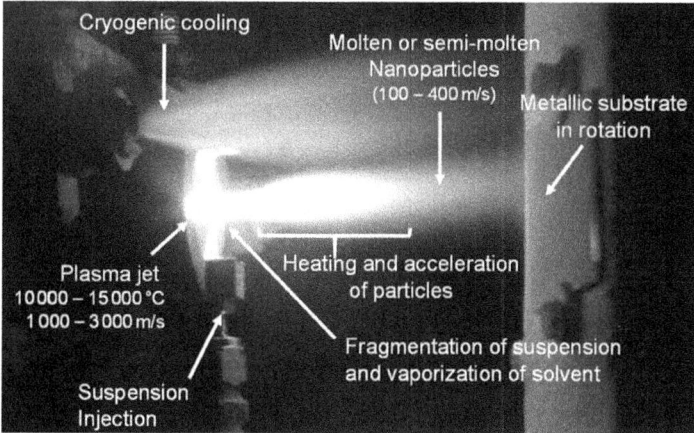

FIGURE 1.18 Schematic of nanoparticles sprayed by suspension plasma spraying [4].

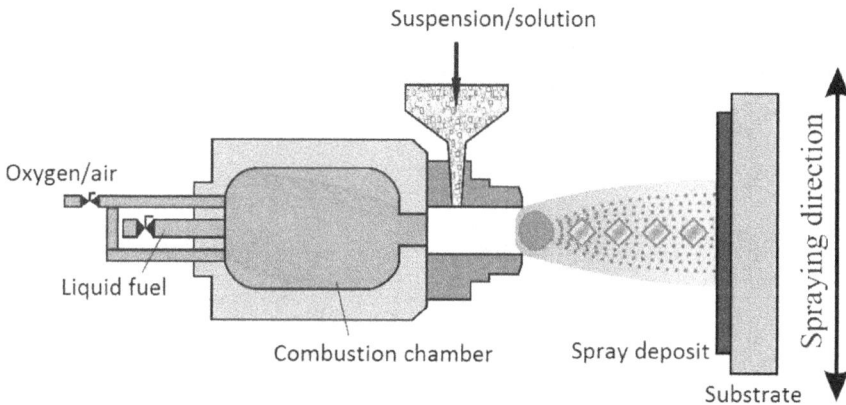

FIGURE 1.19 Schematic diagram showing the basic principle of high-velocity suspension/liquid fuel flame spraying [20].

liquid suspensions. As a result, using liquid feedstock in suspensions or solutions allows for the development of unique fine-microstructured coatings with segmented, columnar, and dense features. Possible applications include TBC with stress-relieving structures and low thermal conductivity for gas turbines and aerospace. Fuel cells, anti-microbial coatings, biomaterials, and photocatalytic are among the other applications. Schematic for both the techniques are shown in Figures 1.18 and 1.19.

These techniques are not yet mature enough to satisfy industry demands or consumer requirements. On the other hand, collaboration among researchers would lead to a better understanding of scientific values and process behavior. This will allow the equipment manufacturers to create processes that are dependable, durable, and simple to implement in an industrial environment.

The following are some of the research priorities listed to ensure a rapid improvement in technological readiness:

1.6.1 FEEDSTOCK

- Development of stable solutions and suspensions.
- Suspensions handling and storage.
- Choice of particle size for the suspension to formulate nanostructured coatings.
- Keeping particles separated.
- Choosing suitable liquid type and solid content.

1.6.2 SPRAY PROCESS

- Design of spray guns that are stable and suitable for liquid feedstock and fine particle production.
- Injection of liquid feedstock into high-energy gas flows in a regulated and stable manner.
- Understanding liquid feedstock/high-energy gas interactions requires both experimental studies and numerical simulations.
- Understanding the mechanism of coating build-up via experimental studies and numerical simulations.
- Improvements in deposition rate and deposition efficiencies.

1.6.3 PROCESS CONTROL TO ENHANCE RELIABILITY OF PROCESS

- Through sensors that control temperature and velocity of fine particles.

1.7 APPLICATION, CHALLENGES, AND FUTURE MARKET OF TS COATINGS

The industry sector uses TS coatings to improve:

- Wear resistance
- Heat resistance (in case of thermal barrier coatings [TBCs])
- Corrosion resistance
- Electrical properties, for instance, resistance and conductivity

However, in some sectors, thermal spraying still requires exploration, such as in electronics, the automobile sector, gas turbines, and biomedical implants. Furthermore, traditional sectors also have to deal with problems like shortage of raw materials and the global pressure of price increase.

1.7.1 ELECTRONICS INDUSTRY

Thermal spraying can further inquire for its application in electronics with the advanced materials used in this industry. The promising future has opened a gateway for advanced applications such as sensors, wiring, antennas, energy storage, and harvesting. Table 1.7 presents the main components of the electronics industry where TS coating is used.

   To be competitive, specifically concerning the high-quality standards of the electronics industry, the coatings deposited by the TS process must exhibit:

TABLE 1.7

Electronics Industry TS Components

Surface Requirement	Components	TS Coatings	TS Process
Radiofrequency shield	Instrument nuts	Cu	CS
Dielectric coating	Heater tube insulation	Al_2O_3-Ti	HVOF
Electrical conductivity	Electrical contacts	Cu	APS and CS
Abrasion-resistant	Magnetic tape heads	Al_2O_3	Controlled APS
Corrosion-resistant	Submarine cable connectors	Al_2O_3	PS

- High electrical conductivity, flexibility for the potential semi-conductor applications, capability for doping in sensing, thermoelectric, and semi-conductor applications [30].
- Properties should remain stable under the conditions of thermal or mechanical cycling, humidity, and aging.
- Repeatable properties in order to reproduce properties of coatings consistently for a given device.
- Consistency between feedstock powder and coating deposited.
- Ability to deposit the patterns without making use of a mask.
- *Challenges to deal with*
  o Ability to deposit a broader range of materials on non-flat surfaces.
  o No call for post-processing.
  o Higher deposition rates, faster processing times, and cost-effectiveness [30].
  o Processing should accomplish at ambient conditions.

1.7.2 AUTOMOBILE INDUSTRY

The automobile industry is developing faster by introducing new and advanced technologies and generating employment for people with excellent academic education. To be competitive in today's global market, advances in the mechanical and automobile sectors is vital for most national economies. Both automotive and industrial manufacturing become the basis for the production of thousands of parts per day. The basis for this mass production is vast and deep knowledge of manufacturing processes is required an inter-disciplinary approach. Good quality and rapid production can be achieved only by process automation equipped with susceptible control systems.

In general, TS technology is still ruled by conventional small- and medium-sized ventures and spray delis. Their primary focus is to add value to their consumers' goods through materials engineering/manufacturing and coating science. Table 1.8 presents the main components of the automobile industry on which TS coating is used.

Woof Thermal Management Technology, based in Bradford, West Yorkshire, UK, uses well-proven plasma ceramic coatings technology to create highly efficient thermal barriers in high-performance automotive applications. The long-lasting plasma ceramic coating on exhausts, turbos, and brake components acts as a thermal shield, allowing high-performance vehicles to operate at lower temperatures. Plasma ceramic coatings work by reducing heat input to other parts, raising power, and preventing heat transfer. A normal drop in under-bonnet temperature of 25°C (45°F)

TABLE 1.8
Automobile Industry TS Components

Surface Requirement	Components	TS Coatings
Fretting resistant	Differential ring gears	Mo
Erosion resistant	Exhaust valve seats	Fused Ni-Co
Resist corrosive gases	Piston heads	Ni-Cr
Oxidizing atmosphere	Exhaust mufflers	Ni-Cr
Cavitation resistant	Cylinder liner	SS–316
Corrosion	Piston head	Fused Ni-Cr
Abrasive grain wear	Cylinder liner	High carbon-Fe-Mo

lowers intake temperature, resulting in a 5% rise in power and a large increase in ancillary reliability.

The availability of a large number of feedstock powder compositions and the capability to manipulate the final deposited coating properties with effective control of the PS parameters means that this method has the ability of being tailored to every cylinder block service requirement. Hence, it gives greater variety than traditional surface treatment methods such as plating or lining methods. Sulzer Metco's Wohlen, Switzerland, plant currently has a manufacturing PS booth that coats various diesel- and gasoline-powered engine blocks (like a few Formula1, and LeMans racing team engines), engine blocks for aircrafts, and some limited-production supercars like the Bugatti Veyron and the recent Aston Martin One-77. In Salzgitter, Germany, the engine plant of Volkswagen has coated more than 3 million engine bores for gasoline-and diesel engines on its production lines over the past 5 years with this technology.

- *The key challenges to meet the needs of the automobile industry are:*
  - o The major challenge in developing a suitable material and process for coating cylinder bores in an industrial environment is transforming the process into a cost-effective, high-volume, fully integrated production coating system [30].
  - o Integration of the crucial analytical instrumentation, better steering, and control systems required for the coating process in order to ensure stability, reproducibility, and dimension tolerances,
  - o Integration of the robotic systems with control systems of the TS with the ability in terms of the tailored spray particle trajectories and release of heat to substrate,
  - o Knowledge achieved through deep scientific exploration must and become basis for the planning and operation of modern industries through virtual reality (VR) engineering.
- *Challenges to deal with:*
  - o Environmental issues (coating technology and products be environmentally friendly),
  - o Advanced tribological systems are re-designed and manufactured with the objective of operational safety [30],
  - o Coatings having functional properties like electrophysical, physicochemical, and optoelectronic, etc.,

o Standardization of TS techniques (suspensions and solution precursors) to achieve better surface functionality and product performance.

1.8 LAND AND AERO-BASED GAS TURBINES

Today, the thermal spraying process is a renowned surface engineering approach in the gas turbine industry, specifically in the production of TBCs, which can tolerate the elevated fuel gas temperatures required for the engine's efficient operation and lesser pollutant emissions.

TBCs for land and aero-based gas turbines are among the current key market sections of thermal spraying. However, it has to face new challenges like increased firing temperature, the improved service life of turbine parts, usage of the non-conventional fuels (like syngas and low-BTU gases), worldwide shortage of key "Rare earth" materials such as Y, Yd, He, and heavier and more complex geometries of the parts [30].

1.8.1 CHALLENGES TO DEAL WITH

- New TS technologies like suspension and solution precursor spraying approach, PS-PVD, and advanced APS including without using helium as a process gas.
- Mechanism-based modeling systems to analyze the failure of TBC under service conditions like CMAS (calcium–magnesium–alumino silicate) and thermal gradient [30].
- Improved concepts for the fabrication of powders to enhance the deposition of material or architecture of coating.
- Introducing multifunctional coatings.
- To develop repair/restoration system for structures having several millimeters thickness.

1.8.2 CRITICAL RESEARCH TO ADDRESS THE CHALLENGES MENTIONED ABOVE

- New compositions and TBC architecture to withstand elevated temperatures and non-conventional fuels by introducing new TS technologies like suspension and solution precursor spraying. But these technologies still require a better understanding of the deposition process.
- Coating systems with enhanced resistance to erosion, vanadium, water vapor, and silicate deposits.
- Higher process deposition efficiency to minimize the wastage of strategic materials.
- Use of robots to deposit coatings on complex parts.

1.9 WEAR- AND CORROSION-RESISTANT COATINGS

Engineering components used in different industrial applications have to operate under high heat loading conditions and deal with high-temperature oxidation and erosion, especially in power plants that use coal as a fuel. High-temperature oxidation and erosion are the two prominent failure modes, which lead to the degradation of materials, resulting in the premature failure of the components. The costs associated with them are estimated to be around 3–5% of the developed Nations' Gross Domestic Product.

The corrosion–erosion related problems can be easily, economically, and effectively resolved with surface treatment techniques. In recent years, TS coatings have increased to overcome the corrosion and erosion of the mechanical components.

However, much room is still available to develop novel, economical coatings that can address erosion–corrosion issues under specific environments. In general, the search for cost-effective, durable, and reliable coatings that enhance the performance and increase the operating range of structural materials has speed-up in-lieu of global concerns for declined energy consumptions, conservation of resources, and reducing the emission levels to the environment [30]. The coatings produced by TS exhibits a microstructure that is entirely different from the coatings developed by other coating routes: microstructure of the coating consists of splats of different sizes, and their shapes are parallel to the surface of the substrate, fully melted/un-melted particles, and some pores which are common structural features of the TS coatings. This morphology has a crucial role in contrast to coatings' performance and behavior against corrosive and erosive wear.

1.9.1   KEY RESEARCH REGIONS WHICH NEED FURTHER DEVELOPMENT OF TS COATINGS

- Characterization of the inter-splat interface, the mechanism involved in bonding formation, and impact of physical, chemical, metallurgical, thermal, and physical reactions of the impacting particle droplets with the underlying coating layer,
- Relationship among the process parameters, microstructure, and performance of the coating,
- Modeling the as-sprayed TS coating microstructure and properties, taking into consideration of interconnected porosity,
- Relationship among wear performance, microstructure and splat properties, spraying processing parameters for low-stress and high-stress wear conditions,
- A database of thermal sprayed coatings applicable to different service environments will be summarized.

1.9.2   CHALLENGES TO DEAL WITH

- Develop hard chrome plating solutions to withstand demanding industrial applications [30].
- Development of dense metallic or composite coatings which have non-connected pores with good inter-splat bonding that can restrict pathways of the corrosive fluids.
- Development of coatings to resist slurry erosive wear of turbine blades in the hydroelectric power plants.
- Corrosion-resistant coatings for elevated temperature processes, as the overall efficiencies of various energy generation processes rises with elevated temperatures like waste-to-energy systems, gas turbines, boilers, and nuclear energy plants.

- Microstructure of the coatings to give high wear performance by altering the composition of alloys.
- Development of nanostructured and bimodal coatings by CS or warm spraying having the strength and excellent mechanical properties.
- Develop smart coatings with self-enhancing capabilities according to the service conditions like becoming strong with rising operating temperature and self-healing properties such as restoration of the self-lubricating or self-protecting film.
- Protection against wear for the polymer composites such as leading edges of aircrafts, engine nacelles, antennae, and wind turbine blades.

1.10 BIOMEDICAL COATINGS

Biomaterials have become a growing field, as these materials have the potential to improve the quality and period of human life, and the research and technology associated with this field has now grown into a multibillion-dollar industry. Fundamental properties of a promising biomaterial include improved mechanical and biological compatibility, as well as increased wear and corrosion resistance in the biological system [33]. The materials currently employed for surgical implants include 316LSS, Co–Cr alloys, and Ti and its alloys. Corrosions and biocompatibility are the most significant considerations for human body implants. Thin coatings have improved corrosive resistance, with recent developments creating bio-active coatings that actively work with the biological system after they are implanted. These coatings can form strong interactions with surrounding tissue due to their surface reactivity resulting in an osseoconductive or osseoproductive response. Hydroxyapatite (HAp) is one of the most common bioactive coatings used for implants into bone tissue. The application HAp is completed by a variety of techniques, with PS being widely used. Recent coatings using HAp have also incorporated metals such as titanium improving the bioactive response, reducing crack formation growth and thermal expansion mismatching, all reducing the chance of failure.

1.10.1 KEY RESEARCH REGIONS WHICH NEED FURTHER DEVELOPMENT OF TS COATINGS

- In the case of HVOF and PS, crystalline HA is converted into tri- and tetra-calcium phosphates (TCP,-TCP, and TTCP) and oxy hydroxyapatite (o-HA) phases due to fast cooling of the coated substrate from elevated temperatures [34, 35]. These phases are typically amorphous and susceptible to dissolving in human blood plasma, resulting in bio-implant instability and, eventually, implant failure.
- The general lack of long-term implant survival in body conditions is one of the most pressing questions regarding the employability of PS HA-coated implants.
- To prevent cracking, spalling, or chipping of the HA coating during implant installation, the adhesive strength among the HA coatings and the metallic implant should be maximized.

1.10.2 CHALLENGES TO DEAL WITH

In biomedical coatings, different materials have been used as to match the Young's modulus of human bone (E = 10–30 GPa). However, still many challenges are being faced by the researchers [36, 37]:

- Wrought CoCrMo alloys are costly, which has restricted their percentage of the medical market in contrast with SS.
- Being employed as a short-time device (from months to years) does not assure the absence of a failure in SS, as premature failures of 316LSS have been reported in orthopedics implants.
- In general, Ti alloys cause tissue reactions when the prostheses are fretted, and they have low wear tolerance in articulating situations when compared to Co alloys. Although α-β Ti alloys have better fatigue strength and UTS, they have relatively bad bending ductility, owning to the HCP structure of the α phase.
- Failure of the femoral stem and loosening of the acetabular components due to wear are also major and severe post-implantation problems of complete joint replacement.
- Only comparison experiments are carried out at various loading and atmosphere environments due to the lack of an acceptable procedure for measuring the wear properties of metallic biomedical materials. To build an alloy with improved wear resistance, further studies should be done on developing an effective procedure for calculating the wear property.

1.11 ROBUSTNESS, RELIABILITY, AND PROCESS ECONOMICS

The process reliability, durability, and reproducibility of TS technology are crucial for its successful adoption in the industrial manufacturing sector. The development of (a) spray control systems that precisely and accurately control various operating parameters like gas, electrical, feed-rates, temperature, and cooling requirements and (b) better diagnostic tools for substrate and/or the plume have proven to assist in enhancing process productivity, quality, and confidence among consumers. Diagnostic tools are very well accepted by the TS family and have been helpful, in general, for optimization of spray process parameters, transfer of parameters booth-to-booth, trouble-shooting, validation of powder lot, on- and off-site monitoring, etc. However, a long journey to cover before those diagnostic tools is still systematically employed for daily production [30]. The optimal situation would be if there were a reliable link between coating properties and different sensor parameters: those measuring the particle parameters in-flight and those who characterize the coating during its development (temperature, residual stress, etc.). An online control may be done in this situation.

1.11.1 REQUIREMENT FOR ONLINE SPRAY DIAGNOSTICS

The TS group realized in the mid-1980s (1984–1986) that in order to take TS processes to the next stage, sensors needed to provide real-time, in-flight particle information right before impingement on the component. The particles become hot during TS and emit infrared radiation, which is used to "see" and classify the particles (temperature, velocity, size, flux, and trajectory). CS particles, on the other hand, are too cold (below 1000°C) to be observed using traditional pyrometric techniques. Cold spraymeter sensors can be used to validate numerical models (CFD) that can be used to design CS nozzles that meet specific particle velocity requirements. Cold spraymeter can characterize particles individually and provide complete velocity and size distributions. In the last 20 years, various diagnostics devices have been designed—both in research facilities and on the shop floor. Figure 1.20 [38] shows the corresponding technique for online monitoring, which should be defined for each coating.

1.11.2 WHAT CAN ONLINE SPRAY DIAGNOSTICS DO FOR YOU

- *Save Time*
  - o Rapid development of spray parameters.
  - o Rapid optimization of spray equipment.
  - o Required fewer test coupons.
  - o Flexible parameter transfer—from booth to booth or among spray facilities.
- *Minimize Cost*
  - o Accurate spray at the first time and minimize stripping of coating.
  - o Optimize usage of powder feedstock and minimize the requirement for post-machining (desired thickness of spray is obtained, not more).
  - o Deposit Efficiency will be optimized and control.
- *Guarantee High Quality, and Reproducible Coatings via:*
  - o Advanced online monitoring systems.
  - o Real-time differentiation of as-sprayed material properties with the pre-determined acceptance ranges.

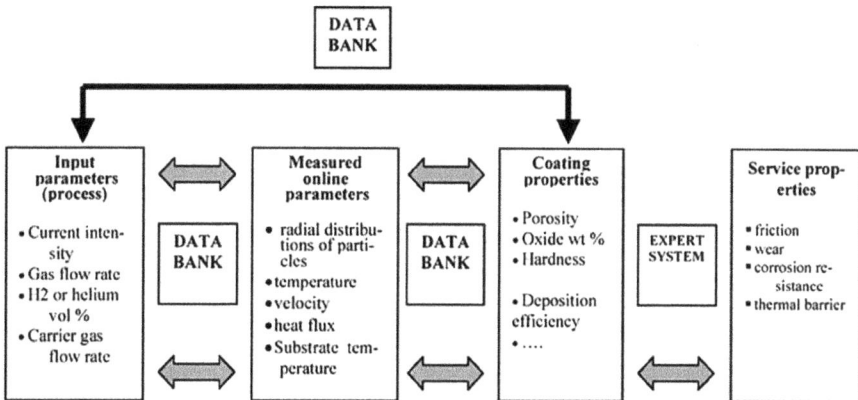

FIGURE 1.20   Possible online control strategy [38].

1.11.3 The Identified Fields of Research Areas

- Adjudicate the admissible process windows for every material and taking necessary steps at the process level if measurements do step on the outside, set tolerance windows (e.g., readjusting a bit the parameters upstream based on the measurements).
- Re-designing standard methods to reflect the outcomes (in terms of mechanical and tribological properties) instead of the procedure.
- Inquest of uncertainty in the value of measuring instrument versus the achievable and preferred coating capability.
- The improved basic science behind the formation of splats and build-up of coating in order to predict the microstructure and various properties of as-sprayed coatings.
- Fully simulated model of plasma torch operation to design the stable and efficient torches.
- Fully simulated model of the thermal spraying methods to optimize the method at a cost-effective approach.
- Fully simulated model of the thermal spraying methods to develop skilled sprayers and engineers.

1.11.4 Challenges to Deal With

- *Hardware (TS)*
  - o Better technology for powder feeder: good stable injection, capability to fluidize and feed nanoscale powder particles without clogging the spraying gun, and higher feeding rate.
  - o Improved gun technology: excellent thermal efficiency, stable jets, higher process deposition efficiency, and less process variability resulting from hardware degradation.
- *Process Control*
  - o Diagnostic tools used on the process spec sheets.
  - o Transparent integration of the process sensors employed during spray with spray controllers.
  - o Development of novel sensors capable of measuring the desired coating thickness and properties (hardness, surface roughness, Young's modulus, and roughness) required online.
  - o Development of novel sensors for the developing TS techniques: suspension and solution spray, LPPS, CGS.
- *TS Processes Greening*
  - o Minimizing the energy usage, resource usage, and reducing the waste and emission, which accounts measurably to the process cost and cutting environmental load.
  - o Efficient energy usage: reducing the total energy usage per gram of the material deposited.
  - o Efficient techniques for the recycling and recovery of the overspray.
  - o Life-cycle assessments of the TS methods and products for declining levels of the process and products' environmental effects.

1.12   FUTURE MARKET FOR TS

The global market of TS coatings was around USD 7.6 billion in the year 2020 and is expected to reach USD 10.7 billion by 2025, at a CAGR (compound annual growth rate) of 7.0%. TS coatings can be applied to parts by two techniques, namely, combustion flame and electrical energy. However, combustion flame is a commercial and widely used technique. Ceramics, cermets, metals and alloys, composites, and polymers are the commonly used materials in TS coating applications. Among these, ceramics alone are accounted for the largest share in the market of thermally sprayed coatings. There is a continuously increasing demand for ceramics due to their low cost and easy availability.

Moreover, growth is also influenced by the growth in industries like aerospace, healthcare, the automobile sector, and other end-use industries. After ceramics, metals and alloys are the most widely preferred materials. Metal and alloys are the second-largest material next to ceramics, consumed in the TS coating market during the forecast period. Due to the easy availability of the raw material in North America and Europe, the cost of materials is significantly less. However, in the Asia Pacific, Middle East, and Africa, the cost is almost 35% more.

Industries like aerospace and automotive are the primary consumers in the market associated with TS coatings. The comprehensive growth is supported by the demand backlog in the aerospace sector, which is nearly up to 10 years mark. North America alone is the highest consumer of this market. This immense growth is related to the rapidly increasing urbanization and modernization, increase in the personal disposable income, and at last, growing demand from the different end-use industries like aerospace. Moreover, one major factor driving the growth of the TS market in this region is the easy availability of raw materials. North America is dominating in the aerospace sector due to the availability of some big manufacturers. Furthermore, big manufacturers like Boeing and Airbus are also going to step up their production capacity to full throttle in order to meet the increasing demand for the commercial and defense sector aircrafts. In addition to the above discussion, cost-effectiveness approach of TS coating techniques coupled with rapidly increasing demand from the healthcare sector in thermal coated components/parts like biomedical implants are expected to increase TS coatings market's growth.

1.13   FEEDSTOCK MATERIALS FOR SPRAYING

The types of materials commonly deposited by thermal spraying are shown in Figure 1.21 (Table 1.9).

1.14   FUTURE GROWTH AREAS

Essential fields that will determine the future growth in TS coatings include:

- Continuous advancements in process control equipment (such as robotics, control of motion, and real-time sensors, etc.) for better coatings.
- Improved techniques for nondestructive testing and the evaluation of coatings.

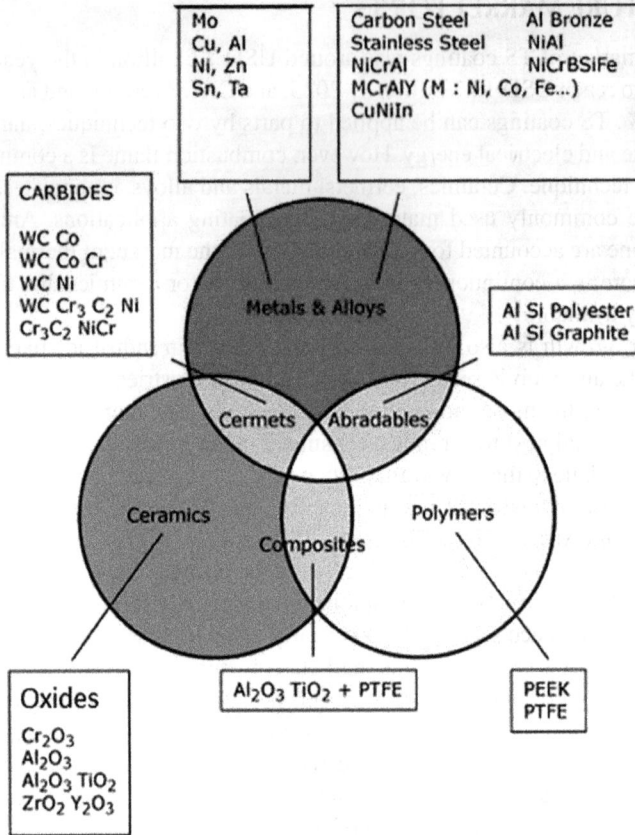

FIGURE 1.21   Materials commonly deposited by TS processes.

- Better optimization of CS technique.
- Thermal spraying of the near-net-shape components.
- Spray forming of the high critical temperature superconducting oxide ceramics.
- CAD and rapid prototyping methods like stereolithography.
- Improved feedstock development method and the quality control.
- New materials and fields (e.g., composites, nanophase materials, and fields such as biomedical composite coatings).
- Even though numerous studies were carried out concerning conventionally produced materials, nowadays, a specific area targeted is additive manufacturing, which is utilizing the TS processes, specifically the CS process. CS additive manufacturing (CSAM) allows the design of parts with complex-shaped geometries that are manufactured with much shorter lead times when compared with the conventional manufacturing approach. In the next few years, CSAM technology is likely to be integrated into the industrial field.

TABLE 1.9

List of Coating materials Used for Industrial Purposes [19]

Industry Sector	Carbides	Self-Fluxing	Iron/Steel	Ni-Alloys	Superalloys	MCrAlY	Cb-Alloys	Nonferrous
Aero gas turbine	✓			✓	✓	✓	✓	✓
Agricultural	✓	✓	✓	✓				✓
Architecture								✓
Automobile engines		✓		✓		✓		✓
Business equipment								
Structural & cement			✓					
Chemical sector		✓	✓	✓	✓		✓	✓
Mills (brass & Cu)			✓	✓			✓	
Defense & aerospace	✓	✓	✓	✓	✓	✓	✓	✓
Engines (diesel)	✓		✓	✓	✓	✓	✓	✓
Electronics & Electr.			✓					✓
Electric usage		✓	✓	✓	✓		✓	✓
Food industry		✓	✓	✓	✓		✓	
Forging industry		✓	✓	✓	✓		✓	
Glass manufacture	✓	✓	✓	✓				✓
Hydro-turbines	✓	✓	✓	✓				✓
Iron & steel casting		✓	✓	✓				✓
Iron & steel manufacturing		✓	✓	✓				✓
Medical								✓
Mining, and dredging		✓	✓	✓				✓
Oil & gas exploration,		✓	✓	✓			✓	✓
Printing equipment								
Pulp & paper			✓					
Railroads			✓	✓				✓
Rock products		✓	✓					✓
Rubber & plastic industry		✓		✓			✓	
Screening		✓						
Ship & boat (repair/manufacture)	✓		✓	✓	✓	✓		✓
Gas turbines (land-based)			✓	✓	✓	✓	✓	✓
Steel rolling mill		✓	✓	✓			✓	✓
Textile			✓				✓	
Transportation, non-engine			✓	✓				✓

1.15   CONCLUSION

The thermal spraying technique is an influencing and economical method to deposit the coatings for tailoring the surface properties of engineering components. As-sprayed coatings can be used in different applications such as automotive systems, coal-fired boiler components, power generation equipment, chemical processes, aircraft engines, orthopedics, dental, and ships. However, industries like aerospace and automotive are the primary consumers in the market associated with TS coatings. TS coatings can be applied to parts by two techniques, based on combustion flame principles and based on electrical energy, and out of these, combustion flame-based is a commercially and widely used technique to get hard, well-bonded, and dense coatings for corrosion and wear resistance applications. Ceramics, cermets, metals and alloys, composites, and polymers are the widely used materials in TS coating applications. Among these, ceramics alone are accounted for the largest share in the market of thermally sprayed coatings. The continuously increasing demand for ceramics is due to its low cost and easy availability. The major drawback of TS is that they can coat only what the spray torch or gun can "see." Moreover, size limitations are also there. It is not possible to deposit a coating on small, deep cavities as the torch or spray gun will not fit inside it properly.

In the near future, the major advancement in the field of thermal spraying would be the use of industrial robots in the development and production of high-quality coatings on complex-shaped components by an offline programming approach. An extensive range of the operating parameters in the thermal spraying process influences the final properties of the deposited coatings. For this cause, extensive research activities have been carried out to clarify the relationships among the process handling parameters and the final coating properties. The other growing sector which is in current demand is biomedical implants. Biomimetic deposition of the composite coating by thermal spraying technique on the polymeric tissue engineering scaffolds will provide a simple and most effortless means of rendering scaffolds osteoconductive. Moreover, biomimetic methods can also be employed to add or improve some other functions for the tissue engineering scaffolds. It is clear that new, advanced, and practical technologies, specifically in the nanotechnology region, will rise and can be used for surface modification of the biomaterials &tissue engineering scaffolds. This study is an effort toward the basic information regarding the TS techniques and their future applications.

REFERENCES

1. Davis, J.R. ed., 2004. *Handbook of Thermal Spray Technology*. ASM International, Park, OH.
2. Babiak, Z., Wenz, T and Engl, L., 2006. Fundamentals of thermal spraying, flame and arc spraying.in: F.-W. Bach, A. Laarmann, and T. Wenz (editors), *Modern Surface Technology*, Wiley VCH, Weinheim, pp. 119–136.
3. Xanthopoulou, G., Marinou, A., Vekinis, G., Lekatou, A. and Vardavoulias, M., 2014. Ni-Al and NiO-Al composite coatings by combustion-assisted flame spraying. *Coatings*, 4(2), pp. 231–252.
4. Billard, A., Maury, F., Aubry, P., Balbaud-Célérier, F., Bernard, B., Lomello, F., Maskrot, H., Meillot, E., Michau, A. and Schuster, F., 2018. Emerging processes for metallurgical coatings and thin films. *ComptesRendus Physique*, 19(8), pp. 755–768.

5. Sadeghi, E., Markocsan, N. and Joshi, S., 2019. Advances in corrosion-resistant thermal spray coatings for renewable energy power plants. Part I: Effect of composition and microstructure. *Journal of Thermal Spray Technology*, 28(8), pp. 1749–1788.

6. Quinlan, F.B. and Grobel, L.P., 1942. "Treatment of metals," US2303869A, 01-Dec-1942.69.

7. Fauchais, P., Montavon, G. and Bertrand, G., 2010. From powders to thermally sprayed coatings. *Journal of Thermal Spray Technology*, 19(1–2), pp. 56–80.

8. Browning, J.A., 1992. Hypervelocity impact fusion—a technical note. *Journal of Thermal Spray Technology*, 1(4), pp. 289–292.

9. N.N.: EN 657: ThermischesSpritzen, Beuth, Berlin, 1994-06 (under revision: prEN 657: 2003–04). Thermal spraying – Terminology, classification

10. Chandra, S. and Fauchais, P., 2009. Formation of solid splats during thermal spray deposition. *Journal of Thermal Spray Technology*, 18(2), pp. 148–180.

11. Xue, M., Chandra, S., Mostaghimi, J. and Salimijazi, H.R., 2007. Formation of pores in thermal spray coatings due to incomplete filling of crevices in patterned surfaces. *Plasma Chemistry and Plasma Processing*, 27(5), pp. 647–657.

12. Haefer, R.A., 1987. *Oberflächen- Und Dünnschichttechnologien*. Springer, Berlin.

13. Fauchais, P., 2004. Understanding plasma spraying. *Journal of Physics D: Applied Physics*, 37(9), p. R86.

14. Klinkov, S.V. and Kosarev, V.F., 2006. Measurements of cold spray deposition efficiency. *Journal of Thermal Spray Technology*, 15(3), pp. 364–371.

15. Morks, M.F., Tsunekawa, Y., Okumiva, M. and Shoeib, M.A., 2002. Splat morphology and microstructure of plasma sprayed cast iron with different preheat substrate temperatures. *Journal of Thermal Spray Technology*, 11(2), pp. 226–232.

16. Chraska, T. and King, A.H., 2002. Effect of different substrate conditions upon interface with plasma sprayed zirconia—a TEM study. *Surface and Coatings Technology*, 157(2–3), pp. 238–246.

17. Mellali, M., Grimaud, A., Leger, A.C., Fauchais, P. and Lu, J., 1997. Alumina grit blasting parameters for surface preparation in the plasma spraying operation. *Journal of Thermal Spray Technology*, 6(2), pp. 217–227.

18. Bahbou, M.F. and Nylén, P., 2005. Relationship between Surface Topgraphy Parameters and Adhesion Strength for Plasma Spraying. *The Material Information Society*, pp. 1027–1031.

19. Fauchais, P.L., Heberlein, J.V.R. and Boulos, M.I., 2014. *Industrial Applications of Thermal Spraying Technology. Thermal Spray Fundamentals*, Springer, New York, pp1401–1566

20. Roata, I.C., Croitoru, C., Pascu, A. and Stanciu, E.M., 2019. Photocatalytic coatings via thermal spraying: A mini-review. *AIMS Materials Science*, 6(3), pp. 335–353.

21. Thakare, J.G., Pandey, C., Mahapatra, M.M. and Mulik, R.S., 2020. Thermal barrier coatings: A state of the art review. *Metals and Materials International*, pp. 1–22.

22. Thorpe, M.L., 1993. Thermal spray industry in transition. *Advanced Materials and Processes*, 143(5), pp. 50–56.

23. Sadeghimeresht, E., Markocsan, N., Nylén, P. and Björklund, S., 2016. Corrosion performance of bi-layer Ni/Cr2C3–NiCr HVAF thermal spray coating. *Applied Surface Science*, 369, pp. 470–481.

24. Pfender, E., 1994. Plasma jet behavior and modeling associated with the plasma spray process. *Thin Solid Films*, 238(2), pp. 228–241.

25. Kang, A.S., Singh, G. and Chawla, V., 2013. Some problems associated with thermal sprayed ha coatings: A review. *International Journal of Surface Engineering and Materials Technology*, 3, pp. 10–20.

26. Heimann, R.B., 1996. Plasma-spray coating. *Principles and Applications*, VCH publishers, New York, p. 2.
27. Moskowitz, L.N., 1993. Application of HVOF thermal spraying to solve corrosion problems in the petroleum industry-an industrial note. *JTST*, 2, pp. 21–29. https://doi.org/10.1007/BF02647419.
28. Kim, G.H., Park, C.K., Ahn, H.J., Kim, H.S., Hong, K.H., Jin, S.W., Lee, H.G., Fukanuma, H., Huang, R., Roh, B.R. and Kim, T.S., 2015. Fabrication feasibility study on copper cold spray in tokamak system. *Fusion Engineering and Design*, 98, pp. 1576–1579.
29. Papyrin, A., Kosarev, V., Klinkov, S., Alkhimov, A. and Fomin, V.M., 2006. *Cold Spray Technology*. Elsevier, Amsterdam.
30. Vardelle, A., Moreau, C., Akedo, J., Ashrafizadeh, H., Berndt, C.C., Berghaus, J.O., Boulos, M., Brogan, J., Bourtsalas, A.C., Dolatabadi, A. and Dorfman, M., 2016. The 2016 thermal spray roadmap. *Journal of Thermal Spray Technology*, 25(8), pp. 1376–1440.
31. Singh, L., Chawla, V. and Grewal, J.S., 2012. A review on detonation gun sprayed coatings. *Journal of Minerals and Materials Characterization and Engineering*, 11(03), p. 243.
32. Vuoristo, P., 2014. Thermal spray coating processes. In *Comprehensive Materials Processing*, Elsevier, (4), pp. 229–276
33. Prashar, G. and Vasudev, H., 2020. Thermal sprayed composite coatings for biomedical implants: A brief review. *Journal of Thermal Spray and Engineering*, 2(1), pp. 50–55.
34. Morks, M.F. and Kobayashi, A., 2007. Effect of gun current on the microstructure and crystallinity of plasma sprayed hydroxyapatite coatings. *Applied Surface Science*, 253(17), pp. 7136–7142.
35. Morks, M.F., Fahim, N.F. and Kobayashi, A., 2008. Structure, mechanical performance and electrochemical characterization of plasmasprayed SiO_2/Ti-reinforced hydroxyapatite biomedical coatings. *Applied Surface Science*, 255(5), pp. 3426–3433.
36. Chen, Q and G. A. Thouas, 2015. Metallic implant biomaterials. *Materials Science and Engineering R: Reports*, (87), pp. 1–57.
37. Geetha, M., Singh, A.K., Asokamani, R. and Gogia, A.K., 2009. Ti based biomaterials, the ultimate choice for orthopaedic implants: A review. *Progress in Materials Science*, 54(3), pp. 397–425.
38. Fauchais, P. and Vardelle, M., 2010. Sensors in spray processes. *Journal of Thermal Spray Technology*, 19(4), pp. 668–694.

2 Cold Spraying of 3D Parts – Challenges

*Hongjian Wu, Rija Nirina Raoelison, Yicha Zhang,
Sihao Deng, and Hanlin Liao*

CONTENTS

2.1 COLD SPRAY TECHNOLOGY: PRINCIPLE AND 3D ADDITIVE MANUFACTURING CHALLENGES

The cold spray additive manufacturing (CSAM) appeared in the mid-1980s, thanks to a technological achievement developed at the Institute of Theoretical and Applied Mechanics of the Russian Academy of Sciences [1,2]. This additive method belongs to the wide family of thermal spray technology that uses a spray of gas flow and micron-sized powders mixture. The gas flow accelerates and heats the powders before they additively collide onto a substrate to produce a coating or on a fixed part to build a thick component. The continuous additive collision is governed by a high velocity impact and also by a thermal softening (or a solid/liquid phase change) where a spray process with a high thermal exposure such HVOF, plasma arc or powder flame process is used (Figure 2.1). Due to the use of low gas temperature and high in-flight powder velocity (Figure 2.1), CS can be distinguished among the thermal spray processes [2,3]. The principle of CS is to provide a high kinetic energy to micron-sized powders so that they can reach supersonic velocities, high enough to generate powders-to-powders bonding during the continuous additive collision onto a substrate or a fixed part. This kinematics relies on a principle of supersonic gas flow produced by a gas expansion across a convergent/divergent nozzle, named De Laval nozzle. The gas does not need to be heated at high temperature such as in other

DOI: 10.1201/9781003213185-2

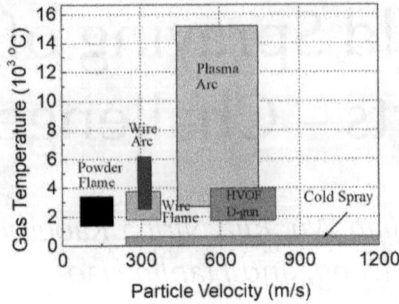

FIGURE 2.1 Thermal spraying processes including CS that makes use of low gas temperature and high velocity of micron-sized powders.

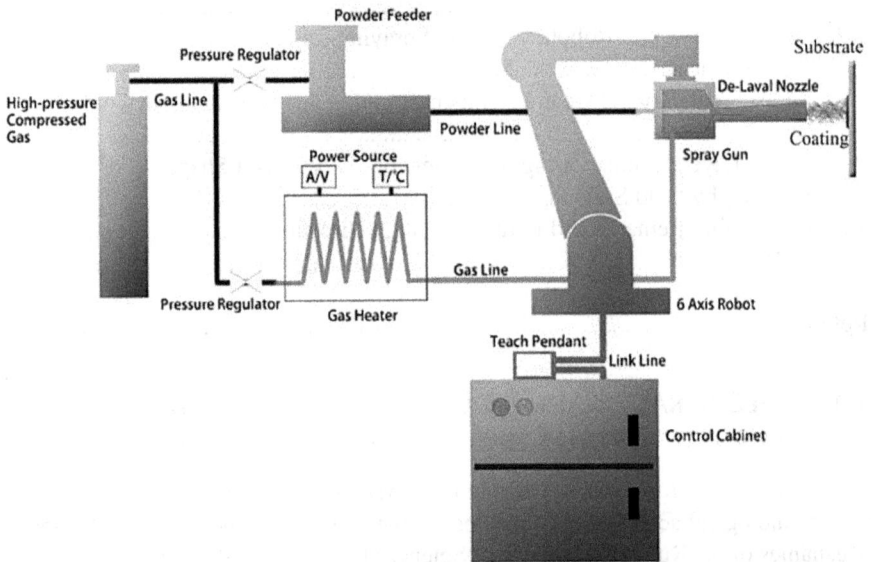

FIGURE 2.2 Typical architecture of a modern CS system.

thermal spray process. In CS, the working temperature is generally less than 500°C which is further reduced down to very low values due the expansion of the gas. This situation has been considered as a major thermal feature of the CS process to endorse the specification "Cold" spraying or "Cold Gas Dynamic" spraying in the literature. Then, a typical CS system is basically composed of a low- or high-pressure compressed gas source, a gas heater, a powder feeder, a spray gun equipped by a De Laval nozzle and an industrial robot (Figure 2.2). The gas source is separated into two different gas lines: one is fed to the gas heater and serves as propellant gas while the other (carrier gas) drives the powders from the powder feeder to the nozzle. The powder feeder can provide a continuous and controllable feeding. During the spraying process, metal powders with a size distribution of 1–50 μm are accelerated at a high velocity (500–1200 m/s) by the gas expansion [2–4], while they remain at a temperature lower than the melting point to form a coating or a 3D part by a ballistic collision [2]. During the deposition process, the spray gun needs to be precisely controlled to achieve the desired coating thickness or the desired final shape of the deposit. Industrial robots are used to perform the motion of the spray gun to achieve controllable, safe and accurate spraying path. They have been widely combined with CS due to their high performance, such as high accuracy, high repeatability, and high flexibility, which directly affects the deposit features [5–8]. The robot assisted spraying strategy plays thereby a major role in manufacturing 3D part by the CS additive process. The CS kinematic parameters are manipulated and controlled by the robot. Via a robot arm, the spraying gun moves with respect to the substrate (Figure 2.3) by monitoring the following kinematic parameters: the spray trajectory, the relative speed between the nozzle and the substrate, the spray distance, the spray angle and the scanning step. The scanning step refers to the interval between two successive scanning tracks when a coating is deposited by a multi-track trajectory (Figure 2.4a). This parameter governs the flatness and the thickness of a coating. Optimal value

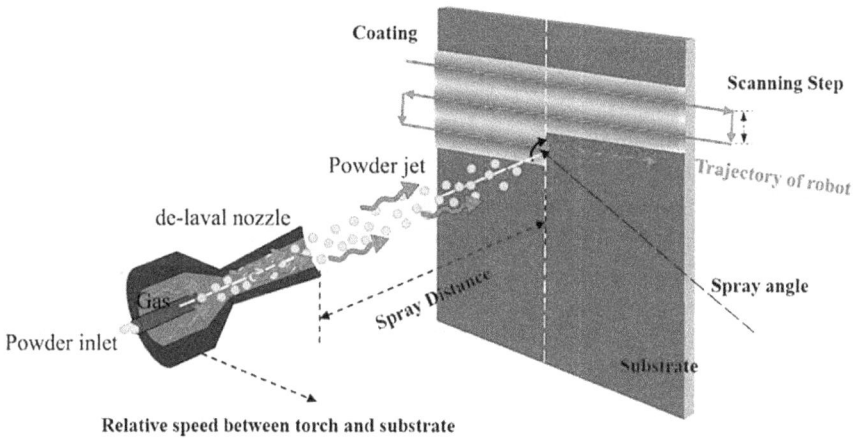

FIGURE 2.3 Kinematic parameters monitored by a robot arm during the cold spraying process.

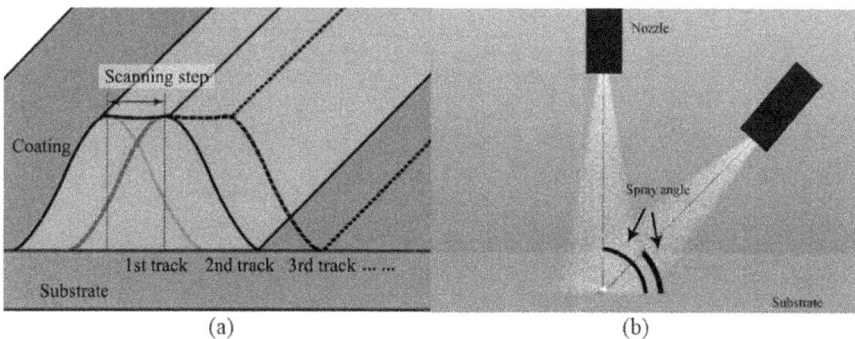

FIGURE 2.4 Scanning step (a) and spray angle (b) in CS.

gives a uniform coating, and a low value reduces the coating surface roughness. If the scan step is too large, the flatness of coating will be decreased, and the coating thickness distribution will become uneven. Besides, different scanning steps will lead to different distribution of the track-to-track profile, thereby affecting the distribution of residual stress and pores [9].

The spray angle is the angle between the nozzle and the surface of the substrate (Figure 2.4b) that should be about 90° to optimize the deposition efficiency[1]. If the angle is oblique, the powders rebound during their collision onto the substrate so that the deposition efficiency decreases, and porosity occurs within the coating. Therefore, the most active manner is that the spray angle should be kept at about 90°, that is easier to achieve a plane surface. However, due to a motion limitation of the robot with respect to the workpiece position and geometry, particularly in case of a complex shape, the robot adapts the spray angle and the scanning speed to ensure a good coating quality and deposition efficiency. This has been proved to be feasible for a spray angle range of 45°–90° [10].

The spray trajectory, also called spray path, is a point set of spraying targets that are connected to each other by robotic programming. The aim of spray trajectory planning is to enable to coat the whole substrate surface as uniform as possible. The spray trajectory influences not only the coating thickness and distribution, but also the property of coating, especially the coating anisotropy [11–14]. Therefore, it is very important to plan the spraying path. With the increase of complex workpiece geometry, path programming has evolved from the initial manual manner to the current computer-aided automatic programming based on different algorithms [5,11,15], especially thanks to the use of robot offline programming technology that improves the capability and reliability of path planning.

In the literature of CS, some efforts investigated the capability of multiple-layer deposition and issues of edge effects, and proposed appropriate manufacturing strategies. A few effective strategies have been suggested [16–19] such as a triangular

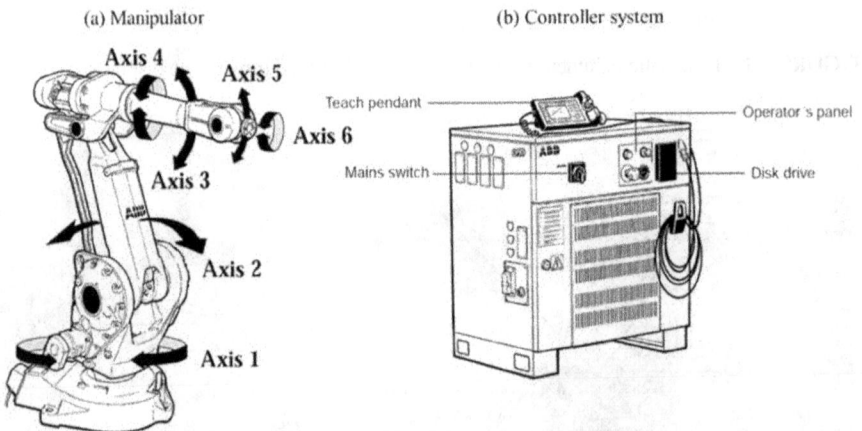

FIGURE 2.5 ABB IRB 2400 robot system including a manipulator with 6 axes (a) and a controller system (b).

tessellation scheme [19] for thick coatings or a vertical wall, or a combination of a CS deposition with a design and topology shape structural optimisation [18]. These methods proposed reliable design guidelines that links the design of the three-dimensional structure to the 3D part via the use of topology optimization technology combined with the control of the nozzle trajectory for various kinds of spray paths. In addition, they also considered the design of removable 3D support which is very consistent with the current 3D printing ideas (Figure 2.6). The notion of removable support has also been used to fabricate an array of 3D structure by additive manufacturing using CS. Cormier et al. [20] used a specially designed mask to obtain a pyramidal fin array used as compact heat exchanger solution.

These 3D manufacturing challenges by CS reveal that improving the accuracy, reliability and ability of CS requires advanced spraying strategies that mainly depend

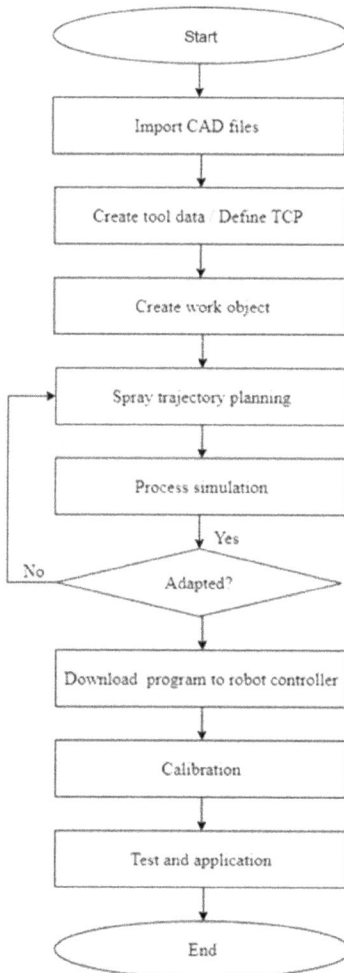

FIGURE 2.6 Procedure of an offline programming.

on a better implementation of robotics. In addition, robot offline programming technology allows to reduce the burden and difficulty of programming, to improve the accuracy and to make CS available for more complex parts. It is necessary to develop an auxiliary system to provide a robust spraying strategy including generating trajectories, process simulation and collision detection. For such a purpose, this chapter suggests viable perspectives based on robotized CS process simulation and planning in order to enhance the ability of CS for 3D metallic additive manufacturing. The implementation of robotics in CS is described in Section 2.2. Then, in Section 2.3, a new approach for 3D coating thickness or shape simulation is presented with its integration into a tool (Profile Kit of Thermal Spray Toolkit [TST]) for anticipating a 3D CS process. Section 2.4 focuses on the development of a novel spraying strategy for developing a freeform 3D objects with acceptable precision by CS. Concluding remarks with future research directions will be also addressed (Section 2.5).

2.2   IMPLEMENTATION OF ROBOTICS IN COLD SPRAYING FOR 3D ADDITIVE MANUFACTURING

Generally, a robot is a machine capable of carrying out a complex series of actions automatically. Nowadays, many kinds of robot have been developed affecting the way people live and work, viz. industrial robots, mobile robots, collaborative robots, biomorphic robots, military robots and so on. Among these categories, industrial robots are currently the most widely used. Popular industrial robots include FUNUC Robots, MOTOMAN Robots, ABB Robots and KUKA Robots. Our CS system is equipped by an ABB IRB2400-16 robot which is a multi-axis robot system (Figure 2.5). This robot is automatically controlled. Generally, a CNC machine has 2 or 3 axes for a basic manipulation and 3–6 for programmable robots [21]. The degree of freedom is equal to the number of axes. Therefore, this multi-axis robot system has greater number of axes to perform complex actions and tasks.

Compared with traditional machines, the main advantage of industrial robots is their programmability. Robots can perform arbitrary sequences of predefined motions. In CS applications, the main task of the industrial robot is to execute the spraying trajectory with high precision. The planning and generation of the spraying trajectories are based on the shape of the workpiece to be sprayed and on the process parameters also. Hence, a generation of trajectories and post-analysis require an efficient and well-structured programming method. Currently, the methods of robotic programming involve the online teaching [22] and offline programming [7,11]. The complexity of the project decides the choice of the programming method. Most robot programming uses an online teaching method which is suitable for certain simple tasks. However, for those complex movements that require higher accuracy, offline programming should be used to define the robot's movement to perform tasks that cannot be accomplished by online teaching method.

For the online programming technique, the industrial robots generally consist of three parts: a controller, a robot arm and a teach pendant [21,23]. The teach pendant is the remote controller of the robot, which is equipped with a user operation interface software and buttons, moving handles, touch screen, etc. The operator controls the robot to complete the specified actions through these human–computer interaction functions. This process is called online programming or teaching programming that is a traditional robotic programming method in most industries. The tool is mounted on the end-effector of the robot (wrist). A Tool Centre Point (TCP) should be defined based on the tool and the task, that is usually located at the end of the tool or at the point of contact. The operator uses the manual control on the teach pendant to control the robot, to move the TCP to the desired position and then to store the information of the position in a series of motion commands (including the position and orientation of the TCP). After storing all target points and robot motion commands, the corresponding trajectory program can be tested. Note that industrial robots have good position repeatability, that is, after teaching, the robot can run the same program repeatedly, and the accuracy of returning to the same spatial position point can reach the level of tens to hundreds of microns. The advantages of this programming method are a low learning cost and an easy use. In the process of online programming, operators are required to perform on-site operations in the working unit of the robot. However, this method can involve a long duration due to several manually operated robot movements that makes the programming process tedious and time-consuming. Depending on the complexity of the task, this may take days or even weeks. In addition, the robot needs to leave the production line during online programming so the production will be interrupted. In CS, the working range of the spray gun, the nozzle travel speed, the spray distance, the spray angle, etc., need to be considered for the programming. However, for tasks that require high complexity with accuracy and repeated modifications, the online programming method may not be suitable. Instead, another programming method, denoted offline programming method, is better since it allows to handle complex programming tasks.

The robot offline programming refers to "virtual" programming of the robot on the computer (PC) through related software tools [24,25]. Programmers need to use CAD models to offer a more scientific programming strategy. This process includes creating CAD models, planning robot path, creating motion programs and robot motion simulation. When using this method of programming, there is no need to stop the robot during a process, so it will not hinder the production operations. With the development of robot technology and with the continuous expansion of its application fields, offline programming technology has gradually become a popular programming method, which is mainly used to complete some complex programming tasks. Currently, this technology has been widely used in welding, laser cutting, thermal spraying, CNC machining and in other fields, to deal with some complex programming tasks. The robotic offline programming technology provides a complete solution for industrial robots, from trajectory generation, parameter selection to process simulation and trajectory optimization. The trajectory of the robot can be generated by using the geometric data of the workpiece to ensure an accuracy, and at the same time, it provides more optimization strategies while also reducing the occurrence of collision accidents. The robot offline programming technology is even more essential in CS application. The robot offline programming technology can provide optimized spraying path and reliable programming [7,24]. In addition, for complex shapes, online programming is difficult to complete with the correct definition of the trajectory point. During the spraying process, it must be guaranteed that the spraying direction is perpendicular to the surface of the workpiece, and at the same time, it must be also guaranteed that the coating can cover the entire surface of the work-piece. Thus, offline programming is an accurate programming method based on the CAD model. This allows the robot to easily adjust the tool operation direction and the starting pose of the robot to avoid the singularity state and obtain the most opti-mized path. Since many applications are no longer just a process of producing simple coatings, Computer-aided design (CAD) and Computer-aided manufacturing (CAM) are needed to provide the ideal spray strategy, including generating trajectories, sim-ulating the process and collision detection. Thereby, the robot offline programming technology can be used as a new production capacity and modern application to provide desirable solutions in the process.

Robot offline programming technology requires the application of related soft-ware systems, called robot offline programming software [25]. Many robot manufac-turers have developed their own robot offline programming systems compatible with their robots. The software operator can create a three-dimensional virtual scene of the entire actual workstation and uses the related functions provided by the software to accomplish many programming tasks that cannot be completed by online pro-gramming. Nowadays, there is various robot offline programming software such as RobotStudio™ which is a powerful offline programming software developed by Asea Brown Bovrie Ltd. (ABB). This software enables very realistic simulations of robot motion [23]. This software is specifically developed for performing robotic tasks including programming, trajectory simulation and collision detection. Robot Studio™ is based on ABB Virtual Controller technology that makes it possible to graphically create a virtual robotic site like the real site. Therefore, users can use the software to specify the robot's motion, automatically form executable code and then to synchronize it to the actual robot. Robot Studio™ provides the ability to design the robot program in a computer with a learning programming method. The Robot Studio ™ software offers also other advantages: (1) all types of robot models of ABB prod-ucts are available in the Robot system library, (2) commonly used geometric models such as STL, IGES, STEP, ASCII, ACIS and CATIA can be imported directly into Robot Studio™ as digital components for modelling, (3) robotic programming and simulation including the detection of collisions in the graphic environment ensure the safety of the operator and the equipment, (4) the program is designed and created in a computer so that it can be prepared in advance even though the robot is still in production, (5) Robot Studio™ provides an interface for users to personalize devel-opment according to specific requirements. Figure 2.6 shows the general process of programming using Robot Studio™ which is suitable for various industrial applica-tions, including CS.

Generally, Robot Studio™ or other offline programming software can meet the needs of general applications. Operators can generate an appropriate trajectory. In CS, the trajectory consists of paths separated by the constant scan step to guarantee that the coating covers the entire surface of the substrate. However, it is not easy to create parallel spray paths on curved substrate surfaces. CSAM can need a creation of a shaping path. The basic function including trajectory generation in Robot Studio™ cannot meet such a specific request. In order to satisfy the specific require-ments of 3D manufacturing by CS, a software based on the offline programming platform is essential to provide the appropriate spray strategy quickly and precisely.

Robot Studio™ allows the developers to develop different kind of customized applications or Add-Ins as a new feature. The laboratory LERMPS (Laboratoired'Etudes et de Recherches sur les Matériaux, les Procédés et les Surfaces) has developed an add-in application program named Thermal Spray Toolkit (TST) in the framework of RobotStudio™ [8,11]. TST is a RobotStudio™-based extension software developed for the generation of trajectory in thermal spray applications, and its whole functions are also suitable for CS. This tool consists of four basic modules as follows: Path Kit, Profile Kit, Monitor Kit, Kinema Kit. The Path Kit module enables the creation of robotic trajectories with respect to the geometry of workpiece. A series of scanning curves can be generated using orthogonal planes that cuts the substrate surface. The normal vector is calculated to define the orientation of the nozzle on every point of the curves. Path Kit uses this method to generate robot trajectories (Figure 2.7) for thermal spraying in the offline programming software Robot Studio™ [26,27]. The Profile Kit module is developed for the purpose of coating deposition simulation. Currently, a coating thickness model based on a numerical method was developed and used to predict the coating thickness and distribution. Combined with the robot kinematic parameters, the coating profile is simulated with an asymmetric Gaussian distribution curve. Thus, the suitable coating thickness can be obtained within the required tolerances. According to the calculation and simulation results, the spraying strategy including the robot trajectory and operating parameters can be adjusted with iteration in the feedback loop to achieve the desired coating thickness distribution. The Monitor Kit module controls both speed and trajectory of the robot and evaluates the path accuracy to optimize the trajectory. Finally, the Kinema Kit module is used to optimize the kinematics parameters of the spraying process by collecting and analysing various signal data of the robot's motion. The signal data mainly include the TCP position and orientation, the position of each axis versus time, etc. Kinematic analysis method is useful for the optimised mounting method of spray nozzle on the robot [28]. Besides, the optimal placement of the workpiece in the robot workspace is selected through robotic kinematic analysis method to optimize the robot's trajectory [29].

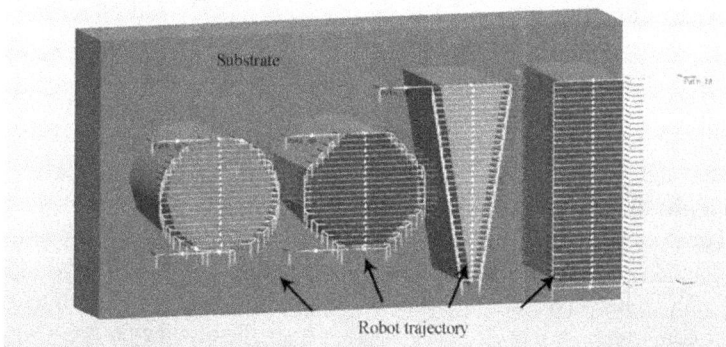

FIGURE 2.7 Spray trajectory generation under Path Kit.

After generating the spraying strategy using the offline programming software, the robot motion program can be created according to the planned CS tool path. The dynamic simulation of the spraying process can be carried out under the offline programming system. Through simulation and analysis, if the expected result is achieved, the generated robot program can be synchronized to the robot controller system to proceed with the next step. If not, the related spraying strategy can be adjusted with iteration in the feedback loop to achieve the desired result. Then, the workpiece will be pre-processed (including blasting or others machining) and fixed at an appropriate position in the work spot. After a series of tests and calibrations, the CS process can be implemented. In addition to depositing materials during this process, a 3D scanning system can also be used to monitor the growth process of the coating in real-time. A complete scanning process includes the following sub-programs: creating a connection, running spray program, sending scan request, receiving returned data, judging the validity of the data, diagnosis and decision-making, returning feedback information and saving data. In practice, after launching the spray program, the robot takes the substrate in front of the nozzle to perform the spraying. When the pre-defined number of cycles is reached, the spray program is interrupted, and the contour scan module is activated. At this stage, the robot arm moves to the measurement window of the 3D profiler, then the robot program sends a trigger signal to the 3D profiler for scanning. The computer will receive and analyse the scanning results so that the decision-making algorithm can compare the real coating profile with the theoretical model to decide the compensation value of the robot motion. The adjustment parameters of the robot trajectory will be sent back via the feedback subroutine to modify the current spray process. This approach is also applicable to many AM processes and enables the closed-loop control as well as the real-time parameter optimization to realize stable and high-accuracy AM.

With the continuous development of CSAM, the sprayed materials, sprayed objects and application fields are also continuously expanding. Faced with different spraying conditions, how to obtain the best spraying strategy and parameters in the shortest time is the key to realize the actual benefits of CSAM. Therefore, from the production and design concept, it is necessary to transform the design and manufacturing concept based on simulation and experience. Through advanced simulation technology, it can provide predictions of spraying results and can optimize the spraying parameters to guide the manufacturing process quickly and effectively and to avoid cost waste caused by repeated trials. During a process simulation, trajectory issues and speed issues are two crucial issues, especially in CSAM. Through the trajectory simulation module, it can be observed whether there is divergence in between the expected and the actual robot trajectories. Then, the spray trajectory will be adjusted and decided. Sometimes, even if the spraying target or spraying trajectory conforms to the robot's motion range due to the problem of the configuration of each robot axis, singular points will appear during the robot's motion resulting in preventing the spraying trajectory program from continuing. The analyser module of Robot Studio™ can be used to record and analyse the position of each axis, to re-adjust the configuration of each axis. But most of all, the accuracy of a CS deposition relies on the dependence of the spray angle, the nozzle traverse speed, the scanning step and the standoff distance on the coating thickness distribution.

2.3 AN APPROACH FOR 3D COATING THICKNESS AND SHAPE SIMULATION IN COLD SPRAYING

The model describing the distribution of powders jet outside the nozzle is important to assess the distribution of the coating thickness. According to the central limit theorem, the average of random variables can be considered as normally distributed when the amount of variable is sufficiently large. Thus, in cold spray process, the coating thickness distribution (coating profile) can be defined using a Gaussian distribution [14,30]:

$$\varphi = \zeta(\theta)\zeta(s)\int_0^T \left[\int \frac{A\zeta(v)}{\sigma\sqrt{2\pi}} e^{-\left[\frac{(x-\mu_x)^2}{2\sigma^2} + \frac{(y-\mu_y)^2}{2\sigma^2}\right]} dxdy \right] dt \tag{2.1}$$

where A is the amplitude factor related to the experimental powders feed rate, σ is the standard deviation of the coating profile, $(\mu x, \mu y)$ is the centre coordinate of the coating profile on the substrate surface. $\zeta(\theta)$, $\zeta(s)$ and $\zeta(v)$ are deposition efficiencies depending on the spray angle, the spray distance and the nozzle traverse speed, respectively.

The single coating profile model is illustrated in Figure 2.8 in a three-dimensional Cartesian coordinate system. The nozzle is defined as a point in the Cartesian coordinate system. The angle of incidence, that is, spray angle, is θ (on X–Y plane). In the YZ plane, the spray cone in the Cartesian coordinate system is divided into a series of rays with a constant interval angle (the dash lines in Figure 2.8). For perpendicular spraying, the coating profile is conical and symmetric with respect to the central line of the nozzle. Thus, the spray length at the same deflection angle is constant. For

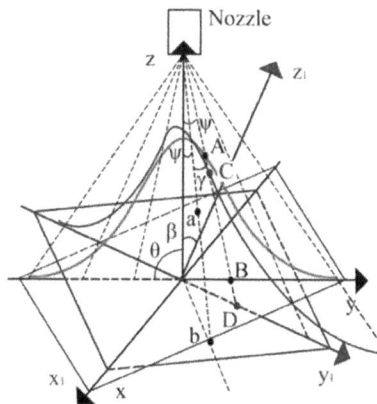

FIGURE 2.8 Schematic of single coating profile model on X–Y plane (red line) and X_1–Y_1 plane (blue line). θ and β are the spray angles on X–Y plane and X_1–Y_1 plane, respectively. a is the angle between Z-axis and Z_1-axis. ψ is the deflection angle (the angle between Z-axis and ab line, as well as AB line). γ is the angle between ab line and AB line.

example, at the deflection angle ψ as indicated in Figure 2.8, the spray length AB has the same value as ab and the angle in between these two rays is γ. Due to the fact that mass distribution out of the nozzle is constant, it can be assumed that the spray length at each deflection angle is constant during inclination. For example, when the XY plane rotates around the x-axis (now the spray angle is β on the $X1$–$Y1$ plane), as indicated in Figure 2.8, the spray length AB has the same value as CD.

In computer graphics and computational physics applications, an object can be represented by a three-element vector which is typically used to describe a position, normal and Euler rotation. In addition, a transformation (rotation and translation) can be described by a matrix from a three-element vector and a quaternion. In order to express the coating profile model in a three-dimensional graphics environment, the coating virtual simulation using the model (Equation 2.1) involves some useful considerations to preserve the characteristics of the spraying. Firstly, the powders leaving the nozzle and deposited on the surface are considered as a finite number of rays sharing the same starting point. As shown in Figure 2.9a,d, a specific point at a certain distance above the surface of a body is created to represent the outlet of the nozzle. Then from this point, many straight rays are created. These rays are obtained by rotation and symmetric operation of the centreline, which is the vertical line from the starting point to the body surface. The rays will intersect the substrate to create intersections. In this way, it will be possible to determine the effective area for generating the coating profile on any shape of the substrate. Secondly, a single coating profile consists of a finite number of cylinders with the same diameter and the height of each cylinder equals the value of the corresponding Gaussian function. As shown in Figure 2.9b,e, based on the intersection, circles will be created and stretched in the opposite direction of the ray to form a cylinder. The distance stretched is equal to the spray length. The smaller the radius of the cylinder is, the higher the accuracy of the simulation results will be. However, the calculation will be time-consuming. This study uses a cylinder with a radius of 0.02 mm. Thirdly, the cylinders are established only where the rays touch. Therefore, the coating profile can be in a myriad of shapes,

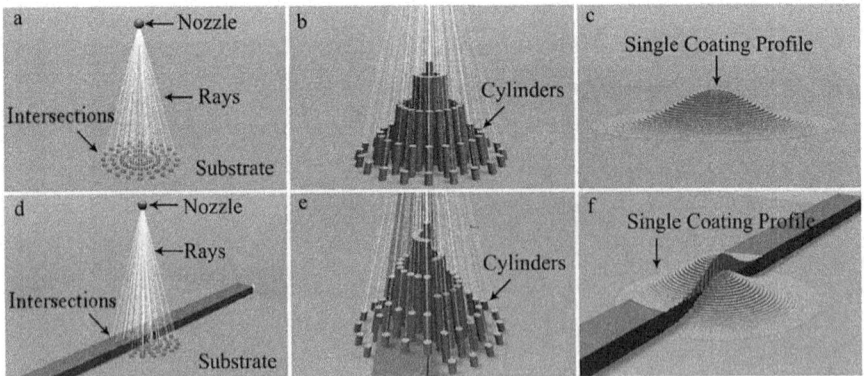

FIGURE 2.9 Creation of rays and intersection on a flat (a); creation of cylinders on a flat (b); single coating profile model on a flat (c); creation of rays and intersection on a non-planar (d); creation of cylinders on a non-planar (e); and single coating profile model on a non-planar (f).

even non-continuous. As shown in Figure 2.9c,f, a number of regular cylinders is composed as a single coating profile model. Although it is created based on the Gaussian model, its final profile is determined by the shape of the substrate.

Normally, a set of continuous random variables can be discretized according to certain rules. In order to simulate the deposited coatings on the substrate during the nozzle motion, it is assumed that the continuous coating profile is a collection of single coating profiles. Figure 2.10 shows the schematic of the coating thickness distribution model. By selecting the appropriate distance between two discrete points, it can create the coating distribution model deposited on the substrate under the traveling of the nozzle. An appropriate distance value is important for the thickness simulation result due to the fact that a large distance value can lead to a less accurate result and a small one can lead to excess computation. Based on the results of trial and error, as long as the distance is less than σ, a continuous coating profile close to the actual situation can be obtained. In this study, the distance between two discrete points was set to $\sigma/2$. In a three-dimensional graphic environment, a continuous coating profile model can be created based on the principle of creating a single coating profile model. The first single coating profile will be created on the substrate (Figure 2.11). Then the second single coating profile will be created on the substrate and overlapped with the first single coating profile and so forth. Finally, by selecting a stacking appropriate distance, continuous coating topography can be created on any

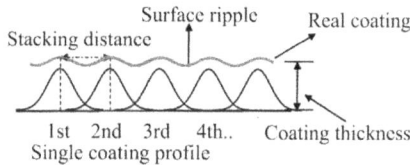

FIGURE 2.10   Illustration of coating thickness distribution model.

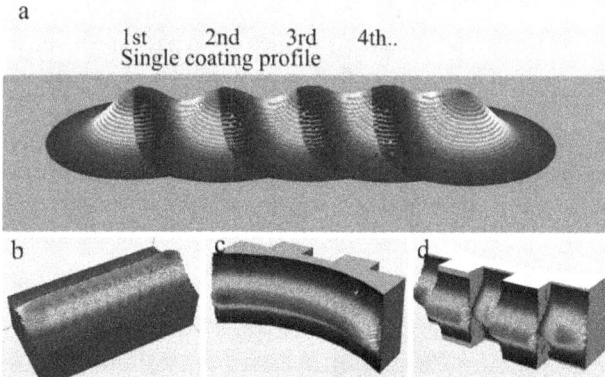

FIGURE 2.11   Discrete single coating profile with overlaps (a); continuous single coating profile on a flat (b); continuous single coating profile on a curved surface (c); continuous single coating profile on a complex surface (d).

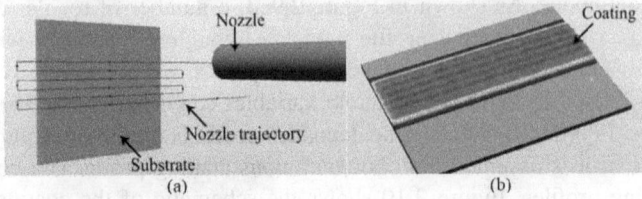

FIGURE 2.12 Coating thickness simulation in Robot Studio™. Generation of trajectory (a) and coating thickness distributed simulation (b).

topographic substrate, such as on a flat surface (Figure 2.11b), curved surface (Figure 2.11c) and complex surface (Figure 2.11d).

A general coating thickness simulation process is displayed in Figure 2.12. First of all, a corresponding CAD file of the component is needed in the process of coating simulation. The formats of CAD models such as IGES, STL, STEP, ACIS and ASCII are available in Robot studio™. After a calibration of the component position and the TCP, the robotic trajectory is generated according to the operating parameters including spray angle, nozzle traverse speed, projection spray distance, scanning step and so on. Here, Path Kit of TST can be used to generate trajectories on the surface of workpieces with different kinds of geometric shapes, including a rectangular surface, a circular surface, a curved surface and a rotation of a workpiece. Figure 2.12a is a simple zigzag path created on a flat substrate. After that, the corresponding robot path program will be created and run. During the robot path simulation in Robot Studio™, the speed, the position and the orientation of the nozzle as well will change with the robot movement. These signals can be recorded to generate discrete points on the path for coating thickness simulation. Finally, according to these discrete points, a 3D coating distribution model can be generated on the surface of the substrate (Figure 2.12b).

In order to evaluate the predictive capabilities of this newly developed approach, both simulation and experimental investigations were carried out. For the experimental verification on a flat surface, the experiment was performed with the same robot program under the same kinematics parameters as predefined in the simulation (Figure 2.12). The simple zigzag path was created. The spray nozzle was always perpendicular to the surface of the substrate through the entire path. The nozzle traverse speed and the spray distance were set to 50 mm/s and 30 mm, respectively. In order to make the coating in the cross-sectional direction easier to observe and to compare, a larger scanning step of 3.8 mm was used. Finally, the experimental result is in good agreement with the simulation (Figure 2.13). The capability of the model (Equation 2.1) to compute a virtual coating development for complex part with a shadow effect has also been demonstrated. Figure 2.14a,b shows the results of the experiment and simulation, respectively. Each zone was affected by shadow effects under experiment settings. The coating in each zone was different: zone 1 had the largest coating thickness and a triangular-like profile was formed; the lower number of particles was deposited in zone 3 so the coating thickness was minimum. This was because the spray angle decreased as the spray distance increased, so that the relative deposition efficiency became smaller. There was no obvious coating on the top

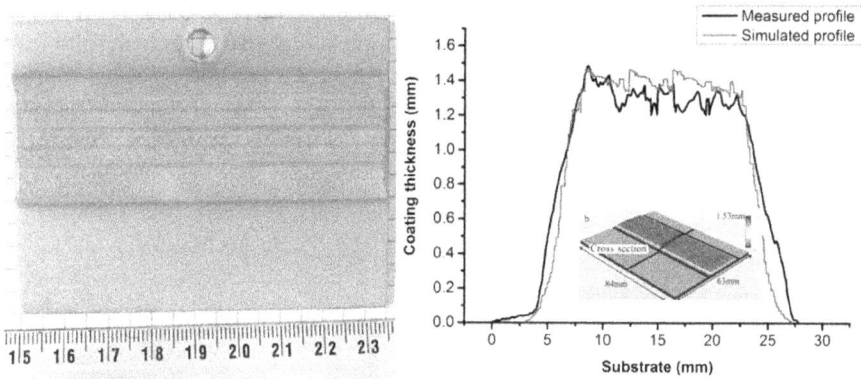

FIGURE 2.13 Comparison of experimental and simulation results of coating thickness.

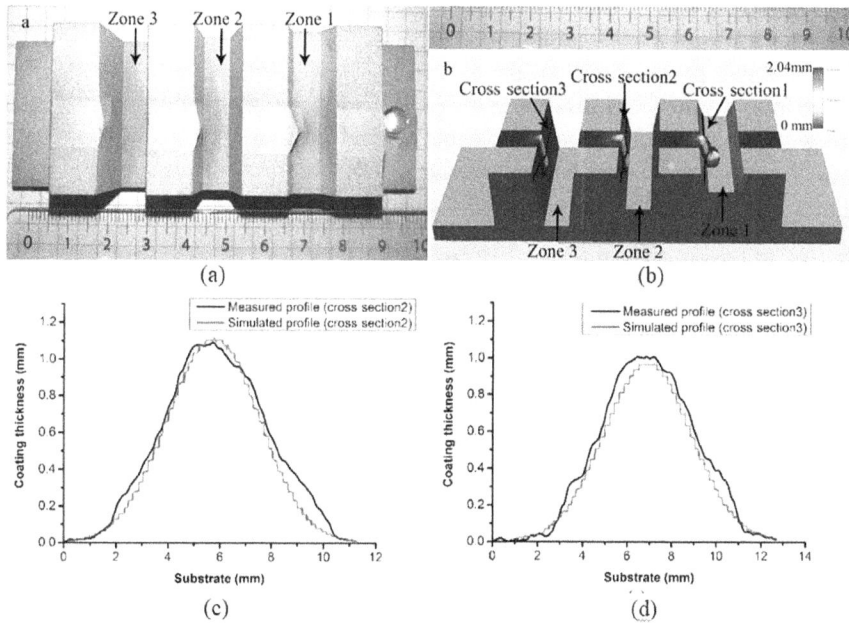

FIGURE 2.14 Experimental (a) and simulation (b) results of CS deposition on workpiece with shadow effect. Comparison of experimental and simulation results of coating thickness at cross section 2 (c) and cross section 3 (d).

surface of the substrate because the deposition efficiency was close to zero. However, the traces washed by the powders could be clearly observed. The cross-sectional comparisons of coating thickness distribution in each zone are displayed in Figure 2.14c,d. It is known that the physical influence factors of deposition are very complicated during cold spraying. The parallel multiple passes in a single track will lead to the triangular-like profiles. This is because increasing coating height leads to a significant decreasing of deposition efficiency (more than 35% according to Kotoban's et al. report [31]) due to the diminishing of effective impact angle (oblique impact on the substrate). Obviously, the coating height of zone 1 increased more quickly than in the other zone owing to its closer spray distance and larger spray angle, that made it the earliest formation of triangular-like profiles. Based on the simulation results; robotic trajectory, operating parameters and spray strategy can be re-adjusted, like changing the spray angle, setting different nozzle travel speeds, until achieving the desired coating thickness distribution.

2.4   STABLE LAYER MANUFACTURING STRATEGIES FOR 3D NEAR-NET SHAPE FORMING BY COLD SPRAYING

Basically, the cold spraying jet can be regarded as a set of continuous random variables. Many investigations used a Gaussian-based numerical modelling to study the feature of spray deposits [14,30,32,33], as done in our previous work [34]. This approach is proved to be feasible and reliable. The Gaussian curve is a standard curve for a normal distribution. Figure 2.15a shows a normal Gaussian curve where μ is the centre position of the peak and σ is the standard deviation. The values less than σ away from the mean account for 68.27% of the set, while the remaining portion is relatively low. It is conceivable that the main weight of a single CS deposited coating is concentrated in the central area. A more uniform coating implies to have a progressive concentration of mass distribution on both sides of the Gaussian distribution. This can be done via a control of the spray angle that shifts the centre of gravity of the sprayed coatings in the direction of the incoming particles. By tilting the nozzle, this method avoids the formation of triangular-like deposits such as during a triangular-tessellation strategy. The oblique spraying can compensate the thickness and weight difference in between the middle and side of the deposit (Figure 2.15a,b). Since then, three main parameters are proposed to control the shape of deposits. As shown in Figure 2.15b, θ is the deflection angle in order to

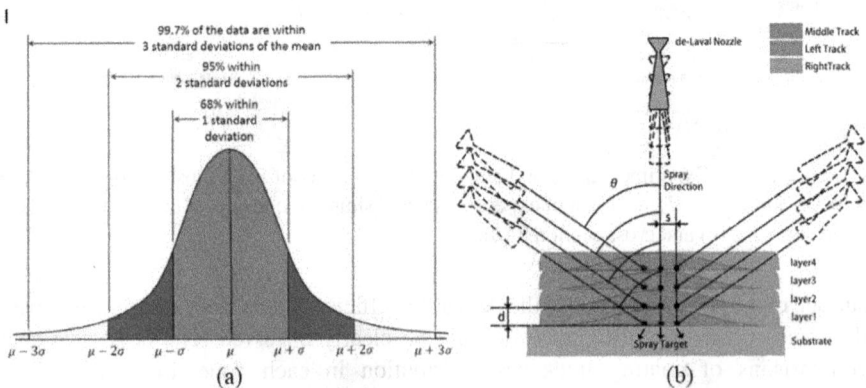

FIGURE 2.15 Gaussian distribution model (a) which is considered for developing a CSAM spray strategy (b) to additively produce thick and vertical walls.

replenish particle depositions at the edge; *s* is the offset that can adjust the track width in horizon; *d* is the value that adjusts a setback distance of the nozzle with respect to the layer's thickening. This setback distance correction provides a consistent spray distance between the top layer and the nozzle during the spraying. The coating profile in perpendicular spray cases is a Gaussian distribution (shown with red blocks), while the coating profile in off-normal spray cases has no longer a normal Gaussian distribution (the left track with green block, the right track with yellow block).

With the previous strategy, a single layer can steadily rise in a track-by-track manner without a formation of a defective triangular-like structure. Meanwhile, this spray strategy is also appropriate to fabricate large bricks or a thick coating (Figure 2.16). The surface of the prescribed area is sprayed in a perpendicular manner by specifying a scanning step of the corrected distribution. Then the edges are sprayed by tilting the nozzle to compensate for the edge loss. A bulk growth occurs layer after layer using this spraying strategy that defines a successful principle of a 3D parts manufacturing by CS. This strategy does allow to avoid triangular-like shapes. Moreover, with the application of high-performance industrial robot, this spraying strategy can be easily realized. There is no doubt that the parameters used in this strategy can be easily controlled and repeated with high accuracy and flexibility. Figure 2.17 shows a simulation procedure based on this strategy. A first track is generated under normal parameters (in case of perpendicular spray) that gives a typical Gaussian profile (Figure 2.17a). Then the left track (Figure 2.17b) is obtained by

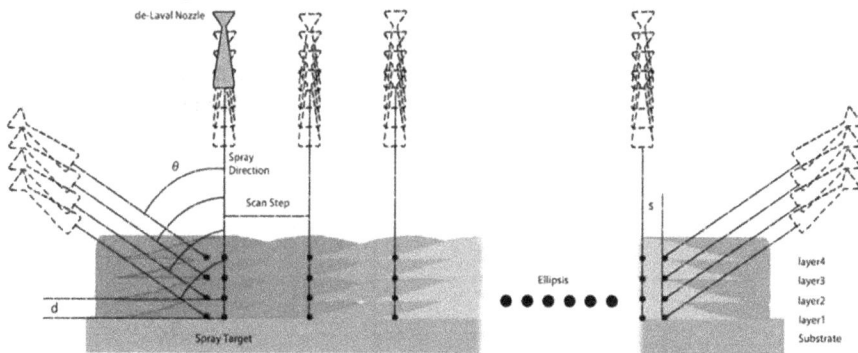

FIGURE 2.16 CS spray strategy for large blocks or thick coatings AM.

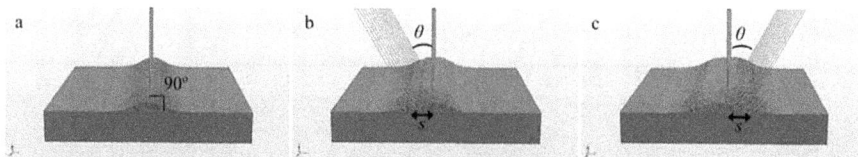

FIGURE 2.17 Illustration of the profile simulation operation based on the strategy of Figure 2.19.

tilting the angle θ and by offsetting the distance s to the left side. Thirdly, the right track is also created by the tilting angle θ and the offsetting distance s to the other side (Figure 2.17c). The coating created by such operations can be considered as one layer. Then, the subsequent layers will be created on top of the previous layer by repeating this procedure. The profile topography strongly depends on the parameters set (θ, s, d) we denote tilting angle, offsetting distance and setback distance, respectively. The flatness and verticality of the formed profile are closely related to the combination of these parameters, while these two attributes are exactly the criteria for judging the quality of the control and for reducing the post-machining treatment.

The experimental implementation of the spray strategy evidences the sensitivity of the profile development to the parameters set (θ, s, d). Figure 2.18a shows the deposits profile curves measured by the 3D profiler with a deflection angle of 30° ($\theta = 30°$) and offset distance of 0, σ, 2σ, 3σ and 4σ mm. When s is set to 0 and σ mm, the deposits profile becomes triangular-like, even though the profile thickness is larger than the other cases. The deposits had a flat top surface and a good efficiency. For s of 3σ and 4σ mm, the efficiency was dissatisfying. Figure 2.18b shows the effect of deflection angle change. With an offset distance of 2σ mm and various deflection angles (10°, 20°, 30° and 40°) and assuming a deposition efficiency of 100% when the deflection angle is set to 90°, the deposition efficiency becomes low when θ is increased up to 40°. The deposits with a deflection angles of 10° and 20° tended to form triangles. Conclusively, these profiles proved that the value of deflection angle and offset distance significantly affect the flatness and the efficiency of the deposits.

This spray strategy is a successful method for building blocks or thick coatings without edge losses while a normal spray strategy leads to edge losses (Figure 2.19a). This means that during the bulk formation, the area of the upper surface shrinks (more than 34% in this case). With the spray strategy we developed, the borders were compensated without edge losses. The bulk thickening does occur layer by layer and

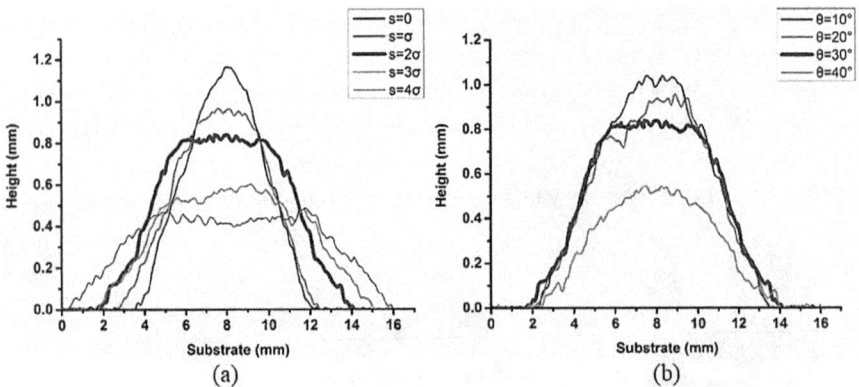

FIGURE 2.18 Experimental results with $\theta = 30°$ and various values of s (a). Experimental results with $s = 2\sigma$ mm and various values of θ (b).

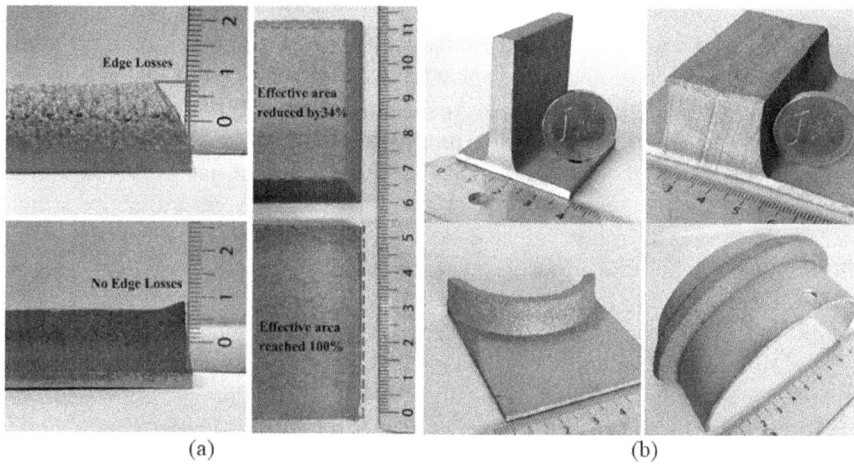

FIGURE 2.19 Creation of thick coatings without edges loss (a) and perspectives of thick and vertical wall created on a various surfaces (b).

the effective area reaches nearly 100%. Furthermore, this spraying strategy can generate different shaped samples. Thick and vertical walls could be built on both flat surface and on curved surface (Figure 2.19b). It proved that this method can be successfully used for a highly accurate shape control and for generating complex shaped components by CSAM.

2.5 CONCLUDING REMARKS AND FUTURE RESEARCH DIRECTIONS

With the continuous development of the CS technology and the expansion of application fields, CS has been widely used to prepare various functional coatings on components surface, to restore damaged metal components, or to fabricate freestanding 3D metal components. Many advantages make the CS process a very competitive additive method. In particular, the combination with high-performance industrial robots improves the accuracy and ability of CS so that it can complete more and more complex tasks. Nowadays, both industrial and academic communities are paying more and more attention to CSAM, especially for the direct 3D manufacturing of metal components. This is a recent development that requires more new methods to improve the manufacturing accuracy, flexibility and reliability. This chapter addressed a brief review on some achievements for a 3D manufacturing by CS and also brings new process implementation to enhance the CSAM of 3D parts. A stable layer-by-layer building strategy and capability of CS is a key preliminary step for forming complex 3D shapes in an additive way. This chapter focused on the development of such a strategy that considers both kinematic factors and coating characteristics in CS. A new approach based on coating process simulate is suggested, using a three-dimensional geometric model based on a Gaussian distribution of the coating virtual profile. The 3D geometric coating thickness model was integrated into the offline programming software Robot Studio™ as a module in the software TST that enables the possibility of a wide application with accurate profile prediction. Both numerical and experimental verifications proved that the proposed method has a reliable prediction accuracy in practice, especially in coating thickness prediction including shadow effects. Three major parameters were proposed to control the shape of deposits: a deflection angle θ, an offset distance s and a nozzle setback distance d. In the current framework of CSAM system, these parameters are directly controlled by a high-performance industrial robot with an easy control and process repetition capabilities. Numerical simulation and physical benchmarking case study results validate the proposed method and show that more complex 3D geometries can be built via this layer-building strategy. It was found that the combination of controlling different parameters played a key role in determining the layer geometry, and thus the component built-up process. Experimental implementation of the stable layer building strategy evidences a real capability for 3D shape forming by a layer-by-layer AM process with a good shape accuracy.

Future work will further investigate the impact of nozzle shape and size on the layer building; and also, the development of a multi-axis robotic tool-path planning strategy for building complex 3D curved layers or freeform 3D objects in near-net-shape with the proposed layer-building method. In addition, to fully exert the potential of CSAM, efforts are still required to integrate more technologies. Future research will focus on the improvements for each component of the framework and on the development of more spray strategies for CSAM. A potential new avenue is to combine this CSAM 3D strategy with conventional machining techniques if the 3D bulk accuracy does not meet the predefined expectations. More efforts should be also made to study the properties and potential defects of the 3D bulk component, such as the porosity, the microstructure, the bonding strength.

ACKNOWLEDGEMENT

The author Hongjian Wu would like to acknowledge the support from the China Scholarship Council.

NOTE

1   The deposition efficiency is defined by the ratio between the deposited powders and the sprayed powders. A DE of 100% means that the whole sprayed powders get adhered on each other during the additive deposition.

REFERENCES

1. R.C. Dykhuizen, M.F. Smith, Gas dynamic principles of cold spray, *Journal of Thermal Spray Technology*. 7 (1998) 205–212.
2. A. Moridi, S.M. Hassani-Gangaraj, M. Guagliano, M. Dao, Cold spray coating: Review of material systems and future perspectives, *Surface Engineering*. 30 (2014) 369–395.
3. V.K. Champagne, *The Cold Spray Materials Deposition Process*, Elsevier, Amsterdam, 2007.

4. S. Yin, M. Meyer, W. Li, H. Liao, R. Lupoi, Gas flow, particle acceleration, and heat transfer in cold spray: A review, *Journal of Thermal Spray Technology*. 25 (2016) 874–896.

5. Z. Cai, Programmationrobotique en utilisant la méthode de maillage et la simulation thermique du procédé de la projection thermique, PhD Thesis, 2014.

6. D. Fang, Diagnostic et adaptation des trajectoiresrobotiques en projection thermique, PhD Thesis, Université de Technologie de Belfort-Montbeliard, 2010.

7. C. Chen, Research and realization of assistant off-line programming system for thermal spraying, PhD Thesis, Belfort-Montbéliard, 2016.

8. S. Deng, Programmationrobotique hors-ligne et contrôle en temps réel des trajectoires: développementd'une extension logicielle de "Robotstudio" pour la projection thermique, PhD Thesis, Besançon, 2006.

9. K. Yang, W. Li, X. Yang, Y. Xu, Anisotropic response of cold sprayed copper deposits, *Surface and Coatings Technology*. 335 (2018) 219–227.

10. G. Wkh, L. Wkh, R. Dqg, *Effect of Spray Angle on Deposition Characteristics in Cold Spraying*, ASM International, Materials Park, OH, 2003, 6.

11. S. Deng, Robot offline programming and real-time monitoring of trajectories: Development of an add-in program of robotstudio™ for the thermal spraying, in: *Graduate School of "Physics for Engineers and Microtechnology"*, University of Technology of Belfort-Montbéliard France, Sevenans, France, 2006.

12. C. Chen, S. Gojon, Y. Xie, S. Yin, C. Verdy, Z. Ren, H. Liao, S. Deng, A novel spiral trajectory for damage component recovery with cold spray, *Surface and Coatings Technology*. 309 (2017) 719–728. doi:10.1016/j.surfcoat.2016.10.096.

13. Z. Cai, T. Chen, C. Zeng, X. Guo, H. Lian, Y. Zheng, X. Wei, A global approach to the optimal trajectory based on an improved ant colony algorithm for cold spray, *Journal of Thermal Spray Technology*.25 (2016) 1631–1637. doi:10.1007/s11666-016-0468-7.

14. Z. Cai, S. Deng, H. Liao, C. Zeng, G. Montavon, The effect of spray distance and scanning step on the coating thickness uniformity in cold spray process, *Journal of Thermal Spray Technology*.23 (2014) 354–362. doi:10.1007/s11666-013-0002-0.

15. A. Frutos, Numerical analysis of the temperature distribution and Offline programming of industrial robot for thermal spraying, 2009. https://repositorio.upct.es/handle/10317/764.

16. R.N. Raoelison, C. Verdy, H. Liao, Cold gas dynamic spray additive manufacturing today: Deposit possibilities, technological solutions and viable applications, *Materials & Design*.133 (2017) 266–287.

17. S. Yin, P. Cavaliere, B. Aldwell, R. Jenkins, H. Liao, W. Li, R. Lupoi, Cold spray additive manufacturing and repair: Fundamentals and applications, *Additive Manufacturing*. 21 (2018) 628–650.

18. M.E. Lynch, W. Gu, T. El-Wardany, A. Hsu, D. Viens, A. Nardi, M. Klecka, Design and topology/shape structural optimisation for additively manufactured cold sprayed components, *Virtual and Physical Prototyping*.8 (2013) 213–231.

19. J. Pattison, S. Celotto, R. Morgan, M. Bray, W. O'Neill, Cold gas dynamic manufacturing: A non-thermal approach to freeform fabrication, *International Journal of Machine Tools and Manufacture*.47 (2007) 627–634.

20. Y. Cormier, P. Dupuis, B. Jodoin, A. Corbeil, Net shape fins for compact heat exchanger produced by cold spray, *Journal of Thermal Spray Technology*.22 (2013) 1210–1221.

21. Unimate - The First Industrial Robot, Robotics Online. (2020). https://www.robotics.org/joseph-engelberger/unimate.cfm. (accessed August 13, 2020).

22. A. León, E.F. Morales, L. Altamirano, J.R. Ruiz, Teaching a robot to perform task through imitation and on-line feedback, in: *Iberoamerican Congress on Pattern Recognition*, Springer, Berlin, Heidelberg, 2011: pp. 549–556.
23. M.M. Dalvand, S. Nahavandi, Teleoperation of ABB industrial robots, *Industrial Robot: An International Journal*, 41 (3), (2014) 286–295.
24. S. Deng, Z. Cai, D. Fang, H. Liao, G. Montavon, Application of robot offline programming in thermal spraying, *Surface and Coatings Technology*. 206 (2012) 3875–3882.
25. Y. Nagatsuka, K. Inoue, Device, method, program and recording medium for robot offline programming, 2012. https://patents.google.com/patent/US8155789.
26. Z. Cai, H. Liang, S. Quan, et al. Computer-aided robot trajectory auto-generation strategy in thermal spraying. *The Journal of Thermal Spray Technology*. 24, 2015) 1235–1245.
27. S. Deng, H. Liao, C. Zeng, C. Coddet, New functions of thermal spray toolkit, asoftware developed for offline and rapid robot programming, *Thermal Spray 2006: Building on 100 Years of Success*. (2006) 15–18. https://www.asminternational.org/home/-/journal_content/56/10192/CP2006ITSC1437/CONFERENCE-PAPER.
28. C. Chen, H. Liao, G. Montavon, S. Deng, Nozzle mounting method optimization based on robot kinematic analysis, *Journal of Thermal Spray Technology*. 25 (2016) 1138–1148. doi:10.1007/s11666-016-0429-1.
29. S. Deng, H. Liang, Z. Cai, H. Liao, G. Montavon, Kinematic optimization of robot trajectories for thermal spray coating application, *Journal of Thermal Spray Technology*. 23 (2014) 1382–1389. doi:10.1007/s11666-014-0137-7.
30. C. Chen, Y. Xie, C. Verdy, H. Liao, S. Deng, Modelling of coating thickness distribution and its application in offline programming software, *Surface and Coatings Technology*. 100 (2017)315–325.
31. D. Kotoban, S. Grigoriev, A. Okunkova, A. Sova, Influence of a shape of single track on deposition efficiency of 316L stainless steel powder in cold spray, *Surface and Coatings Technology*. 309 (2017) 951–958.
32. M.M. Fasching, F.B. Prinz, L.E. Weiss, Planning robotic trajectories for thermal spray shape deposition, *Journal of Thermal Spray Technology*. 2 (1993) 45–57.
33. C. Chen, Y. Xie, C. Verdy, R. Huang, H. Liao, Z. Ren, S. Deng, Numerical investigation of transient coating build-up and heat transfer in cold spray, *Surface and Coatings Technology*. 326 (2017) 355–365.
34. H. Wu, X. Xie, M. Liu, C. Chen, H. Liao, Y. Zhang, S. Deng, A new approach to simulate coating thickness in cold spray, *Surface and Coatings Technology*. 382 (2020) 125151.

3 Instant Tuning of Wettability of Metallic Coating

Anup Kumar Keshri and Swati Sharma

CONTENTS

3.1   INTRODUCTION TO WETTABILITY AND ITS APPLICATIONS

Wettability refers to how a liquid spreads out when deposited on any solid (or liquid) substrate. In other words, wettability is a response related to surface tension and contact angle (CA) of the liquid on the substrate [1–3]. A drop of water deposited on a clean glass of sheet spreads completely on the surface. By contrast, the same drop of water on lotus leaf does not wet the leaf surface. This exemplifies the two extremes of wetting. This chapter will highlight the fundamental concepts of wettability and contact angle, which play crucial roles in determining the surface properties of materials. There are two regimes of wettability and can be characterized as follows [4]:

(i)   Total wetting
(ii) Partial or nearly no wetting

When we rest a small fluid on any solid substrate, there exists a surface tension between the two interfaces. A liquid vapor interface is formed at a certain contact angle with the solid substrate. This contact angle gives the quantitative measure of wettability of the solid by any fluid. Figure 3.1 shows a liquid droplet on the solid

DOI: 10.1201/9781003213185-3

59

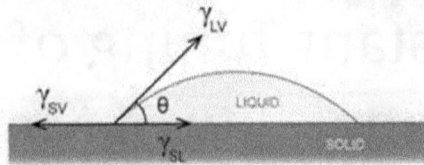

FIGURE 3.1   Illustration of a liquid droplet on the solid substrate.

substrate. Equilibrium consideration of the respective surface tensions acting upon different interfaces allows us to calculate the wetting (or dihedral) angle θ using Equation 3.1 [5].

$$\gamma_{LV} \cos\theta = \gamma_{SV} - \gamma_{SL} \tag{3.1}$$

where γ_{LV} is the interfacial tension between liquid and vapor, γ_{SV} is the interfacial tension between solid and vapor, and γ_{SL} is the interfacial tension between solid and liquid.

Hence, the contact angle can be written as Equation 3.2 [5]:

$$\cos\theta = \frac{\gamma_{SV} - \gamma_{SL}}{\gamma_{LV}} \tag{3.2}$$

The contact angle can range from 0° to 180°. If the droplet is strongly attached to the solid substrate, it spreads out completely on the solid surface and the solid substrate is hydrophilic in nature. Intensely hydrophilic substrates have contact angles in the range of 0°–10° and can be categorized under superhydrophilic material. Lesser hydrophilic substrates have contact angles of around 90°. In other case, when the droplets are weakly attached to the solid substrate, the contact angle is around 120°and the surface is categorized as hydrophobic. However, if the contact angle ranges at around 150° the substrate is superhydrophobic [1–3]. Figure 3.2 shows

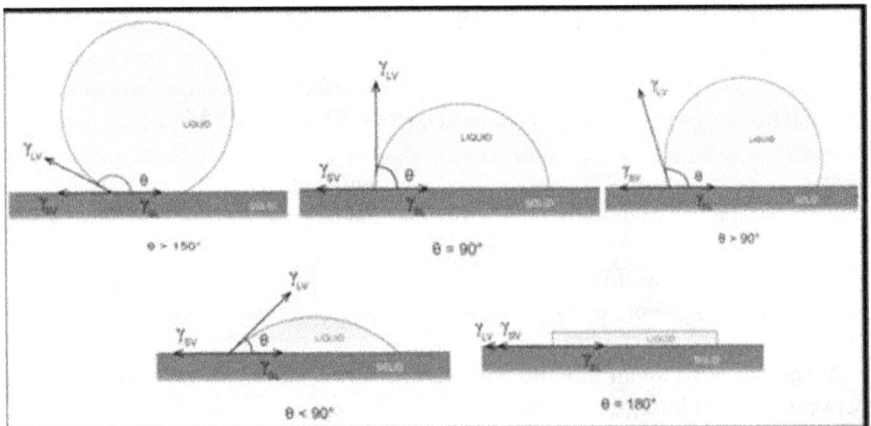

FIGURE 3.2   Different contact angles and different wetting properties.

different wetting conditions with different contact angles and different wetting properties.

Wetting abilities of different hydrophilic and hydrophobic surfaces have attracted extensive research attention for both practical and fundamental applications [1–5]. Some of the practical applications of wettability have been discussed further.

In case of biomaterials, the wettability becomes a prerequisite property to ensure the desired response [6]. One of the best examples is the contact lenses worn by a huge portion of the total world population [6]. Biocompatibility of such biomaterials is of great concern as any type of unfavorable response when exposed to the human tissues cannot be accepted. For such applications, the surface wettability significantly influences the biocompatibility and this characteristic is determined by measuring the contact angle of lens material. Biocompatibility of dental implants, blood contacting devices, tissue engineering surfaces contact lenses, etc., are dependent on the wettability characteristics of the surfaces. The surface of the biomaterials gets exposed to biological cells or fluids. Thus, the biocompatibility gets strongly influenced by the wetting characteristics of the surface leading to extensive research in the field of effect of contact angle on the biocompatibility of the biomaterials [6].

The self-cleaning property is one of the most widely sought-after properties in recent years. It can be obtained by manipulating the surface properties. Both hydrophilic and hydrophobic surfaces offer self-cleaning properties by either rolling droplet of water or by sheeting water that will carry away the dirt [3–5]. A lot of work has also been done to develop contamination-resistant coatings. Self-cleaning coatings and anti-contamination coatings find a wide range of application like window glasses, cement, textiles, etc. Titanium dioxide (TiO_2) is one of the largely investigated materials for its wetting properties for self-cleaning, anti-contamination and antifogging properties [7, 8].

Wettability characteristics of modified wood substrates in cellulose fiber-based composites with high strength also directs the adhesion process. The wetting properties can strongly influence the adhesion and gluing process. Hence, the wetting characteristics of modified wood have also been investigated to study the effect of thermal modification process of hot pressing on its wettability and surface roughness.

Superhydrophobic surfaces have aroused great interest due to diverse and wide range of applications like anti-corrosion, antifouling, self-cleaning, anti-icing, etc. Efforts are directed toward achieving such surfaces mimicking the surface structures available in nature, such as lotus leaf, butterfly wings, etc.

The wettability of any surface depends on its chemical composition and topographical features. In this chapter, we have tried to highlight the effect of chemical composition and surface topological features which are instrumental in altering the wetting characteristics of different surfaces.

3.2 CONVENTIONAL MATERIALS FOR TUNING THE WETTABILITY

Out of the three broad spectrums of materials, polymeric materials have been used mostly for tuning the wettability. Figure 3.3 shows the pie chart which is based upon the last 10 years of research papers (≥2000 research articles) from the Scopus. This clearly indicates that 80% of the work used polymer for tailoring the wettability,

■ **Polymer** ■ **Metal** ■ **Ceramics**

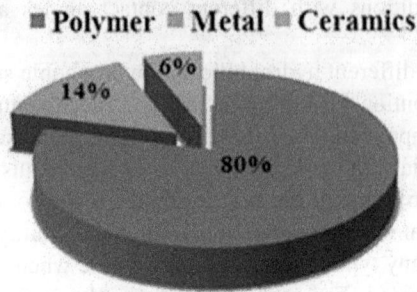

FIGURE 3.3   Pie chart based on the last 10 years work on wettability of materials.

while 14% focused on metallic materials and rest 6% on ceramics. The next section discusses the wettability of different kinds of materials.

(a) **Polymer-Based Materials**

Polymeric materials (natural or synthetic) play a vital role in everyday life. Nowadays, they are being extensively used because of their wide range of applicability and properties like high elasticity, chemical stability and light weight. Several polymeric materials like High-Density Polyethylene (HDPE), Low-Density Polyethylene, Silicone, Polypropylene, Polytetrafluroethylene (PTFE), polyimide (PI), etc., are widely used in automotive industries, as bio-materials, in tissue engineering and many other industries [9]. These polymeric materials and their derivatives offer a low surface energy resulting in poor wettability [9]. Their chemical composition, morphology and structure influence the surface-free energy to a great extent and consequently its wettability. There has been extensive research in the area of tailoring the surface roughness and chemical composition of these materials in order to engineer the wetting properties. Polymeric surfaces provide non-polar characteristics which results in low affinity for a wide range of compounds and fluids [9]. The nonpolar nature provides a non-wetting surface specifically to the water-based and polar compounds.

Most of the work strategies have aimed toward surface modification to exploit the desired property requirements, for example, laser surface treatment and different types of plasma treatment. Though these treatments provide excellent wetting characteristics, they are associated with highly sophisticated surface modification techniques. In addition, these techniques also involve multiple step processing and majority of these strategies involve surface modifiers. This limits their application only to laboratory-scale preparation and the benefits cannot be realized at a larger scale.

Polymeric materials are again limited in their usability because of the poor mechanical properties (except the elastic property) and thermal durability [9]. This suggests that the microscopic roughness can be easily destroyed, resulting in a decline of the hydrophilic or hydrophobic properties upon mechanical or thermal exposure. Thus, the tailored wetting properties of such polymeric materials are quite unstable and are liable to decline. Consequently, commodities prepared by such approaches are not very readily available for practical applications. Hence, scalable method for production of robust polymeric coatings with desired properties remains a challenge. It becomes a challenging task to manipulate the wetting behavior of such materials without the addition of any surface modifiers or application of UV light or lasers. Hence, the sophisticated processing routes and requirement for surface modifiers makes the process uneconomical as it involves high cost [9].

Despite the attractive hydrophobicity achieved from polymeric and organic materials, their use is limited in engineering applications due to the lack of mechanical robustness and inherent thermal instability. Therefore, development of superhydrophobic materials with good mechanical properties and thermal properties to withstand high temperature and harsh environments are very much required.

(b) **Ceramic-Based Materials**

While superhydrophobic surfaces successfully incorporated on various substrates have involved the use of low surface energy materials like polymers, the major drawback lies in the sophisticated multi-step process and use of modifiers. Moreover, these processes are suitable for rapid and large-scale production. Besides, these surfaces are also vulnerable to easy degradation and damage when exposed to harsh environments of mechanical forces or thermal energy. Therefore, a simple, rapid process/route to produce a stable superhydrophobic surface at a large scale becomes highly significant.

In an attempt to address these issues, ceramic powders were looked upon as potential candidates owing to their high hardness, corrosion- and thermal resistance and excellent chemical stability. However, ceramics are intrinsically hydrophilic in nature because of the high-energy surfaces arising from numerous polar sites [10, 11]. The polar sites arise due to unsaturated coordination structures of the atoms in the surface. Ceramic oxide cations have affinity toward hydrogen and tend to form water molecules at the interface [10, 11]. Thus, the ceramic surfaces are usually hydrophilic in nature. However, for superhydrophobicity of ceramic materials, it is important to attain a low surface energy by incorporating materials like SiO_2, Ta_2O_5, CF_x, etc., by adopting various fabrication techniques [10, 11]. Azimi et al. were first to report hydrophobic rare earth oxides from the Lanthanide series [12]. Their hydrophobicity was attributed to the specific electron structures. Thereafter, the hydrophobicity of the REOs have been extensively investigated over this behavior [12]. However, two schools of thoughts have been illustrated [13, 14]. One of them believes that the hydrophobicity can be attributed to the oxygen-to-metal ratio at the surface while the other believes that the hydrophobicity results from the absorption of hydrocarbons from the surface of the REOs [13, 14]. However, the basis of hydrophobicity of REOs still remains unclear and efforts to develop fabricating processes for hydrophobic REOs are continuing.

Ceramics have the advantage of mechanical robustness and thermal stability. However, the challenge remains in their fabrication. Work done so far report multi-step processing techniques which are very time-consuming and have scalability concerns. Some processes necessitate the creation of micro-textured surfaces before any hydrophobic ceramic can be deposited. Many recent works have attempted toward obtaining superhydrophobicity by tailoring the surface roughness. Some of the reported works include growing ceria (CeO_2) nanorods and their cathodic electrodeposition [15], sputtering of ceria on nanograss, leaf-like Y_2O_3 surface fabricated on porous fiber-board substrate [16], pulse-laser deposition (PLD) processes, boron ink, etc. All these processes have focused upon altering the typographical features utilizing the role of liquid vapor surface energy in improving the contact angle. Attempt has also been made to achieve hierarchical structures of micro- to nanoscaled grains to achieve superhydrophobicity and adequate bonding of the coating. This has been attained by adopting high-velocity oxy-fuel (HVOF) technique.

Although the ceramic coatings provide excellent mechanical properties and chemical stability, the routes of synthesis are complex, time-consuming, expensive and not robust enough for industrial scale-up.

(c) **Metal-Based Materials**

In recent years, the focus of obtaining superhydrophobicity has shifted toward exploring and understating the potential of metallic surface. Bhattacharya et al. have attempted to fabricate clusters of copper nanowires by cathodic electrode-position on anodic porous alumina template [17]. However, the processes of fabricating such surfaces are usually quite complicated and time-consuming. In addition, it is also important to maintain the integrity of such surfaces in harsh mechanical, chemical and thermal environment.

It is well known that metals are intrinsically hydrophilic in nature because of their high-energy surfaces. A significant advancement has already taken place in tailoring the wetting characteristics of the surfaces and researchers are in a pursuit to find a single-step scalable fabrication technique with an emphasis to achieve mechanically robust and thermally and chemically stable surfaces. Liang et al. have developed two extremities of wetting of Ti by a one-step anodization and fluorination [18]. They considered the process to be quite facile and simplified. However, the work did not produce relevant and sufficient data regarding the robustness and durability of the fabricated surface. Further, despite the authors' claim for the fabrication to be a single step process, the process involved several conventional multi-steps, for example, tailoring the roughness of the surface and consequent use of surface modifiers. Moreover, the use of fluorinated chemicals like fluoroalkyl-silanes impose hazardous health effects and are a potential threat for the environment and human health. In another work by Bae et al., intrinsically hydrophobic Al 7075 alloy surface was fabricated to offer superhydrophobicity by a one-step process using wire electrical discharge machining (WEDM) technique [19]. The process was scalable, and the authors also achieved a high static contact angle (CA) of nearly 156° and a low water contact angle hysteresis or CAH (3°). However, the process becomes strenuous if the wetting characteristics had to be manipulated from an intrinsically hydrophilic (<90°) surface just by altering the gas environment from O_2 to H_2 or Ar during its laser processing. Some advanced materials like bulk metallic glasses have also attracted much attention in recent years as they offer unique combination of various properties like high strength and hardness with significantly large elasticity, and excellent wear and corrosion resistance. It has been reported that this class of materials have shown the potential to exhibit superhydrophobicity after proper etching and surface modification by materials with low surface energy.

Table 3.1 demonstrates different materials from different classes of polymers, ceramics and metals [20–23]. The contact angle of water on their surface has also been given to understand their wetting characteristics. Additionally, their applications as given in the Table 3.1 also emphasizes on the utilization of these materials based on their wetting behavior.

TABLE 3.1
Materials and the Contact Angle of Water on Their Surface with Their Most-Specific Applications

S. No.	Material	Contact Angle(s) of Water on the Material Surface	Applications
1	Polyvinyl alcohol (PVOH)	51 [20]	Hydrophilic water-soluble synthetic polymer for biomedical applications
2	Polyvinyl acetate (PVA)	60.6 [20]	Adhesive, thickeners for paint, plasticizers
3	Polysulfone (PSU)	70.5 [20]	Used in microfiltration, ultrafiltration, etc.
4	Polymethyl methacrylate (PMMA)	70.9 [20]	Biomaterial applications such as bone, bone substitutes, cement, lenses and drug delivery systems
5	Polyethylene terephthalate (PET)	72.5 [20]	Tape application, microwave containers, filter oil, packaging applications, mesh fabrics in printing, cosmetic jars and agricultural applications
6	Epoxies	76.3 [20]	Primers, plastics, paints, adhesives, coatings, sealers, flooring and other products and materials that are used in building and construction applications
7	Polyoxymethylene (POM, polyacetal, polymethylene oxide)	76.8 [20]	Eyeglass frames, gear wheels, ball bearings, fasteners, guns, ski bindings, knife handles and lock systems, consumer electronics and automotive industry
8	Polyvinylidene chloride (PVDC, Saran)	80 [20]	Impermeable plastic food wrap
9	Polyphenylene sulfide (PPS)	80.3 [20]	Material for automobile and electronic parts
10	Acrylonitrile butadiene styrene (ABS)	80.9 [20]	Molded products, protective head gear, pipe, automotive body parts, wheel covers and enclosures
11	Polycarbonate (PC)	82 [20]	Medical equipment

(Continued)

TABLE 3.1 (Continued)

S. No.	Material	Contact Angle(s) of Water on the Material Surface	Applications
12	Polyvinyl fluoride (PVF)	84.5 [20]	Coatings for lowering flammability of airplane interiors and photovoltaic module back-sheets. Also used in raincoats and metal sheeting
13	Polyvinyl chloride (PVC)	85.6 [20]	Construction and buildings, electronics, health care, automobile, piping, blood bags and tubing, cable insulation and wires
14	Polystyrene (PS)	87.4 [20]	Electronic appliances, toys, automobile parts, gardening pots
15	Polyvinylidene fluoride (PVDF)	89 [20]	Insulation electrical wires
16	Polybutadiene	96 [20]	Tires
17	Polyethylene (PE)	96 [20]	Grocery bags, packaging film, trash, agricultural mulch, wire and cable insulation, squeeze bottles, toys and housewares
18	Polypropylene (PP)	102.1 [20]	Crates, bottles, food packaging and pots
19	Polydimethylsiloxane (PDMS)	107.2 [20]	Used as a stamp resin in the procedure of soft lithography, used for flow delivery in microfluidics chips
21	Fluorinated ethylene propylene (FEP)	108.5 [20]	Medical and chemical equipment, chemical apparatus, wire coverings, extruded insulation and glazing film for solar panels and coatings and protective linings
22	Paraffin	108.9 [20]	Skin-softening salon and spa. It can also be used for pain relief to sore joints and muscles
23	Polytetrafluoroethylene (PTFE)	109.2 [20]	Non-stick coating in kitchen cookware (pans, baking trays, etc.)
24	Polyisobutylene (PIB, butyl rubber)	112.1 [20]	Thickening agent, bladders in basketballs, footballs, soccer balls and other inflatable balls to provide a tough, airtight inner compartment
25	LDPE	86 [20]	Films for both packaging and non-packaging applications, extrusion coatings, sheathing in cables and injection molding applications. It is used where high-clarity films
26	HDPE	87 [20]	Shampoo bottles, toys, milk jugs, cereal box liners, chemical containers, recycling bins, grocery bags, pipe systems
27	PP	88 [20]	Automotive industry, special devices like living hinges and textiles
28	Silicone	104 [21]	Microfluidics, seals, gaskets, shrouds and other applications requiring high biocompatibility. Additionally, the gel form is used in bandages and dressings, breast implants, testicle implants, pectoral implants, contact lenses and a variety of other medical uses

(Continued)

TABLE 3.1 (Continued)

S. No.	Material	Contact Angle(s) of Water on the Material Surface	Applications
29	Bare Zr foil	58.7 [22]	Used to make high-temperature superconductive materials, electrodes components, flanges bolts, foils, rods and tubes for special applications
30	Nanoporous Ta_2O_5	26 [22]	Optical glass, fiber and other instruments, used as sputtering target for production of semiconductor, high-frequency integrated circuits and flash memory
31	Bare tantalum	47.9 [22]	Biomaterial applications
32	Transparent Ta_2O_5 nanostructured	155 [22]	Self-cleaning and anti-fogging optical coatings
33	Copper	70 and 90.8 [23]	Electronic and communication industry, pipelines for domestic and industrial water utilities including seawater, heat conductors and heat exchangers
34	Aluminum	62.1 and 85.6 [23]	Various industrial fields as basic material, Aerospace industry, food and packaging industry, etc.
35	Stainless steel	63.8 and 89.7 [23]	Biomedical implants and surgical instruments, marine industry, etc.

3.3 CONVENTIONAL PROCESSES EMPLOYED

There are a number of techniques available for tuning the wettability of the surface and each one of them has their own merits and demerits. Figure 3.4 shows the few techniques which are most commonly exploited for manipulating the wettability. Scientific merits and challenges about these techniques are discussed as below.

(a) **Physical Vapor Deposition**

Physical vapor deposition (PVD) is an excellent vacuum coating process known for over 100 years. PVD is a thin-film deposition technique in which the coating develops atom by atom on the substrate [24]. This process involves the vaporization or atomization of a solid material which is usually termed as the target. The atomic deposition of the metal vapor produced is made on electrically conductive material and can be done in gaseous, vacuum, electrolytic or plasma environment. As the process of deposition grows atom by atom and layer by layer, the thickness of the coating can range from a few atomic layers to several microns. The process is carried out in a high-vacuum chamber (10^{-6} torr) using a cathodic arc source. In PVD process, the coating is deposited over the entire substrate simultaneously [24]. Some of the primary PVD methods include sputtering, ion implantation, ion plating and laser surface alloying [24].

(b) **Chemical Vapor Deposition**

Chemical vapor deposition (CVD) is another widely used materials processing technology. CVD is primarily used for thin-film coatings [25]. However, it is

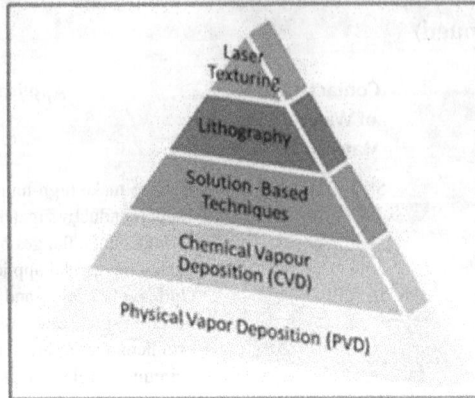

FIGURE 3.4 Most commonly available technique for manipulating the wettability.

also used to produce high-purity bulk powders and for fabricating composites as well. In its simplest form, CVD entails precursor gases flowing into a chamber where the substrate is placed at a predefined temperature. Chemical interaction takes place on and near the hot surface leading to thin-film deposition on the surface. However, such a process is accompanied by the production of reaction by-products and unreacted precursor gases. The temperature range for CVD is typically 200–1600°C. Some of the advanced CVD processes include the use of ions, plasma, lasers, combustion reactions or hot filaments [25]. This increases the deposition rate or lowers the deposition temperature.

(c) **Solution-Based Technique**

The disadvantages with the conventional coating techniques include the problems associated with control over composition, porosity, doping, thickness and most importantly, 3D-substrate coating. Complex composite systems and multi-layered systems are furthermore difficult to achieve by the route of CVD or atomic layer deposition [26].

Devices and technologies for the fabrication of metal-oxide coatings require improved deposition methods. More than this, there is a need to develop technologies for intricate design coatings. The application of solution-based techniques has become very important with the advent of curved, flexible and stretchable electronic devices. Solution-based techniques serve as the best coating method for large-scale thin-film deposition. In the simplest form of processing, a solution is prepared by mixing the precursor materials with a suitable solvent. The viscosity and compositions are also tailored. Catalysts are also used for altering the viscosity, wetting characteristics or for better dissolution of the precursor in the solvent [26].

(d) **Lithography**

Lithography is a process similar to printing of one- and two-dimensional structures in which at least one of the dimensions is in the range of nanometers. The introduction of lithography entails a mention of IC engine and MEMS in the first step as the success of modern IC design is essentially due to lithography. The solution-processed techniques may not be competitive in applications where precise architecture design is required. There are various types of lithography techniques. To name a few, photolithography, UV lithography, soft lithography, nanoimprint lithography, scanning probe and atomic probe lithography, etc. We have two different approach in this process – top down and bottom-up approach. Etching process is a top-down approach and growth of nanolayers is a bottom-up approach. The current top-down lithographic process is a more viable process for the fabrication of electronic materials and devices. The process of lithography requires a hard copy of the pattern, known as reticle or mask, to be generated first through exposure to light, ions or lasers. This design is transferred to a wafer in a 1:1 ratio or in a smaller ratio. The wafer surface is exposed after the pattern is formed. This wafer surface is etched. The material can also be deposited on the exposed surface.

(e) **Laser Texturing**
Laser texturing is a process in which a mark is created on the material by eliminating an undesired section where surface coated layer is not desired. The marked section appears bright or has a contrast with the coated surface. In recent years, this process has emerged as a sustainable process for biomedical applications and for enhancing the tribological properties in terms of reduction of friction of coefficient and damping of wear. This surface coating technique offers many advantages; in particular, most important is the surface roughness and chemistry modification in just one single step and avoiding the use of any toxic materials. Laser surface texturing is widely used for modifying polymeric and biomaterials in particular [27]. Since the process involves a non-contact technique, the contamination of the work piece is easily avoided. This is an important advantage specifically for biomaterials where clean environment is a primary requirement. Another advantage offered by this process is that the processing speed is high and the technique involves easy automation [27].

The advantages and drawbacks of different coating techniques are summarized in Table 3.2.

3.4   GAP IN THE LITERATURE

The concept of achieving superhydrophobicity by introducing either a low surface energy and/or by manipulating the surface roughness in nano- or microscale was first pointed out by Barthlott and Neinhuis [28] in 1997. Since then, there have been tremendous efforts to tailor the surface energies and surface roughness owing to the fascinating wetting properties. While superhydrophobic surfaces find their potential application as self-cleaning, anti-icing, water repellent, corrosion-resistant antibacterial surface, etc., superhydrophilic surface has a huge demand in

TABLE 3.2
Common Advantages and Drawbacks for Different Coating Techniques [28–33]

S. No.	Process	Advantages	Drawbacks
1.	PVD	Improved surface strength and durability obtained, diverse group of substrates can be coated and vast group of coating materials can be used.	Higher cost, high vacuum required, line of sight technique
2.	CVD	Relatively simple apparatus used, excellent uniform coating, high deposition rate, industrially scalable, multi-directional type of deposition	Vacuum required, safety hazards, difficult to deposit multi-component materials with well-controlled stoichiometry, higher cost involved
3.	Solution-based technique	Intricate design coatings, large-scale thin-film deposition, lower cost than other similar techniques, no extreme heating or vacuum required	Uniform thickness is a challenge, coating failures in multi-layered coating structures, not cost effective for higher productivity, not very precise architecture design
4.	Lithography	Cost effective for high production rates, high quality coating	Tedious coating method
5.	Laser texturing	Sustainable process for biomedical applications with improved tribological properties, surface roughness and chemistry modification in single step, non-contact technique, high productivity, easy automation	Low efficiency

applications like antifogging, heat exchanger, in biological systems like cell proliferation, cell activity, etc. Currently several methods such as PVD, CVD, inkjet printing, sol-gel, lithography, etching and electrodeposition have already been established for the fabrication of superhydrophobic and superhydrophilic surfaces. While most of these processes render excellent wetting characteristics as desired, the synthesis of such a surface by these processes is quite complex and challenging as they are highly sophisticated involving multi-step processing limiting their production only to laboratory scale. Furthermore, these approaches involve polymeric surface modifiers.

It is well known that the polymers have limited mechanical and thermal stability which could damage of the microscopic roughness or the hydrophilic/hydrophobic surfaces upon any mechanical or thermal exposure. Consequently, the wetting characteristics of the surfaces deteriorate. Hence, only a few commodities synthesized by these approaches are available for practical use. These processes are therefore not well instituted and suffer from the demerits of low-scalability, high cost and poor robustness of the products. Furthermore, it is a challenging task to fabricate such hydrophobic/hydrophilic surfaces without the use of any modifiers or application of UV light, which further limits the application.

Hence, it is very important to design and fabricate efficient superhydrophilic/ superhydrophobic surfaces which can be achieved by simple, one-step and scalable technique. Additionally, significant interest has also been drawn toward fabricating these surfaces to maintain their integrity even in harsh environments.

Though a lot of advancements have already taken place in manipulating a robust surface that can render the desired wetting characteristics, the pursuit of developing a one-step and scalable technique is still going on with an emphasis on producing mechanically and thermally robust surface.

3.5 PLASMA SPRAYING TECHNIQUE

Plasma is the fourth state of matter in which an ionized gas becomes highly electrically conductive. In this technique, an arc is struck between a tungsten cathode and a copper anode to generate the plasma. The temperature at the core of the plasma reaches more than 10,000 K [30]. A schematic representation of the plasma spraying technique is presented in Figure 3.5. During the plasma spraying, powder is infused into the high-temperature jet. The powder is melted in this high-temperature jet and is accelerated toward a substrate. Molten/semi-molten particles exit from the plasma jet and impact the substrate, gets flattened and rapid solidification takes place at a very high cooling rate of approximately 10^6 K/s [30]. Successive deposition of the particles leads to increasing thickness layer by layer and subsequently the coating is developed.

Plasma spraying technique is extensively used to fabricate metallic, intermetallic, ceramic and composite coating to protect the substrate from wear, corrosion and high-temperature environment. However, this technique involves a large number of variables such as equipment, environment, process and powder parameters which impact the coating properties. Optimization of these properties leads to development a vast range of coating properties.

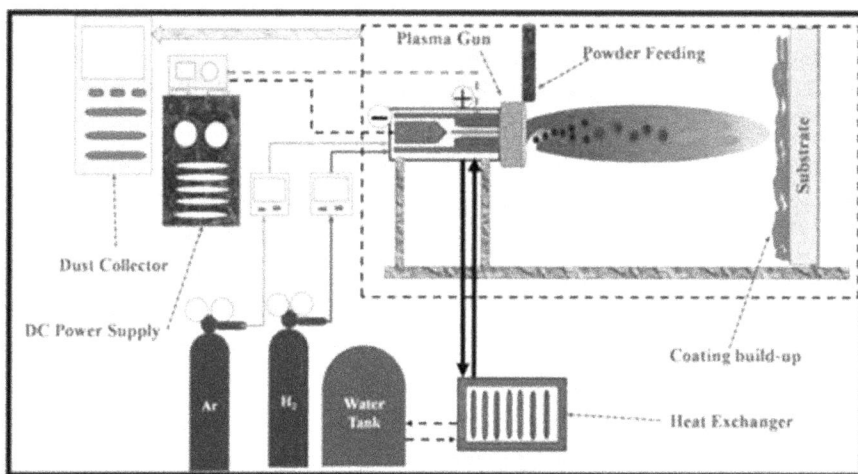

FIGURE 3.5   Schematic representation of the complete plasma spray setup for the deposition of coating over substrate.

Many studies have been conducted using empirical approach to study the influence of plasma process parameters on the coating properties [31, 33]. The empirical approach has involved the changing of one process parameter at a time and observing its effect on the final coating property. However, such an approach requires a large set of data generated through a large number of experimental trials followed by systematic identification of the significant parameters affecting the coating characteristics.

3.6 FABRICATION OF ALUMINUM COATING BY PLASMA SPRAYING

Lightweight aluminum powder (density: 2.6 g/cm^3) of 99.99% purity was procured from Trixotech Pvt. Ltd., India, for synthesizing the coating powder. The Al powder was spray-dried to improve its flowability for efficient plasma spraying. The size of the precursor powder for spray drying was 0.7 ± 0.2 µm. Figure 3.6a illustrates low-magnification FE-SEM image of the spray-dried Al powder. Corresponding high-magnification image of a single powder is illustrated in Figure 3.6b. It can clearly be

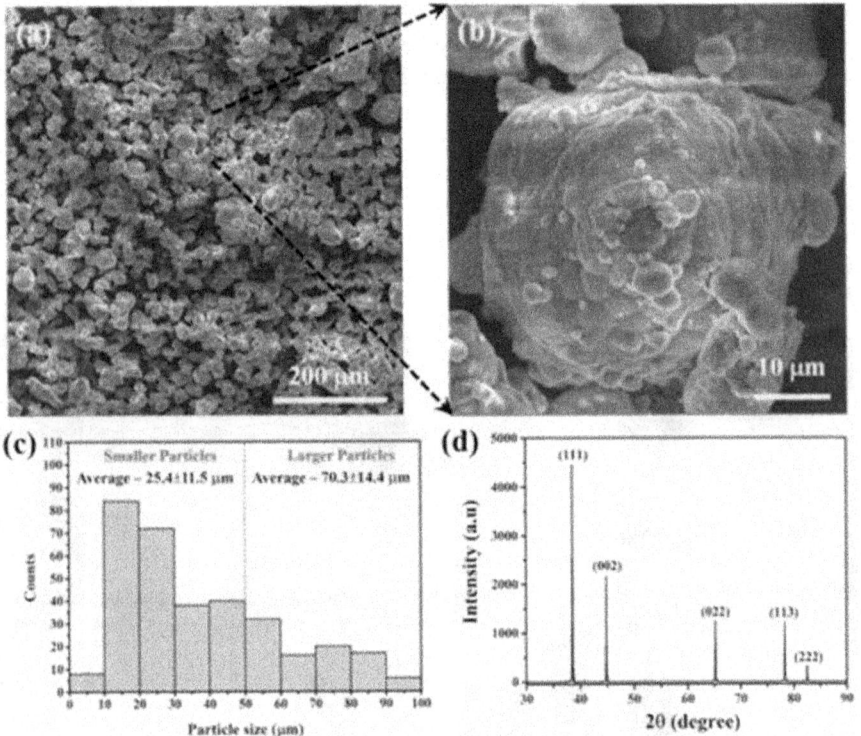

FIGURE 3.6   (a) The low-magnification FE-SEM images of the spray-dried Al particle and (b) its corresponding highly magnified images of the single spray-dried Al particle. (c) Particle size distribution graph of the spray-dried Al Powder and (d) XRD spectra of the spray-dried Al Powder [32].

observed that the particles are spherical in shape. The particle size distribution of this powder is also presented in the form of the histogram illustrated in Figure 3.6c. The histogram reveals that the particles have a wide range of size variation (10–100 μm). The cumulative powder particle size gives a value for d_{50} to be 26 μm and d_{50} is 72 μm. This indicates that the powder is of dual size. Figure 3.6d shows the X-ray diffraction pattern of the spray-dried powder revealing sharp crystalline peaks of FCC Al (JCPDS Card No. 98-004-3423). The absence of any other peak other than that for FCC Al suggests that powder was highly pure and free of contamination.

Tuning of the metallic surface was done by atmospheric plasma spray (Oerlikon Metco, Switzerland) which was equipped with a 9 MB gun. The setup design was further modified by attaching a conventional inert atmosphere shroud in front of the plasma gun as shown in Figure 3.7.

This shroud could provide a controlled atmosphere to the plasma plume. A coaxial gas shroud (dia.: 40 mm) was used to supply inert gas, nitrogen (N_2) around the plasma plume at a constant rate of 40 psi. The function of the shroud was to affect the rate and degree of melting of the particles in the plasma and to prevent the plasma plume from atmospheric oxygen. The surface roughness for coating was tuned by tailoring two key process parameters which were dependent on the plasma gas and application of shroud during the process of spraying. Different wetting characteristics were achieved by applying different combinations of gas and shroud, which are as follows: (i) superhydrophilic surface: N_2 as primary gas and without shroud, (ii) hydrophilic surface: Ar as primary gas and without shroud, (iii) hydrophobic surface: Ar as primary gas and with shroud and (iv) superhydrophobic surface: N_2 as primary gas and with shroud.

For easy reference and convenience, the process conditions can be referred to as P1, P2, P3 and P4 and the fabricated surfaces under these conditions can be designated as W1, W2, W3 and W4, respectively. Token samples were prepared by plasma

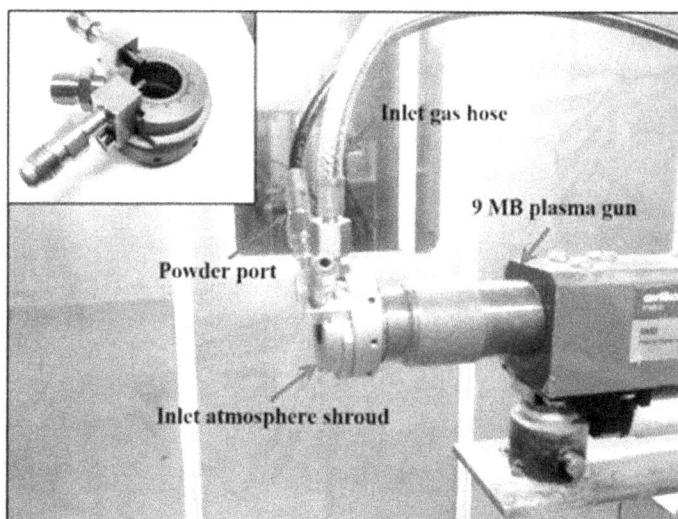

FIGURE 3.7 Illustration of the 9 MB plasma gun integrated with an inert atmosphere shroud. Inset shows the top view of the shroud [32].

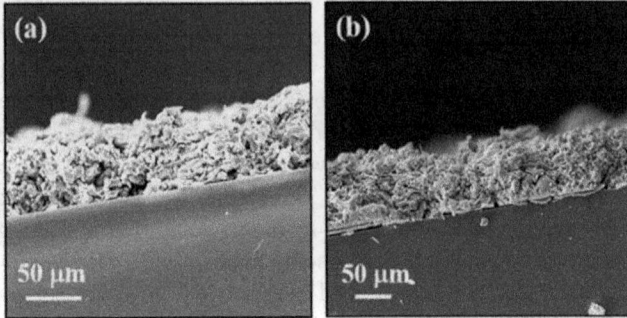

FIGURE 3.8 SEM images of the cross-sectional surfaces of (a) W1 and (b) W4 plasma-sprayed coatings [32].

spray coating of the fabricated powder on a glass substrate of dimension 60 × 20 × 2 mm. Prior to deposition, the glass substrates were cleaned carefully in a water bath ultrasonicator using acetone. The coatings on the glass substrate were found to be well adhered and the range of the thickness was measured to be 50–70 μm as shown in Figure 3.8 (a, b). These samples rendered the desired wetting properties in terms of wetting and de-wetting.

3.7 SUPERHYDROPHILICITY TO SUPERHYDROPHOBICITY

As mentioned before, the static contact angles give a measure of the wetting properties. A surface having static CA less than 90° is hydrophilic and static CA less than 10° is superhydrophilic. By contrast, if the static CA is higher than 90°, the surface is hydrophobic and CA being higher than 150° is referred to as superhydrophobic.

Figure 3.9 shows the static CA for all the glass substrates prepared under different processing conditions as mentioned in Section 3.6 and their respective CA hysteresis

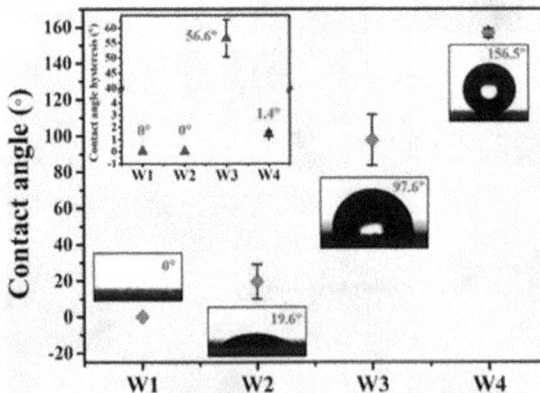

FIGURE 3.9 Static contact angle of the four different surfaces, i.e., W1, W2, W3 and W4, while inset shows their respective contact angle hysteresis [32].

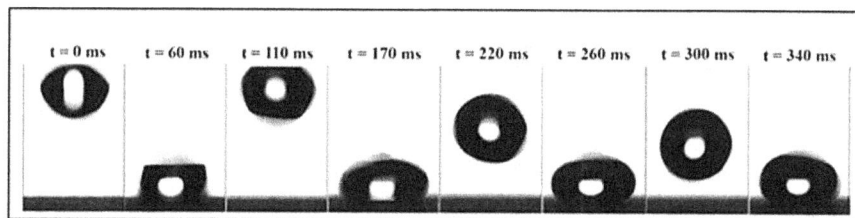

FIGURE 3.10 Series of snapshots depicting a water droplet (5μL) bouncing on the W4 surface with function of time [32].

is illustrated in the inset. The fabricated surfaces of W1, W2, W3 and W4 show a mean CA of 0°, 19.6°, 97.6° and 156.4°, respectively. This mean CA value of the water droplets was obtained by measuring contact angles at different positions of the coated surface. CA of 0° obtained on the coated surface of W1 prepared using the process conditions P1 confirms the superhydrophilic wetting characteristics. A slight increment in the CA value (19.6°) obtained for W2 surface using process parameter P2 exhibited hydrophilic wetting characteristic. In contrast, W3 surface that was tuned under P3 process parameter showed significantly higher CA of 97.6° that indicates hydrophobic nature of the W3 surface. Additionally, this surface also displayed higher CA hysteresis of 56.6°, indicating hydrophobic characteristics of the surface. Finally, when the process parameter P4 was used for the surface W4, the CA reached a significantly high value of 156.4°. The surface also showed very low CA hysteresis 1.4° and sliding angle ~5° which is representative of excellent superhydrophobicity. Thus, instant tuning of wetting characteristics in all the four surfaces by varying the process conditions of primary gas used around the plasma plume and the usage of shroud resulted in different wetting characteristics.

Water bouncing experiments in which the water droplets were dropped from a height of 8 mm over the tilted surface at 3° reconfirmed the wetting characteristics. On the one hand, the water droplet falling on the surface W3 adhered immediately on to the surface with a CA greater than 90° representing hydrophobic nature, a typical Wenzel state where the droplet impregnates the solid surface. On the other hand, the water droplet bounced over the W4 surface exhibiting excellent water repellence and the step-by-step bouncing as a function of time is demonstrated in Figure 3.10. The water droplet bounces off elastically and no residual droplets could be traced on the surface and ultimately the water droplet rests, maintaining a CA of more than 150°.

3.8 SELF-CLEANING COATING

Self-cleaning coatings are classified into two categories: superhydrophilic and superhydrophobic. Superhydrophobic surfaces use the concept of rolling droplets of water to clean the surface and superhydrophilic surfaces clean themselves by forming sheeting water that will remove the dirt from the surface. TiO_2 is one of the most widely used coatings for self-cleaning applications. Some novel self-cleaning glass have also been produced by coating SiC nanowires.

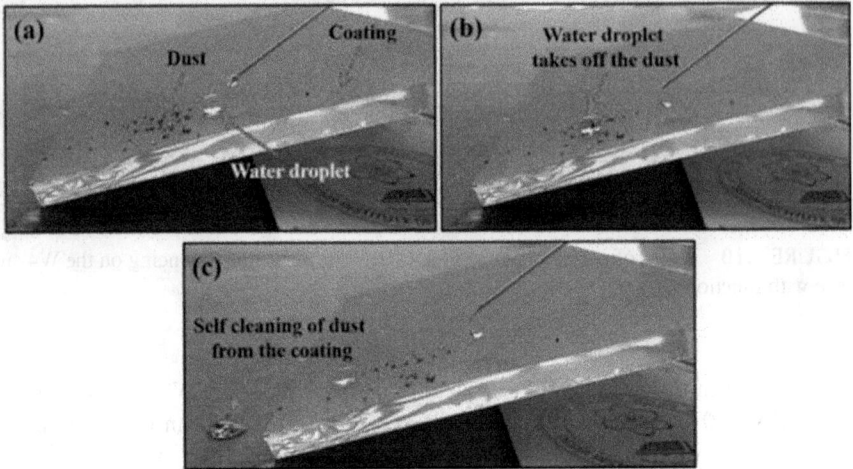

FIGURE 3.11   Sequential photographs of the droplets showing the self-cleaning of dust from the superhydrophobic surface (W4, 32).

The investigation discussed in the previous section has demonstrated and established that different wetting characteristics and self-cleaning property of the surfaces can be attained by tailoring the plasma processing parameters. The ability of this superhydrophobic surface to serve as a potent self-cleaning surface is remarkable. Figure 3.11 (a–c) demonstrates a sequential image of the droplets exhibiting the cleaning of nearly all the dust particles coming in the path of the rolling droplet. Subsequent water droplets rolling from behind clean the surface thoroughly, without leaving any trace of dust or water droplet in its path. Such kind of a surface with superhydrophobicity exhibits excellent self-cleaning behavior and such coatings can be used for a wide range of applications such as self-cleaning coatings in textile industries, wind turbine blades, condenser tubes, cements, steam turbines of hydro and thermal power plants, etc.

3.9   MECHANISM BEHIND DIFFERENT STATES OF WETTABILITY

It can be recollected that the wetting characteristics of any surface is governed by its surface roughness and/or its surface energy, which in turn is decided by its chemistry. Metals have been intrinsically hydrophilic in nature interminably. However, subsequent studies on the wetting characteristics of metallic surfaces have proven that they can rapidly absorb hydrocarbon from its atmosphere resulting in an increase in CA as a function of time. Li et al. have demonstrated this behavior of metals in nickel–chromium (Ni-20Cr) plasma-sprayed coating [34]. The coating exhibited superhydrophilic nature immediately after the fabrication of the coated surface. However, this coating when exposed to an ambient air condition for 60 days, changed its wetting characteristics from superhydrophilic to superhydrophobic. They suggested that the surface adsorbs the atmospheric hydrocarbons and eventually the surface transforms from superhydrophilic to superhydrophobic in nature. Interestingly, in our case, the manipulated surfaces have shown different wetting characteristics immediately after their fabrication. However, an XPS analysis performed to investigate any contribution of surface contamination or hydrocarbons could throw some light on the wetting behavior of the tuned surfaces. Figure 3.12 shows the representative XPS spectra of all the tuned surfaces. Photoelectron peaks of Al, C and O are detected on these surfaces. The XPS spectrum also shows C peak on all the surfaces. This peak of C is often referred to as "adventitious carbon" peak and is originating due to adsorption of hydrocarbons on the surface. This adventitious carbon is considered to be the probable source of the hydrophobicity of the hydrophilic surface when exposed to the atmosphere for an extended duration of time.

Preston et al. have investigated the wetting characteristics of gold. They have shown that gold which is intrinsically hydrophilic in nature contained ~25 atom% carbon after and exposure of 1 hr to laboratory atmosphere [34]. Although, the accumulation of hydrocarbons on the surface of the gold did not alter the CA and hence the wetting characteristics of gold. This suggested that the surface must be exposed for prolonged duration to exhibit hydrophobic nature.

However, in the present case, the manipulated surfaces demonstrated instant tuning of the wetting characteristics after the fabrication of the coatings. Additionally, these surfaces also maintained their wetting characteristics (no change in CA) at an elevated temperature (773 K) for 30 minutes. It should also be mentioned here that when a hydrocarbon-contaminated surface is heated in an oxidizing surface the adsorbed hydrocarbons gets depleted without affecting the surface roughness. This suggests that wetting characteristics should again get reverted to hydrophilicity. Therefore, it can

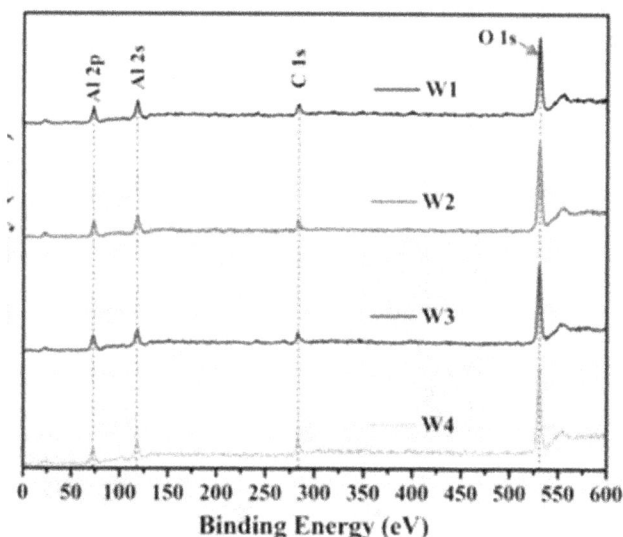

FIGURE 3.12 XPS spectra of all the tuned surfaces showing the photoelectron peaks of Al 2p, Al 2s, C 1s and O 1s [32].

TABLE 3.3

Percentage of Al-N, Al-O and Al in All Four Coatings

Samples	Al (%)	Al-O (%)	Al-N (%)
W1	79.72	9.48	10.78
W2	83.23	16.76	--
W3	84.33	2.13	13.52
W4	73.48		26.51

easily be understood that the wetting behavior of the tuned surface in our case is not governed by the presence of adsorbed hydrocarbons on the surface. Additionally, the Al 2p spectra were also deconvoluted and the percent content of Al, Al-O and Al-N were quantified (Table 3.3) to understand the effect of the inert environment shroud on the surface chemistry. The theory behind this has been discussed later in this section.

Meanwhile, the optical profiles and the respective FE-SEM images for the tuned W1 surface are also presented in Figure 3.13a, b, respectively. Two distinct features, (i) smooth region and (ii) randomly oriented micropillars, can clearly be differentiated in the optical profile of W1 (Figure 3.13a) which was fabricated under P1 processing condition. Such features are responsible for providing the desired multi-scale roughness on the surface. The root-mean-square (rms) roughness of this surface was measured to be in the range of 0.915 ± 0.40 µm, which was attributed to the presence of both the features. Similar topographical features were also observed in the FE-SEM images as seen in Figure 3.13b. The presence of smooth regions can be explained on the basis of presence of melted particles in the plasma while the micropillars (17%) have resulted from the un-melted particles. As it was mentioned earlier that powder particles are of dual size, melting of the powders in the size range of 10–50 µm is relatively easier and it results in the smoother region. And the larger

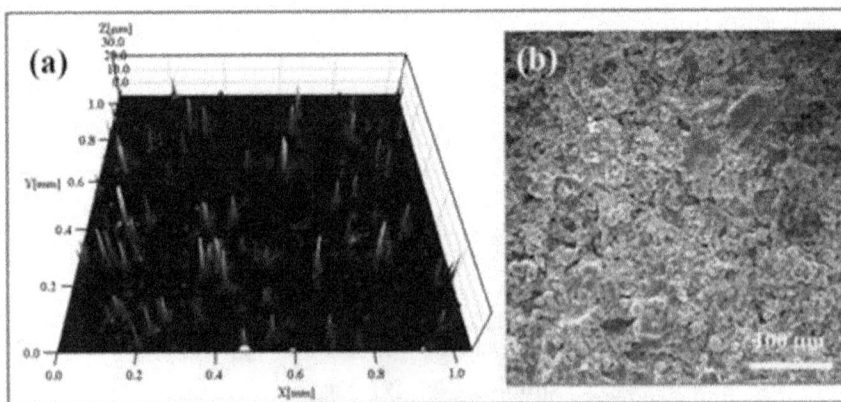

FIGURE 3.13 The 3D optical profiler images of W1 surface and its corresponding FE-SEM [32].

particles (50–90 μm) undergo partial melting due to the temperature gradient within the particle resulting in the micropillars. As plasma spray is a layer-by-layer deposition technique, the un-melted particles keep on depositing one over the other and results in the formation of micropillars or hillocks. Also, the population density of these micropillars is directly dependent on the fraction of larger particles or, in other way, the increased amount of the unmelted particles.

The extremely low CA of the W1 surface can be explained based on Wenzel model. This model predicts that the wetting characteristics (hydrophilicity and hydrophobicity) of the surface are a function of surface roughness according to the following empirical relation 3.3.

$$\cos\theta_{rough} = r\cos\theta_{flat} \tag{3.3}$$

where θ_{rough} is the apparent CA on a rough surface, θ_{flat} is the ideal CA on a flat homogeneous surface and r is the roughness factor. r is defined by the ratio of the actual surface area over projected area. It is evident that for an intrinsically hydrophilic material, insertion of surface roughness will lead to the decrease in apparent CA of the surface. To verify this theory, the CA of grinded and polished Al surface was measured and the value obtained was 51.6° which is θ_{flat} in this case. The details are shown in Figure 3.14. Apart from this, the value obtained from deconvoluted Al 2p XPS revealed the existence of high surface energy compounds Al_2O_3 ($\gamma \approx 970$ mJ/m^2) and AlN ($\gamma \approx 1000$mJ/m^2) as shown in Figure 3.15. This probably would be another reason for obtaining a very low CA (close to 0°).

Figure 3.16 a shows the optical profiling image of the W2 surface processed using process parameters P2. In this process condition, Ar was used as the primary gas instead of N_2 and the other parameters remained same as for the P1 condition. This W2 surface also showed two distinct features: a smooth region and some randomly oriented micropillars. The rms value for this surface was calculated to be 0.826 ± 0.04 μm. Similar characteristics of topographical features were also observed in the corresponding FE-SEM image as shown in Figure 3.13b. However, there is a slight decrease in the roughness which can be attributed to the use of Ar as primary gas. Ar is a monoatomic gas and requires less energy to ionize and form the plasma.

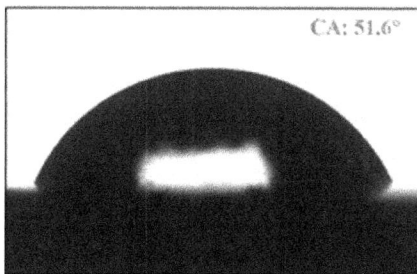

FIGURE 3.14 Water contact angle on the smooth surface of Al pellet [32].

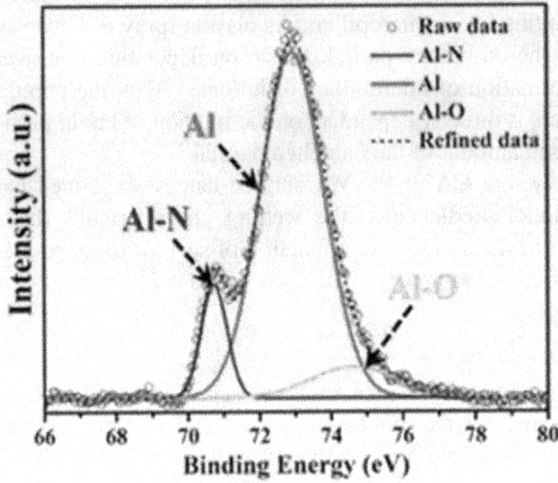

FIGURE 3.15 Illustration of the deconvoluted Al 2p peak of the W1 surface showing the Al-N, Al and Al-O peaks [32].

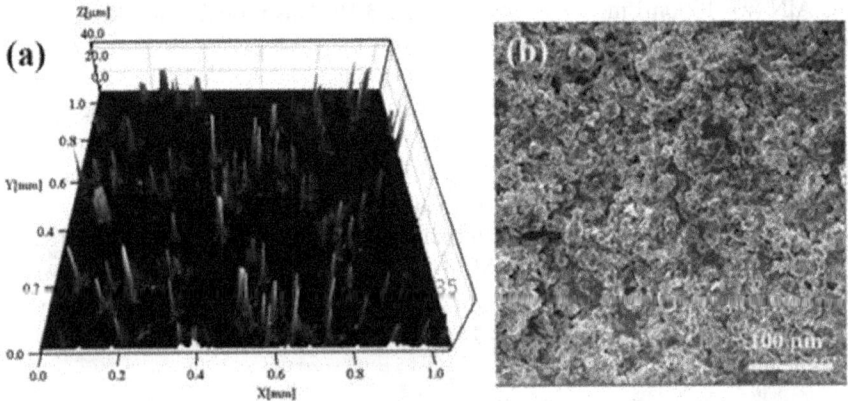

FIGURE 3.16 3D optical profiler images of W2 surface and its corresponding FE-SEM [32].

Hence, in this case, a large amount of thermal energy is available with the plasma as compared to that in case of N2, as the primary gas. Consequently, it melts the powder particle, which further assists the formation of smooth surface with increased density of flat regions and lower density of micropillars (12%) and, hence, lesser surface roughness. However, the superhydrophilic W1 surface, the CA of W2 (~19.6°) falls in the hydrophilic regime which can be attributed to the decrease in the surface roughness. Also, from Figure 3.17, the presence of Al_2O_3 is confirmed and there is a complete absence of high-energy compound, AlN. This could be an additional reason for the hydrophilic nature of W2 surface. Now, the third case of W3

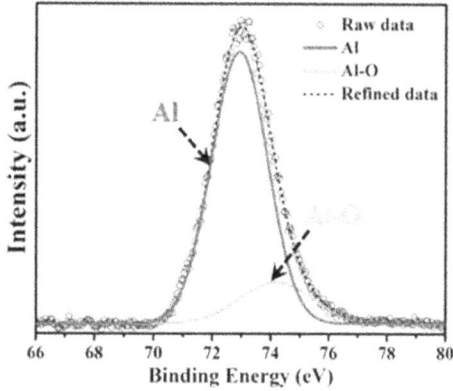

FIGURE 3.17 Illustration of the deconvoluted Al 2p peak of the W2 surface showing the Al and Al-O peaks [32].

surface where Ar was used as the primary gas along with an inert atmosphere shroud, resulted in exponential increment in the density of micropillars.

Figure 3.18 demonstrates the OP image of W3 surface indicating higher density of micropillars (40%) compared to the W1 and W2 surfaces. As a result, the surface roughness of W3 was relatively higher and the rms value obtained was 1.03 ± 0.01 μm. The increase in the number of micropillars can be attributed to the use of (i) the inert atmosphere shroud and (ii) the dual-size powder particles. The inert atmosphere shroud significantly lowers the plasma plume temperature. Additionally, the design of the shroud was such that the powder particles are fed perpendicular to the plasma plume and approximately 25 mm away from the nozzle exit, where the temperature is around 2000–3000 K less than that at the exit pint of the nozzle.

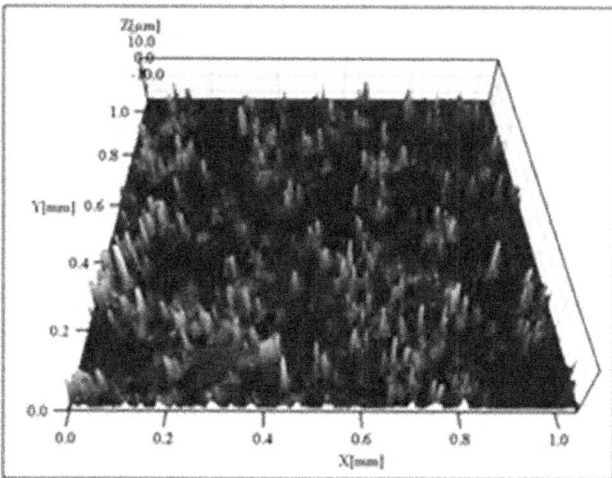

FIGURE 3.18 3D optical profiler images of W3 surface [32].

This drop in temperature by the usage of shroud has been explained schematically in Figure 3.19.

Even for this case, it is anticipated that the powder particles with larger size will not melt uniformly resulting in a dense population of micropillars. These micropillars induce a hierarchical roughness, as shown in Figure 3.20.

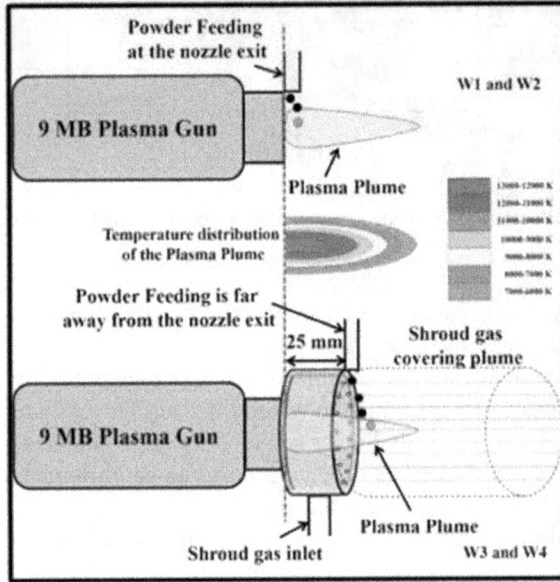

FIGURE 3.19 Temperature distribution schematic of plasma plume both with and without shroud plasma-sprayed coatings [32].

FIGURE 3.20 FE-SEM images of W3 surface [32].

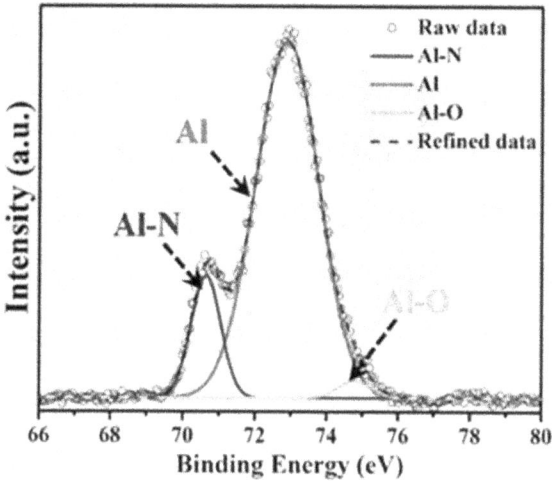

FIGURE 3.21 Deconvoluted Al 2p peak of the W3 surface showing the Al-N, Al and Al-O peaks [32].

Furthermore, the deconvoluted al 2p XPS data on W3 surface revealed the presence of AlN which possesses high surface energy and the lower intensity peak of Al_2O_3 as shown in Figure 3.21. Despite the presence of high surface energy compounds (AlN) on the W3 surface, it presented a hydrophobic (WCA > 90°) wetting characteristic which was attributed to the presence of hierarchical micropillars, where the water droplets were expected to get pinned. It is also assumed that the liquid fills the voids between the protruded micropillars and demonstrates a high CAH of 56.6°.

Figure 3.22a shows the OP image of the W4 surface, which was fabricated under the P4 processing parameters, where the primary gas used was N_2 along with the use of shroud under inert atmosphere. The remaining parameters were kept identical. In this case, a very high population density of micropillars (82%) was obtained which was significantly higher than all the three cases. Also, a higher rms roughness (1.33 ± 0.24 μm) was obtained. Formation of this highly dense population of micropillars could be attributed to the following parallel phenomena: (i) use of N2 as primary gas, (ii) inert atmosphere of the shroud and (iii) dual size of the powder particles. It can also be observed from the FE-SEM images (Figure 3.22b) that the surface has a large density of micropillars with no smooth region. These micropillars are hierarchical and resemble the microstructural pattern of the naturally existing superhydrophobic material, lotus leaf. Further, the investigation of Al 2p peak revealed the presence of high surface energy compound, AlN (Figure 9.11c). Though the presence of high surface energy material on the W4 surface is expected to provide hydrophilicity, the W4 displayed a very high CA (i.e., 156.5°) in the superhydrophobic regime (>150°).

The superhydrophobicity of this surface was understood through the topographical features. High-magnification FE-SEM images of these micropillars (Figure 3.23 (a–c)) revealed an interesting set of microstructures which

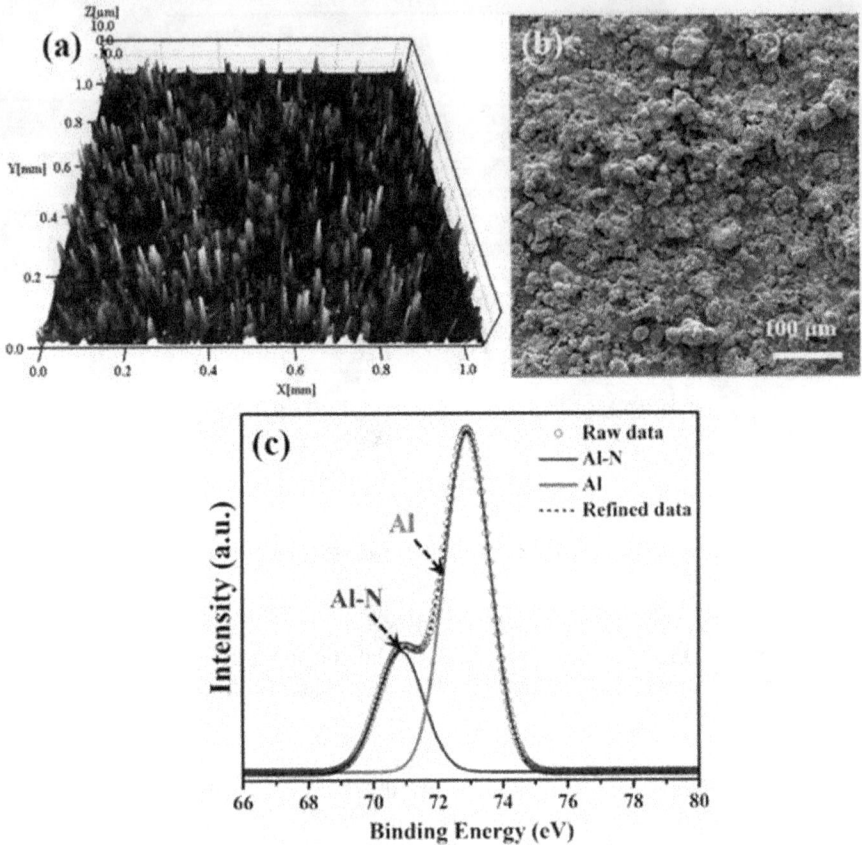

FIGURE 3.22 (a–b) The 3D optical profiler images of W4 surface and its corresponding FE-SEM and (c) deconvoluted Al 2p peak of the W4 surface showing the Al-N and Al peaks [32].

resembled features similar to mushroom, cauliflower and cornets. Tiny irregular particles (nanograins) in the range of nanometers were identified on each micro-pillar as shown in Figure 3.23. These nanograins over the micropillars are respon-sible for a hierarchical microstructure which is known to be favoring a superhydrophobic surface. It should be noted that these hierarchical structures were absent in case of the W1 surface as manifested by Figure 3.24. For the W4 surface, fabricated under P4 processing condition, the powder is fed far from the nozzle exit which results in a drop in the temperature. Moreover, the shroud N_2 gas also lowered the flame temperature which led to the increase in the percent-age of unmelted particles in the plasma. The formation of special nanograins could be attributed to the splashing of the fully melted particles over the island of unmelted microparticles. This distinct microstructure was fabricated by some other researcher and was named as "re-entrant geometry" which renders

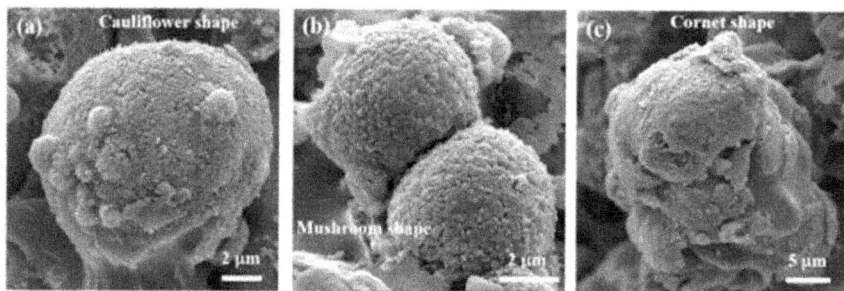

FIGURE 3.23 High-magnification FE-SEM image showing the presence of Re-entrant geometry structures with hierarchical micropillars [32].

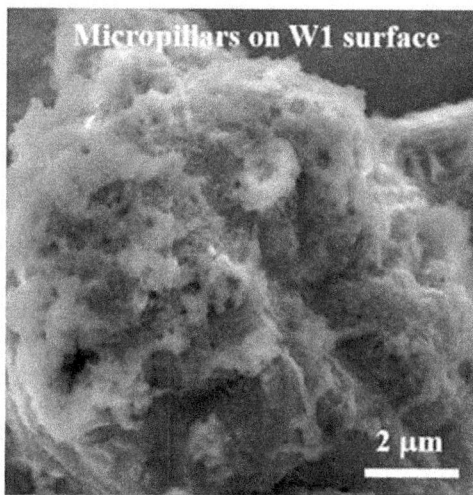

FIGURE 3.24 High-magnification FE-SEM images of the W1 surface shows the presence of melted micropillars [32].

superhydrophobic nature to a hydrophilic surface. The mechanism behind this behavior is explained by the hanging water droplet between the nano-protrudes.

It worth pointing out that these re-entrant structures were produced by sophisticated multi-step techniques. However, in the present case, this exciting microstructure was obtained just by altering a few process parameters that prove to be an industry-benign technique. The specific role of these re-entrants with hierarchical structures in rendering superhydrophobicity to an intrinsically superhydrophilic surface can be elucidated by the schematic (Figure 3.25).There are several re-entrant structures (cauliflower, mushroom, cornet, etc.) on which the water droplet is hanging. These re-entrant structures primarily act as solid overhangs thereby assisting the water droplet to remain suspended without getting pierced by the overhangs. Secondly, the nanograins also act as air-trapping pockets between the solid

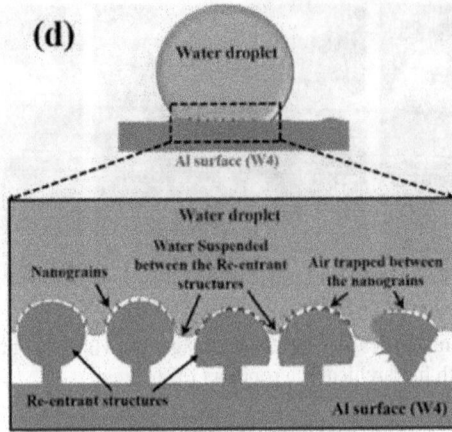

FIGURE 3.25 Schematic illustration of the drop of water in contact with these special kinds of micropillars [32].

overhangs and the droplets. This replicates a scenario of water droplets sitting on a composite surface of air and the coating material and hence, the Cassie–Baxter model becomes valid for these structures.

$$\cos\theta_{rough} = \varnothing_s \cos\theta_{flat} - \left(1 - \varnothing_s\right)$$

where $\varnothing_s$ is the area fraction of the solid surface in contact with liquid, θ_{rough} is the apparent CA on a rough surface and θ_{flat} is the ideal CA of water on smooth/flat homogeneous surface. These hierarchical structures, known as re-entrants in this context, have been beneficial in rendering superhydrophobic surface with excellent water-repelling characteristics.

This work has presented instant tuning of coated metallic surface to render different wetting characteristics by direct, simple, one-step and industrially scalable plasma spray technique. The main mechanism behind the wetting characteristics can be attributed to the re-entrant surface curvature and topography in conjunction with the surface roughness and its chemical composition.

3.10 THERMAL AND MECHANICAL ROBUSTNESS

To establish the robustness of the manipulated superhydrophobic W4 surface, they were exposed to three different temperatures in the range between room temperature (RT) to 773 K. The results are displayed in Figure 3.26 (where the superhydrophobic surface invariantly displays approximately constant CA even after a long exposure to elevated temperature (773 K). The inset image displays the digital images, illustrating the water droplet on the surface immediately after the surfaces were exposed to different temperatures. This ascertains that the coating can be used for all practical purposes and as it can retain integral wetting characteristics even in very harsh environments, which otherwise cannot be expected from polymeric or ceramic coatings.

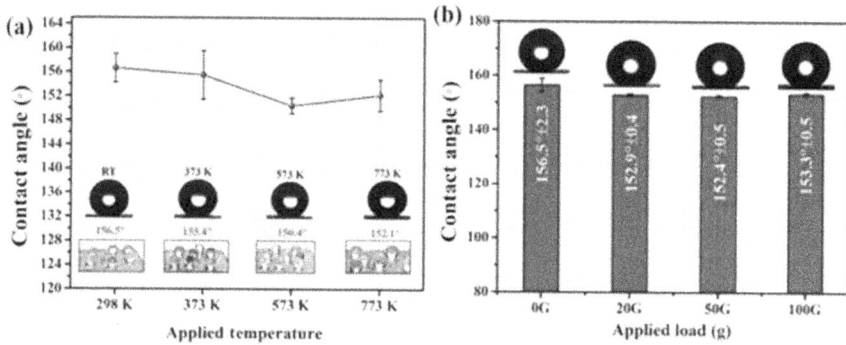

FIGURE 3.26 Static contact angle after (a) heat treatment of W4 surface at varying temperature, i.e., RT, 373, 573 and 773K and (b) abrasion testing of W4 surface by varying the load (0–100 g) using SiC P800 grit paper [32].

Thus, the process can also claims industrial acceptance and can be looked upon for scale-up.

Mechanical durability was also investigated and analyzed by performing abrasion test. The water droplets on the surface exhibited nearly constant CA after mechanical abrasion, and spherical water droplets could still be observed on the abraded surface as presented in Figure 3.26. This suggests that this plasma-coated surface has the potential to retain its superhydrophobicity or, in other words, the water repellency even after an exposure to harsh mechanical and thermal environments.

3.11   SCOPE FOR FUTURE WORK

These wettability-tuned metallic surfaces have immense applications in several industries. In particular, the superhydrophilic and superhydrophobic surfaces have potential in boiling and condensation fields, respectively. The high water affinity of superhydrophilic surfaces enhances capillary water transport, prevents fogging and facilitates boiling. Hence, this surface could be used to check the performance of anyone of the above-mentioned applications in future. Condensation is a common phenomenon in our daily life and is widely applicable in electronics cooling and industrial systems. Mostly the superhydrophobic surface contains the Cassie droplets that lead to drop-wise condensation and offers a superior heat transfer rate than film-wise condensation. In future, drop-wise condensation experiment should be done on this surface to check the heat transfer performance. Further, the dynamic contact angle of the superhydrophilic and superhydrophobic surfaces should show stability for a longer time.

3.12   CONCLUSIONS

Surface coatings rendering tunable wetting characteristics could be fabricated using plasma spraying technique. The process is industrially benign as it offers a simple, one-step and scalable technique to obtain a mechanically robust and thermally stable coated surface. Different wetting characteristics could be attained by the varying the process parameters. Use of Ar as the primary gas offered a superhydrophilic surface where the contact angle was measured to be around 0°, which switched to 19.6° when the primary gas used was N_2 instead of Ar. Further, on application of inert N_2 shroud and Ar as primary gas, the surface became hydrophobic and the measured CA was 97.6°. Again, the use of N_2 as primary gas and the application of inert nitrogen shroud rendered a superhydrophobic surface with a CA of 156.5°. Different wetting characteristics were attributed to the presence of smooth/flat regions and micropillars. Such surfaces resulted in hierarchical structures which offered different wetting characteristics. Special re-entrant geometries like mushroom, cauliflower and cornet shape added to the excellent superhydrophobicity obtained in such surfaces. The superhydrophobic surface showed excellent self-cleaning properties. These surfaces were also tested for their mechanical and thermal robustness and they showed no change in their CA upon exposure to mechanical abrasion, elevated temperatures or harsh environment. Further, it is envisioned that this unique and simple technique will find potential application in preparing metallic surfaces that can offer exciting wetting characteristics, which can have diverse applications in harsh environments as well.

REFERENCES

1. Kesong Liuand Lei Jiang,Metallic surfaces with special wettability. *Nanoscale*, 2011, 3, 825–838.
2. Zhangxin Wang, Menachem Elimelech and Shihong Lin, Environmental applications of interfacial materials with special wettability. *Environ. Sci. Technol*, 2016, 50, 2132–2150.
3. Shuhui Li, Jianying Huang, Zhong Chen, Guoqiang Chen and Yuekun Lai, A review on special wettability textiles: Theoretical models, fabrication technologies and multifunctional applications. *J. Mater. Chem. A*, 2017, 5, 31–55.
4. C. G. Jothi Prakash and R. Prasanth, Approaches to design a surface with tunablewettability: A review on surface properties. *J. Mater. Sci.*, 2021,56,108–113.
5. Yong-Ming Liu, Zi-Qing Wu, Sheng Bao, Wei-Hong Guo, Da-Wei Li, Jin He, Xiang-Bin Zeng, Lin-Jun Huang, Qin-Qin Lu, Yun-Zhu Guo, Rui-Qing Chen, Ya-Jing Yeet al.,The possibility of changing the wettability of material surface by adjusting gravity AAAS. *Research*, 2020, 2020, 11, ArticleID2640834.
6. Kara L. Menzies and Lyndon Jones, The impact of contact angle on the biocompatibility of biomaterials. *Optom. Vis. Sci.*, 2010, 87, 387–399.
7. Jian-Ying Huang and Yue-Kun Lai, TiO_2-based surfaces with special wettability–from nature to biomimetic application.48–84. doi:10.5772/60826.
8. Chien-Sheng Kuo, Yao-Hsuan Tseng and Yuan-Yao Li, Wettability and superhydrophilic TiO_2 film formed by chemical vapor deposition. *Chem. Lett.*, 2006, 35(4), 356–357.
9. N. Encinas, M. Pantoja, J. Abenojar, M.A. Martínez,Control of wettabilityof polymers bysurface roughness modification. *J. Adhes. Sci. Technol.*, 2010, 24, 1869–1883.
10. Feifei Zhang, Ben W. Robinson, Heidi de Villiers-Lovelock, Robert J. K. Wood and Shun Cai Wang, Wettability of hierarchically-textured ceramic coatings produced by suspension HVOF spraying. *J. Mater. Chem. A*, 2015, 3, 13864–13873.
11. Pengyun Xu, Thomas W. Coyle, Larry Pershinand Javad Mostaghimi, Fabrication of micro-/nano-structured superhydrophobic ceramic coating with reversible wettability via a novel solution precursor vacuum plasma spray process. *Mater. Des.*, 2018,160, 974–984.

12. Gisele Azimi, Rajeev Dhiman, Hyuk-Min Kwon, Adam T. Paxson and Kripa K. Varanasi, Hydrophobicity of rare-earth oxide ceramics. *Nat. Mater.*, 2013, 12, 315–320.
13. Sami Khan, Gisele Azimi, Bilge Yildizand Kripa K. Varanasi, Role of surface oxygen-to-metal ratio on the wettability of rare-earth oxide. *Appl. Phys. Lett.*, 2015, 106(061601), 1–4.
14. F. Pedraza, S. A. Mahadik and B. Bouchaud, Synthesis of ceria based superhydrophobic coating on Ni20Cr substrate via cathodic electrodeposition. *Phys. Chem. Chem.Phys.*, 2015, 17, 31750–31757
15. Young Jun Cho, Hanmin Jang, Kwan-Soo Leeand Dong Rip Kim, Direct growth of cerium oxide nanorods on diverse substrates for superhydrophobicity and corrosion resistance. *Appl. Surf. Sci.*, 2015, 340, 96–101
16. Christopher A. Traina, Jeffrey Schwartz, Surface modification of Y_2O_3 nanoparticles. *Langmuir*, 2007 Aug 28, 23(18), 9158–9161
17. Kapil Manoharan and Shantanu Bhattacharya, Superhydrophobic surfaces review: Functional application, fabrication techniques and limitations. *J. Micromanufactu.*, 2019, 2, 59–78
18. Junsheng Liang,Kuanyao Liu,Dazhi Wang,Hao Li,Pengfei Li,Shouzuo Li,Shijie Su,Shuangchao Xuand Ying Luo, Facile fabrication of superhydrophilic/superhydrophobic surface on titanium substrate by single-step anodization and fluorination. *Appl. Surf. Sci.*, 2015, 338, 126–136.
19. Won Gyu Bae,Ki Young Song,Yudi Rahmawan,Chong Nam Chu,Dookon Kim,Do Kwan Chungand Kahp Y. Suh,One-step process for superhydrophobic metallic surfaces by wire electrical discharge machining. *ACS Appl. Mater. Interf.*, 2012, 4(7), 3685–3691.Retrieved from: https://www.accudynetest.com/polytable_03.html?sortby=contact_angle
20. Junping Zhangand Stefan Seeger,Polyester materials with superwetting silicone nanofilaments for oil/water separation and selective oil absorption. *Adv. Funct. Mater.*, 2011, 21, 4699–4704.
21. Sepideh Minagar, Christopher C. Berndt and Cuie Wen,Fabrication and characterization of nanoporous niobia, and nanotubular tantala, titania and zirconia via anodization. *J. Funct. Biomater.*, 2015, 6, 153–170.
22. Donald J. Trevoy and Jr Hollister Johnson,The water wettability of metal surfaces. *J. Phys. Chem.*, 1958, 62(7), 833–837.
23. Andresa Baptista, Francisco Silva, Jacobo Porteiro, José Míguez and Gustavo Pinto, Sputtering physical vapour deposition (PVD) coatings: A critical review on process improvement and market trend demands. *Coatings*, 2018, 8(402), 1–22.
24. J. R. Creighton and P. Ho. Chapter 1 *Introduction to Chemical Vapor Deposition*, Chem. Vap. Depos.2001ASM International, Materials park, OH, 1–13.
25. Colm Glynn and Colm O'Dwyer, Solution processable metal oxide thin film deposition and material growth for electronic and photonic devices. *Adv. Mater. Interfaces*, 2017, 1600610, 1–36.
26. Rhander Viana, Milton Sérgio Fernandes de Lima, Wisley Falco Sales, Washington Martins da Silva Jr. Álisson Rocha Machado,Laser texturing of substrate of coated tools: Performance during machining and in adhesion tests. *Surf. Coat. Technol.*, 2015, 276,485–501.
27. Behzad Fotovvati, Navid Namdari and Amir Dehghanghadikolaei,On coating techniques for surface protection: A review. *J. Manuf. Mater. Proc.*, 2019, 3, 28.
28. Wilhelm Barthlott and Christoph Neinhuis, Purity of the sacred lotus, or escape from contamination in biological surfaces. *Planta*, 1997, 202(1), 1–8
29. Biswajyoti Mukherjee, O. S. Asiq Rahman, Aminul Islam, Krishna Kant Pandey, and Anup Kumar Keshri,Deposition of multiscale thickness graphene coating by harnessing extreme heat and rapid quenching: Towardscommercialization. *ACS Appl. Mater. Interfaces*, 2019, 11, 25500–25507

30. Anup Kumar Keshri, Riken Pateland Arvind Agarwal, Comprehensive process maps to synthesize high density plasma sprayed aluminum oxide composite coatings with varying carbon nanotube content. *Surf. Coat. Technol.*, 2010, 205,690–702

31. J.F. Li, H.L. Liao, C.X. Ding, C. Coddet, Optimizing the plasma spray process parameters of yttria stabilized zirconia coatings using a uniform design of experiments. *J. Mater. Proc. Technol.*, 2005, 160,34–42

32. O. S. Asiq Rahman, Biswajyoti Mukherjee, Aminul Islam and Anup Kumar Keshri, Instant tuning of superhydrophilic to robust superhydrophobic and self-cleaning metallic coating: Simple, direct, one-step, and scalable technique, *ACS Appl. Mater. Interfaces*, 2019, 11, 4616–4624

33. Qiang Zhang, Chang-Jiu Li, Cheng-Xin Li, Guan-Jun Yang and Siu-Ching Lui,Effect of vapor depositionin shrouded plasma spraying on morphology and wettability of the metallic Ni20Cr coating surface. *J. Alloys Compd.*, 2018, 735, 430–440

34. Daniel J. Preston, Nenad Miljkovic, Jean Sack, Ryan Enright, John Queeney and Evelyn N. Wang, Effect of hydrocarbon adsorption on the wettability ofrare earth oxide ceramics. *Appl. Phys. Lett.*, 2014, 105, 011601.

4 Tribological Behaviour of Surface Coatings for Hot Forming Die Steels

Manpreet Kaur, Pankaj Chhabra, and Gagandeep Singh

CONTENTS

DOI: 10.1201/9781003213185-4

4.1 INTRODUCTION

Tribology-a study of friction, wear, and lubrication, is the branch of science which deals with the study of interacting surfaces in relative motion [1]. It is derived from the Greek word "tribos" which means to rub. The term "Tribology" was coined by Jost in 1966. Tribology is aimed at improving product life and performance, lowering costs, saving raw material and energy. The tribological behavior of a system is governed by four factors i.e., elemental composition of the two mating surfaces, lubricants used, and surrounding environment [2]. All these elements together determine the friction and wear characteristics of the particular system.

4.1.1 FRICTION

Friction is a resistive force that opposes relative motion between surfaces in contact. The coefficient of friction μ is a dimensionless entity. It is the ratio of frictional force to the normal force that presses the two bodies together. The main types of friction are static friction, sliding friction, rolling friction, and fluid friction. Amonton Laws define the elementary properties of sliding friction. According to Amonton, the frictional force exerted by one surface on the other surface is proportional to the applied load and is independent of the apparent contact area. Based on Amonton Laws, Coulomb developed an approximate model to measure friction written as Equation 4.1:

$$F \leq \mu N \tag{4.1}$$

where F is the frictional force acting parallel to the surface and in a direction opposite to the applied force, μ is the coefficient of friction, and N is the normal load acting perpendicular to the surface.

For surfaces at rest $\mu = \mu_s$, μ_s is the coefficient of static friction

For surfaces in relative motion, $\mu = \mu_k$, μ_k is the coefficient of kinetic friction

Usually for the same surfaces in contact, the coefficient of kinetic friction has a lesser value than the coefficient of static friction.

4.2 WEAR

Wear is characterized as the removal of material from mating surfaces. The material removal may be from one or both surfaces in the form of debris. It may be a transfer of material from a softer surface to a harder surface. Wear is frequently expressed in terms of mass loss per unit distance travelled. Archard developed a reliable mathematical model for the quantification of wear using a series of pin-on-ring tests. According to Archard, wear results from plastic deformation and fracture of asperities due to localized pressure at contact. He concluded that wear is directly proportional to the applied load and independent of the apparent contact area. These inferences were combined into Archard equation given by Equation 4.2:

$$W = K \frac{P}{p_m} s \tag{4.2}$$

where W is the volume of worn material, K is the wear constant, P is the applied pressure, p_m is the yield strength of the softest material of the pair, and s is the total sliding distance.

K is used for comparing the tribological behavior of materials tested under similar conditions (Equation 4.3):

$$K = \frac{W}{sP} \tag{4.3}$$

where K, in [mm³/Nm], is the wear constant or wear ratio, W [mm³] is the worn volume, s [m] is the total sliding distance, and P [N] is the normal load. The mechanisms relevant to wear of hot forming dies are briefly explained below:

4.2.1 ADHESIVE WEAR

Adhesive wear results from the shearing of plastically deformed and cold-welded asperities formed on mating surfaces (Figure 4.1). When surfaces slide against each other under load, the high contact pressure is generated at asperities junctions. These asperities deform plastically under extreme load and produce wear particles. This form of wear damages the sliding surfaces causing cavitations and ultimately failure of the component.

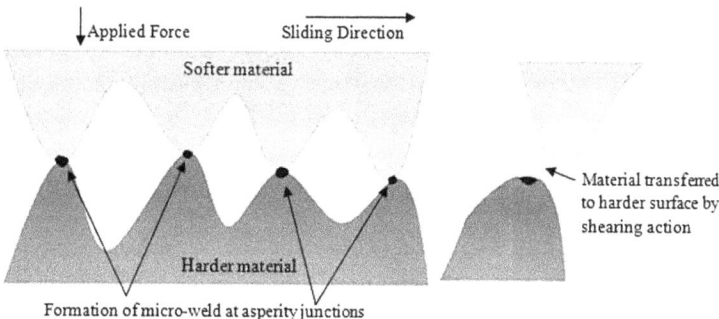

FIGURE 4.1 Adhesive wear mechanisms.

4.2.2 ABRASIVE WEAR

Abrasive wear is caused by the removal of material from the softer surface by the harder material (Figure 4.2). It is also known as scratching, scoring, or grooving depending upon the severity of the wear. Abrasive wear maybe two-body abrasion, in this the harder material continuously removes particles from softer material, e.g., grinding, cutting, and machining. The second type is the three-body abrasion in which the removed particles are trapped in between the mating surfaces and cause wear of both surfaces.

4.2.3 FATIGUE WEAR

This wear happens when a surface is loaded and unloaded periodically (Figure 4.3). These loading–unloading cycles produce compressive and shear stresses resulting in micro-cracks on weakened points. With time, the cracks propagate along and beneath the surface. Upon reaching critical size, the cracked part is detached from the surface. This phenomenon is also known as delamination or spalling. In hot forming processes, thermal fatigue also occurs along with mechanical fatigue due to heating and cooling of the exposed surface. This produces additional stresses and speeds up the formation of micro-cracks. Fatigue may cause sudden tool failure.

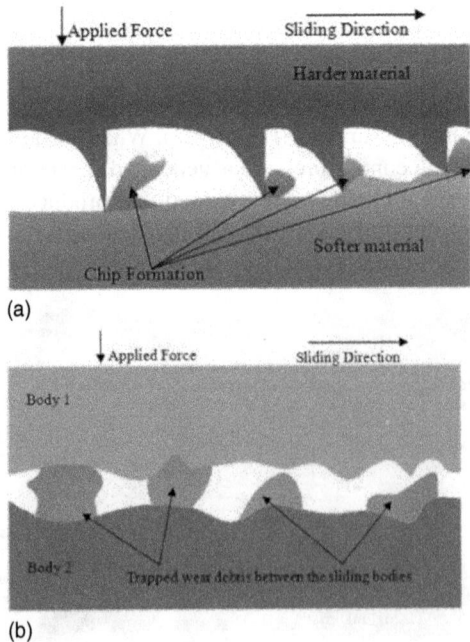

FIGURE 4.2 (a) Two-body abrasive wear mechanism, (b) three-body abrasive wear mechanism [3].

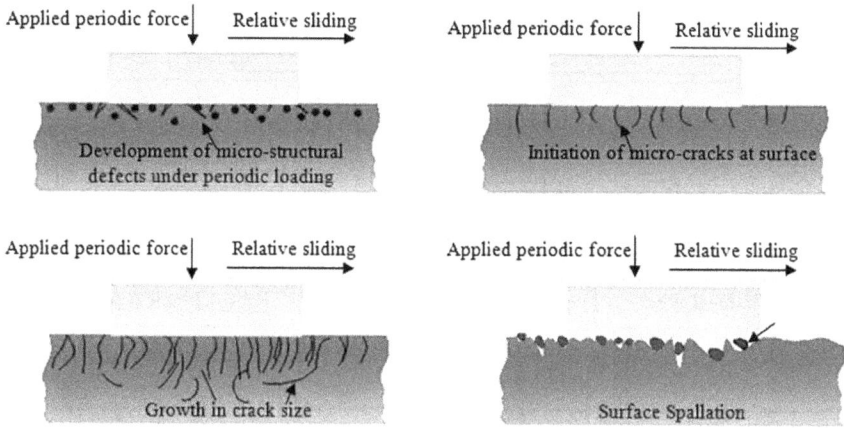

FIGURE 4.3 Fatigue wear mechanism.

4.2.4 TRIBO-CHEMICAL WEAR

In tribo-chemical wear, chemical reactions happening on the surface such as oxidation or corrosion causes the formation of undesirable and weak byproducts on the surface which, in some cases, are easily removed from the surface (Figure 4.4). This exposes fresh surfaces for further degradation. This mechanism thus acts in combination with mechanical wear mechanisms.

Many of the wear mechanisms interrelate each other and are very similar to each other in concepts [4]. Figure 4.5 summarizes the interrelations of wear mechanism in terms of counterparts, type of motion, deformation process, working environment for a better understanding of the wear the mechanism required to control them.

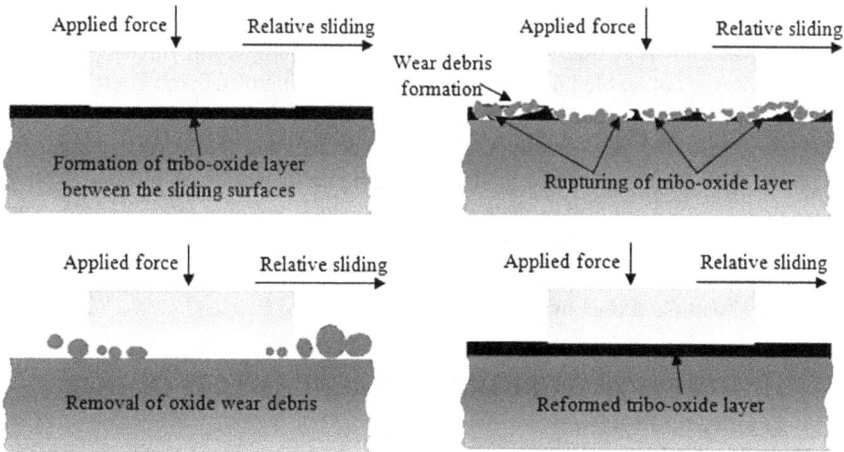

FIGURE 4.4 Tribo-oxidative wear mechanisms.

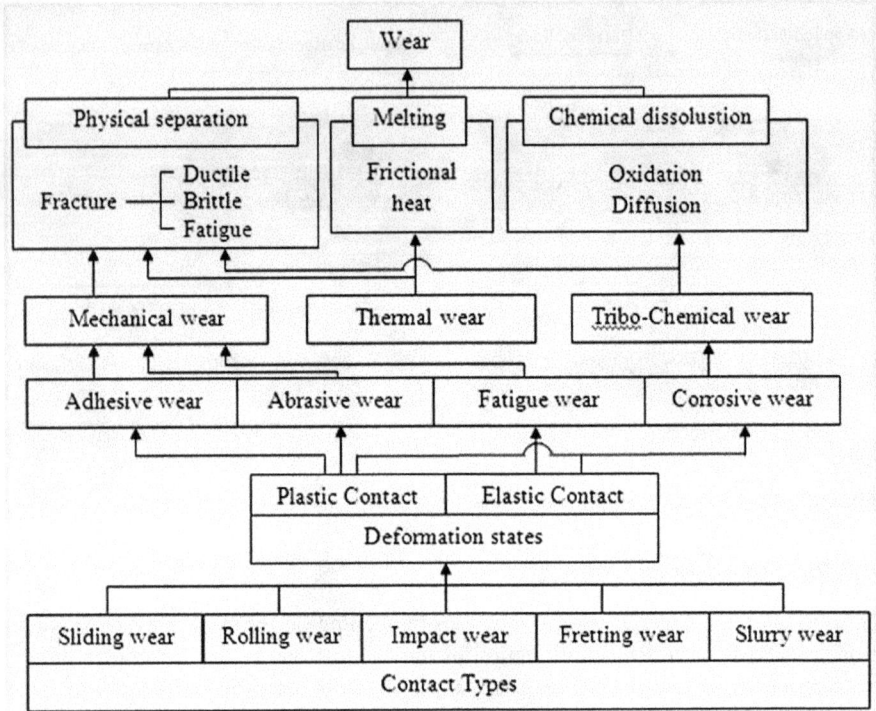

FIGURE 4.5 Description of wear and its interrelations.

4.3 LUBRICATION

The primary purpose of lubrication is to reduce both friction and wear. Lubricants also provide cooling, lower vibrations, and inhibit corrosion. There are two main ways by which lubricants lower friction. One is the method in which a stable lubricant layer is formed between the surfaces. This causes complete separation of the surfaces and friction is due to the shear resistance of lubricant. In other cases, the lubricant film is not stable but still results in the lowering of friction coefficients and temperatures. Wear is also lowered extensively due to lower flash temperature and removal of any worn particles by a lubricant. According to the working parameters, i.e., load, working temperature, and surrounding environment, a suitable lubricant either fluid or solid could be selected. Metals, oxides, sulfides, graphite, etc., could also act as lubricants and they need not be necessarily oil or grease. In certain cases, the lubricant residues could be problematic. The disposal of these residues may be harmful to health and the environment.

4.4 HIGH-TEMPERATURE TRIBOLOGICAL ASPECTS OF HOT FORMING DIES

The moving machine assemblies that operate at high temperatures are subjected to friction and wear. The problem can be seen in power generation, aerospace, and met-alworking industries. At temperatures above 300°C, the semisolid and liquid lubricants rapidly decompose and deteriorate, enhancing wear losses. Also, with an increase in temperature, the friction value changes and in due course affects the functionality of dies in the forming operation.

Hot forming of high-strength steel sheets has become the need of industry to overcome various problems such as poor formability to complex shapes, high spring back, and inclination to work-harden are generally experienced during the cold metal forming operation [5, 6]. The hot forming/forging operations produce components with better mechanical properties and less wastage of materials. In actual industry, the blank (workpiece) is heated to temperatures more than 1000°C and is shifted between the dies. The dies close and the forming of the job takes place. In few cases, the dies are preheated to 250°C temperature. Also, during the forming process, the die (tool) surface temperature may exceed 500°C. Figure 4.6 shows the schematic of the hot forming process. The hot forming process becomes more complicated due to the presence of increased oxidation, thermal fatigue, changes in mechanical properties, and the occurrence of tribo-chemical reactions. The dies are exposed to high tempering due to high mechanical and thermal loading. This causes a reduction in hardness at die surfaces. Further, due to the plastic deformation, the catastrophic fracture, and an intolerable amount of wear, the dies fail. Die failure by surface wear is reported in 70% of the cases [7, 8]. In spite of such importance, very little attention is being given to the high-temperature tribology.

The production expenditures are directly influenced by costs related to die. This is mainly due to high initial costs and the costs involved in the refurbishing of die and loss in production time. A successful and cost-effective production can be achieved

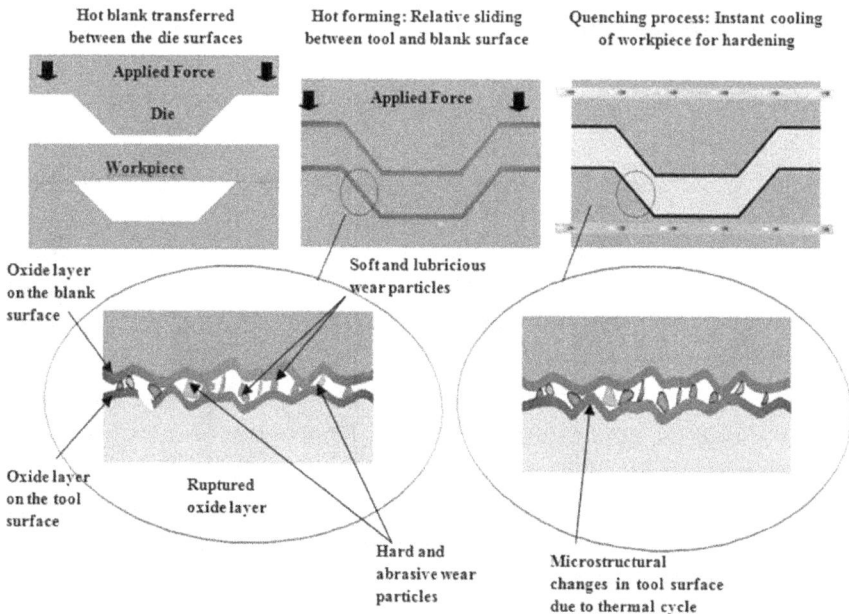

FIGURE 4.6   Schematic of hot forming process.

by using high-quality dies [9]. The dies must withstand the stresses induced during forming, and retain their hardness and adequate strength in the most severe conditions.

Different surveys conducted in many countries report that friction and wear can lead to huge economic losses that can amount to 1–2% of the total Gross Domestic Product (GDP) of the country [10]. In India, bad tribological practices have led to an estimated loss of Rs. 20,000 crores. It has become very important to find ways to control the concerns arising due to friction and wear. Moreover, to ensure the quality of the end products, it very important to select the accurate die material with adequate die life and reasonable costs.

The mechanical properties like toughness, hardness, and ductility of die material are very important to control the wear. AISI H13 and AISI H11 steels are the most popular steels owing to their high toughness, machinability, ability to attain and retain hardness [11–13]. These materials are being used in the industries to make casting dies, hot forging dies, and punches. However, the die materials must possess resistance to wear, corrosion, fatigue, and other mechanical failures in severe conditions such as higher-repetitive loads, elevated temperatures, and corrosive environments.

4.4.1 Methods to Increase the Life and Performance of Hot Forming Dies

The various methods employed to improve the life and performance of hot forming dies are listed below and shown in Figure 4.7.

> **Selection of Die Material:** This is the most crucial factor relating to die life. The material selected for the manufacturing hot forming die must have high hot hardness, toughness, and yield strength. The material should have

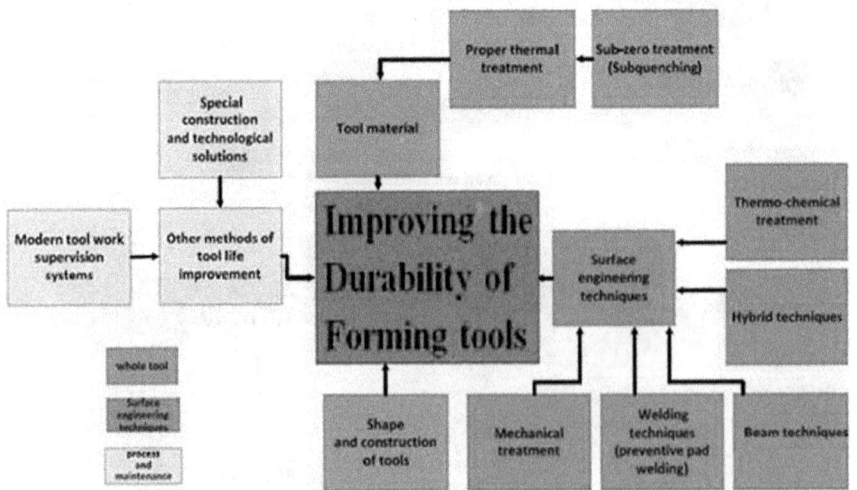

FIGURE 4.7 Methods for increasing the die life [17].

high thermal conductivity and lower thermal expansions. Further, the material should have the ability to be treated with surface engineering techniques for attaining desired properties.

Shape and Design of Die: For increasing the durability of the hot forming die, various design factors such as fillet radii, draft angle, and flash thickness should be optimally selected [14]. The use of mathematical and CAD/CAM techniques can be utilized to obtain the best design. Also, the surface roughness plays a major role in controlling die life. The increase in roughness increases the die wear. On the other hand, the ultra-smooth surface also tends to increase wear by adhesion mechanism and thus the cost of manufacturing goes higher. Thus, an optimum value needs to be selected.

Process Parameters: The process parameters also play an important role in increasing die life. The parameters such as workpiece temperature, the load applied, loading time, production rate, and die cooling time and method, etc., need to be selected appropriately.

Workpiece Properties: The workpiece properties such as hardness, oxidation behavior, strength, oxide scale nature, the thermal effect, and weight are the factors which on selecting judiciously improves die life [15–16].

Lubrication: Proper lubrication reduces the workpiece and die surface contact causing a reduction in wear rate but also reduces frictional force causing energy savings. Lubrication also transfers less heat to the die surface, reducing thermal stresses. The amount of lubrication is also needed to be checked as excess lubrication also causes die failure and manufacturing defects, as it makes a buildup in the die cavity and leads to pre-mature filling and excess die pressure.

Surface engineering techniques: As the die life is largely affected through die surfaces. Thus, the techniques employed for raising surface properties can enhance the die life appreciably. In general, the methods employed for enhancing surface properties include thermochemical treatment (thermal treatments, nitriding or nitrocarburizing, boriding, etc.), mechanical techniques (burnishing, shot-peening, etc.), surface coatings (thermal spray coatings, PVD and CVD, etc.), beam techniques (ion implantation, laser processing), and hybrid techniques (a combination of two or more techniques).

4.4.2 SURFACE ENGINEERING

Surface engineering plays a vital role in these demanding applications. Surface Engineering can be defined as "changing the properties of the surface of a material to obtain properties which cannot be achieved by the bulk material alone" [6]. To attain higher wear resistance, surface treatments and coatings have shown better results. Some important considerations are coating thickness, thermal and chemical stability, adhesive strength, spalling behavior, and cost. Exotic coatings such as ceramic coatings have found use in high-temperature applications, such as hot extrusion, and in drawing dies [18]. This means that by utilizing surface treatments and/or surface coatings, it is possible to modify the material to get the desired properties where it is most needed.

The bulk material can provide strength and toughness for the component. The surface (which is provided with treatment or coating) imparts improved wear resistance, low friction properties, corrosion protection, and so on. This also has an economic aspect. Cheaper bulk material can be utilized and materials that are more expensive can be used for the coating since very small amounts are required [6].

Researchers from all over the world have reported significant developments in surface coatings and treatments and found that they have the potential in high-temperature applications. Several studies have been undertaken for investigating their friction and wear behaviour at room temperature, but only a few studies have been reported about their elevated temperature performance. The surface engineering techniques not only help in the retention of substrate mechanical properties but also help in providing hardness and/or corrosion-resistant layers at the die surface. Traditional surface modification methods such as nitriding, boriding, or carburizing suffer from disadvantages such as long treatment time, uneven treatment depths, and higher costs. They could also damage the environment by releasing a sizeable amount of CO_2. Currently, chemical vapour deposition (CVD), physical vapour deposition (PVD), and thermal spraying techniques are extensively used for different wear and corrosion applications. But in comparison, thermal spray coatings are more environment-friendly, possess better properties, higher coating thickness, and lower application costs [19].

The thermal spraying techniques can spray the widest variety of materials with a large range of coating thickness and coating properties. Due to these advantages, thermal spray techniques have been used in many applications such as automobile, mining, aviation, thermal power plant, printing, orthopedics, and dental implants. However, there is a lack of research in the field of application and performance of thermal spray techniques in hot forming processes. From a scientific point of view, our knowledge-base regarding the use of thermal spray coatings on tool-workpiece materials used in hot forming processes in various automotive industries is still lacking. The potential of thermal spray coatings in reducing friction and wear must be explored.

The authors of this chapter have studied the tribological behavior of various thermal spray coatings developed onto the surface of AISI H11 and AISI H13 tool die steels. The experimental techniques, types of equipment used, and the results have been described in the following sections.

4.5 TRIBOLOGICAL BEHAVIOUR OF THERMAL SPRAY COATED HOT FORMING DIE STEELS

4.5.1 Materials for the Study

The hot forming tool steels namely AISI H11 and AISI H13 were selected as the pin (tool) material in the investigation. The pins with 8 mm in diameter and 50 mm in length were prepared.

Figure 4.8 shows the camera macrographs of the prepared cylindrical pins. DIN 20MnCr5 steel was chosen as the counter-face disc. It is a high-strength low-alloy steel. The chemical composition of the pin and the disc steel is given in Table 4.1. The discs with 100 mm diameter and 8 mm thickness were prepared. The discs were heat-treated and plasma nitrided and the hardness was increased to ~80 HRC.

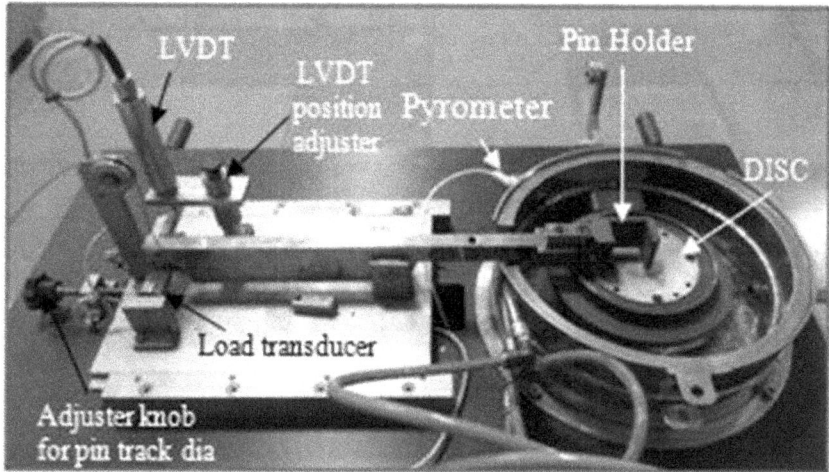

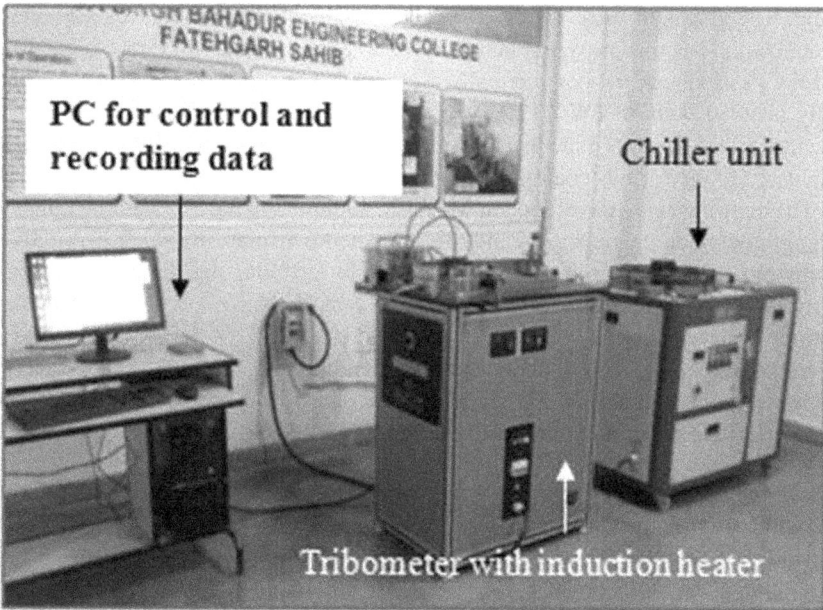

FIGURE 4.8 Descriptive view of high-temperature tribometer employed for tribological testing.

This was done according to the ASTM standard for studying sliding wear behaviour using a pin-on-disc tribometer.

4.5.2 DEVELOPMENT OF THE COATINGS

Three coating compositions: Cr_3C_2-25(NiCr), NiCrBSi, and NiCrBSi blended with 35%(88WC-12Co) were developed on the selected die steels by the HVOF spraying and atmospheric plasma spraying processes. Cr_3C_2-25(80Ni20Cr) powder, NiCrBSi

TABLE 4.1

Chemical Composition of the Pin and Disc Material

Tool Steels	Elemental Composition (wt %)										
	Cr	Mo	Si	S	C	Mn	V	Cu	P	Ni	Fe
H11	4.75–5.50	1.10–1.60	0.80–1.20	<0.03	0.33–0.43	0.20–0.50	0.3–0.6	<0.25	<0.03	<0.3	Bal.
H13	4.75–5.50	1.10–1.75	0.80–1.20	<0.03	0.32–0.45	0.20–0.50	0.8–1.2	<0.25	<0.03	<0.3	Bal.
Disc Material DIN 20MnCr5	P Max 0.025	S Max 0.03	Si Max 0.4	Cr 1.0–1.3	C 0.17–0.22		Fe Bal	—	—	—	—

powder, and 88WC-12Co powder with particle sizes of −45+15 μm were commercially available at the Metallizing Equipment Company, Pvt. Ltd. (MECPL), Jodhpur, India. The powders were procured from Powder Alloy Corporation, Cincinnati, Ohio, USA. The powders were manufactured by agglomeration and sintering. Further at MECPL, Jodhpur, the commercially available NiCrSiFeBC powder was mechanically blended with 88%WC-12%Co powder in the ratio of 65:35 by weight. The mechanical mixing was done in a ball mill for 8 hours. The composition and designation used for the various coatings are given in Table 4.2.

The coatings were developed at MECPL, Jodhpur, India. To raise the adhesion strength between the substrate and the projected particles, the surface roughness of the pin surface was raised. This was done by grit blasting the Al_2O_3 (Grit 60) particles prior to the coating development and Ra value of 6–8 microns was obtained.

In the case of the HVOF spray coating, liquid petroleum gas (LPG) and oxygen were used as fuel gases and nitrogen was used as a powder carrier gas. The coating was performed using a commercially available HIPOJET-2700 torch. Atmospheric

TABLE 4.2

Composition, Particle Size, Designation, and Process of the Coating Powders

Coating Powder	Composition (in wt%)	Particle Size	Coating Method	Designation
Cr_3C_2-25(NiCr)	75%Cr_3C_2-25%(80Ni-20Cr)	−45+15 μm	HVOF	[S1]
			APS	[S2]
NiCrBSi	C-0.81, B-3.35, Fe-4.17, Si-4.27, Cr-15.7, Ni-71.7	−45+15 μm	HVOF	[C1]
			APS	[C2]
65%(NiCrSiFeBC)-35%(WC-Co)	65 %(Ni-71.7, Cr-15.7, Si-4.27, Fe-4.17, B-3.35, C-0.81)-35%(88%WC-12%Co)	−45+15 μm	HVOF	[C4]
			APS	[C5]

plasma spray coatings were developed using argon as the primary plasma gas and hydrogen as the secondary gas. MSG-100 spray gun was used to perform the coating process. The coating guns were mounted on the robotic arm of KUKA make, made in Germany, for automatic spray. For APS [C5] coatings, a bond coat of 80Ni-20Cr with approximately 50 µm thickness on the substrate steels was developed for achieving better adhesion of the prepared composite coating.

The particle velocity and the powder temperature during spraying were recorded by the Spray Watch 2S system by OSEIR, Ltd., Finland. The flame temperature was measured by the thermocouple. The traverse speed of the gun was 2.5 mm/s for both techniques. The powder feed rate was 25 g/min and 35 g/min for the HVOF spray and the APS techniques, respectively. To maintain the temperature of the substrate during and after the spraying process, cooling was done with the compressed air jets. The parameters employed during the HVOF spray and APS coating deposition are shown in Tables 4.3 and 4.4, respectively.

TABLE 4.3
Spray Parameters Used during the Deposition of the HVOF Coating

Particle velocity (m/s)	450–500
Powder feed rate (g/min)	25
Powder pressure (kg/cm²)	3
Particle temperature (°C)	2200
Flame temperature (°C)	2800–3200
Flow rate of Oxygen (SLPM)	300–350
Oxygen pressure (kg/cm²)	9
Fuel (LPG) flow rate (SLPM)	60–80
N_2 flow rate (SLPM)	15–20
Fuel pressure (kg/cm²)	5.6
Spray distance (mm)	250

TABLE 4.4
Spray Parameters Used during the Deposition of the Atmospheric Plasma Spraying

Particle velocity (m/s)	200–250
Powder feed rate (g/min)	35
Flame temperature (°C)	17000–18000
Particle temperature (°C)	2500–2700
Spray distance (mm)	75
Powder carrier gas, Argon (SLPM)	41
Fuel gas, Hydrogen (SLPM)	9

4.5.3 CHARACTERIZATION OF THE AS-SPRAYED SPECIMENS

The surface roughness measurements, porosity measurements, bond strength evalua-tion, and microhardness evaluation of the as-sprayed coated pin specimens were per-formed. The surface roughness tester (Mitutoyo Digital SJ-301, Make Japan), image analyzer software, bond strength universal testing equipment (Instron make, Tamil Nadu, India), and the digital Vickers microhardness tester were used to measure the properties, respectively. The roughness values (Ra) were evaluated at three different positions and the average of three values were evaluated before and after polishing of the coated specimens. The coating bond strength was evaluated with HTK Ultra bond 100 bonding agent. The ASTM Standard C-633 was followed. The microhardness tester was equipped with a diamond indenter of square pyramidal in shape. A dwell time of 10 s and a load of 2.45 N were applied for indentation. The average value of hardness obtained at three different places has been reported. Further, the XRD anal-ysis of the as-sprayed pin specimens was conducted using X'Pert MPD diffractom-eter (make: PANalytical Company, Netherlands).

Scanning electron microscope (SEM) equipped with energy dispersive spectros-copy (EDS) (make: JEOL JSM 6610LV) was used to analyze the composition and microstructure of as-sprayed coatings. To evaluate the coating thickness and compo-sition of elements along with coating depth the specimens were examined along the cross section with SEM/EDS technique. For this testing, the specimens were cut at slow speed, mounted, and mirror-polished.

4.5.4 HIGH-TEMPERATURE WEAR TESTS

A high-temperature pin-on-disc tribometer (make: Ducom, Bangalore) (Figure 4.8) was used to perform the tribological studies. Figure 4.9 shows the schematic of Pin-on-Disc wear test. The tests were conducted as per ASTM G99-04 standard for the unidi-rectional dry sliding conditions. The tribological testing was performed at room temperature (RT), 400°C, and 800°C, and at loads of 25 N and 50 N. The sliding speed was maintained at 0.5 ms^{-1} for a sliding distance of 1500 m. A software (WINDUCOM™) provided by the manufacturer was used to control and acquire data during testing.

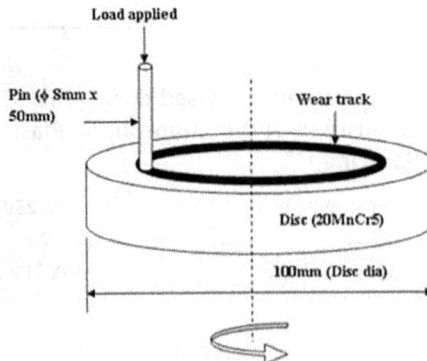

FIGURE 4.9 Schematic of pin-on-disc friction and wear test.

4.5.5   ANALYSIS OF WORN-OUT SPECIMENS

The worn-out specimens were carefully examined. Camera macrographs were taken to keep the record of any changes in macrostructure, color, spalling tendency, and coating adherence.

The weight loss in grams from the pin specimens was recorded carefully by evaluating weight change before and after the testing.

Archimedes principle mentioned in Equation 4.4 was used to calculate the coating density. Volume loss (mm³) from the samples was evaluated as mentioned in Equation 4.5. The specific wear rates (mm³/Nm) for all the specimens were evaluated using Equation 4.6. For obtaining good reproducibility, three tests were performed under each experimental condition.

$$\text{Density of coating}\ (\rho)\ \text{g/cm}^3 = \rho_a \left(\frac{m_{cw}}{m_{ca}} - m_{cw} \right) \tag{4.4}$$

$$\text{Volume loss}\left(\text{mm}^3\right) = \frac{\text{Mass loss}}{\rho} \times 1000 \tag{4.5}$$

where ρ_a is the water density (1 g/cm³), m_{ca} is the coating weight in the air (g), and m_{cw} is the coating weight in water (g).

$$\text{Specific wear rate}\left(\text{mm}^3/\text{Nm}\right) = \text{Volume loss/Sliding distance} \\ \times \text{Applied load} \tag{4.6}$$

Further, average values of COF vs. temperature graphs were plotted demonstrating the average values of coefficients of friction (COF) for the uncoated and coated steel specimens evaluated at two applied loads. The COF graphs with time were also plotted for each coating at all testing temperatures and applied loads. These graphs showed a variation of COF values throughout the test duration of 50 minutes. The WINDUCOM™ software continuously records the value of COF after every second. The SEM/EDS analysis of the worn-out samples was done at critical areas of interest to understand the wear mechanisms, surface deformations, and nature of oxide layers formed on the worn-out surface. Point analysis by EDS was carried out on various locations to identify the composition (wt %) of different phases appearing on the surface. Although these compositions correspond to selected points on the surfaces, still the data could be useful to support the formation of various phases in the oxide scales, in general. For identification of different phases formed in the worn-out specimen, an X-ray diffraction analysis of the specimens was conducted and the results have been reported.

4.6   CHARACTERIZATION RESULTS

4.6.1   EVALUATION OF SURFACE ROUGHNESS, COATING THICKNESS, POROSITY, AND BOND STRENGTH

Table 4.5 shows the average surface roughness values obtained. The surface roughness values of all the developed coatings were found to be in the range of ~4–8 μm. The Ra values after polishing were recorded to be less than 1 μm.

TABLE 4.5

Surface Roughness of all Coated Samples

Sr. No.	Coating	Value of Average Surface Roughness before Polishing (Ra)		Value of Average Surface Roughness after Polishing (Ra)	
		on H11	on H13	on H11	on H13
1	HVOF-sprayed Cr_3C_2-25 (NiCr) [S1]	4.68	4.92	0.83	0.87
2	AP sprayed Cr_3C_2-25 (NiCr)APS [S2]	4.29	4.12	0.49	0.55
3	HVOF sprayed NiCrBSi [C1]	7.71	6.62	0.11	0.13
4	APS NiCrBSi [C2]	4.73	4.42	0.25	0.15
5	HVOF sprayed 65%(NiCrSiFeBC)+ 35%(WC-12Co)[C4]	6.68	6.45	0.24	0.14
6	APS 65%(NiCrSiFeBC)+ 35%(WC-12Co)[C5]	4.27	3.83	0.18	0.26

Table 4.6 shows the recorded values of the coating thickness, porosity, and bond strength of the developed coatings. The cross-sectional SEM morphologies were analyzed to obtain the average thickness of the coatings. The coating thicknesses of all the developed coatings were above 100 microns. The porosity values for the coating compositions were evaluated in the present study. Porosity lies in the range of 1.7–2.91%. Most likely, the AP-sprayed coatings showed higher values of porosity. This was mainly attributed to the air pockets entrapped in the coating and the pores created by solidification and shrinking of splats in the coatings.

The coating bond strength is another significant characteristic that impacts the performance of the coating. The bond strength of the coating is usually determined by adhesive failure (failure at substrate/coating interface), cohesive failure (failure in between the coating), and glue failure (failure occurs at coating/glue interface) during experimentation. The various factors which affect the bond strength of the coating are substrate roughness, diffusion at the interface, surface area, chemical and thermal properties of substrate and coating material, and spraying parameters such as velocity at impact, spraying angle, coating temperature, and many other factors [20–22]. The bond strength values for the various coatings developed by the authors were found in the range of 42–73 MPa. The HVOF-sprayed coating showed higher bond strength values than the AP-sprayed coatings. Among all the deposited coatings, the HVOF-sprayed Cr_3C_2-25(NiCr) coatings [S1] showed a higher bond strength value. The bond strength of the HVOF-sprayed coating was measured by glue failure (failure at coating/glue interface). Higher values of bond strength in HVOF-sprayed coatings are mainly due to the partially melted coating powder particles impinging at higher

TABLE 4.6
Coating Bond Strength, Thickness, and Porosity

Sr. No.	Coating	Substrate Steel	Bond Strength (MPa)	Coating Thickness (μm)	Porosity (%)
1	HVOF-sprayed Cr_3C_2-25(NiCr) [S1]	H11	73	175	2.26
2	HVOF-sprayed Cr_3C_2-25(NiCr) [S1]	H13	73	235	2.06
3	AP-sprayed Cr_3C_2-25(NiCr)APS [S2]	H11	63	250	1.98
4	AP-sprayed Cr_3C_2-25(NiCr)APS [S2]	H13	64	130	1.7
5	HVOF-sprayed NiCrBSi [C1]	H11	61	280	1.64
6	HVOF-sprayed NiCrBSi [C1]	H13	57	300	1.48
7	APS NiCrBSi [C2]	H11	42	200	2.20
8	APS NiCrBSi [C2]	H13	49	190	1.96
9	HVOF-sprayed 65%(NiCrSiFeBC)+35%(WC-12Co)[C4]	H11	67	190	2.69
10	HVOF-sprayed 65%(NiCrSiFeBC)+35%(WC-12Co)[C4]	H13	69	240	1.71
11	APS 65%(NiCrSiFeBC)+35%(WC-12Co)[C5]	H11	44	230	2.91
12	APS 65%(NiCrSiFeBC)+35%(WC-12Co)[C5]	H13	53	240	2.36

velocities causing the anchoring effect. The particles get plastically deformed and not flatten out completely. This causes impact pressure and the coating comprises compressive residual stresses resulting in good adhesive strength. In the APS spray process, the high temperature of the plasma leads to a higher percentage of fully molten powder particles. This led to fully molten splats and also the disintegration of splats reducing the anchoring effect. Also, higher temperature causes oxidation of powders and entrapment of some gases between splats which increases porosity. This causes cohesive failure of some coatings during the testing.

4.6.2 EVALUATION OF MICROHARDNESS

Microhardness of thermal-sprayed coatings directly impacts the wear performance and life of the coatings. Higher hardness is generally associated with higher wear resistance and is the most frequently quoted mechanical property of the coatings [22]. The substrate steel H11 and H13 showed an average microhardness value of 250±15 HV. The

TABLE 4.7

Microhardness Values for the As-Sprayed Coatings

Sr. No.	Coating	Substrate Steel	Microhardness Range (HV)	Average Microhardness (HV)
1	HVOF-sprayed Cr_3C_2-25(NiCr) [S1]	H11	450±23 to 825±41	712±36
2	HVOF-sprayed Cr_3C_2-25(NiCr) [S1]	H13	425±29 to 800±40	700±35
3	AP-sprayed Cr_3C_2-25(NiCr)APS [S2]	H11	520±26 to 827±41	735±37.
4	AP-sprayed Cr_3C_2-25(NiCr)APS [S2]	H13	430±22 to 830±42	734±37
5	HVOF-sprayed NiCrBSi [C1]	H11	510±25 to 717±36	638±34
6	HVOF-sprayed NiCrBSi [C1]	H13	633±31 to 810±45	695±38
7	AP-sprayed NiCrBSi [C2]	H11	484±24 to 626±31	565±29
8	AP-sprayed NiCrBSi [C2]	H13	411±25 to 669±38	588±32
9	HVOF-sprayed 65%(NiCrSiFeBC)+35%(WC-12Co)[C4]	H11	691±36 to 1185±83	933±69
10	HVOF-sprayed 65%(NiCrSiFeBC)+35%(WC-12Co)[C4]	H13	688±43 to 1150±90	916±78
11	AP-sprayed 65%(NiCrSiFeBC)+35%(WC-12Co)[C5]	H11	630±48 to 1150±90	829±67
12	AP-sprayed 65%(NiCrSiFeBC)+35%(WC-12Co)[C5]	H13	507±30 to 990±84	808±64

microhardness values for all the as-sprayed samples were evaluated and are given in Table 4.7. A sharp increase in the hardness values was observed after the development of coatings on the steel specimens. Among all the coated samples, the HVOF-sprayed [C4] coatings on H11 steel showed the highest average value of microhardness. The higher hardness of the coated specimens indicated the retention of harder carbides during the spraying process. Also, excellent cohesion between the inter splat layers and between the matrix and reinforced phases led to higher hardness in the coated specimens.

4.6.3 XRD Analysis

The various phases present on the strong-, medium-, and weak-intensity peaks were identified from the XRD diffractograms. Table 4.8 shows the various phases present on the as-spayed coated H11 specimens. Similar phases were observed on the H13 substrate material.

TABLE 4.8

Various Phases Present on the As-Spayed Coated Specimens

Sr. No.	Coating	Substrate Steel	Strong Intensity Phases	Medium Intensity Phases	Weak Intensity Phases
1	HVOF-sprayed Cr_3C_2-25 (NiCr) [S1]	H11	Cr_3C_2 and Cr_7C_3	NiCr	Ni_3Cr and C
2	AP-sprayed Cr_3C_2-25 NiCrAPS [S2]	H11	Cr_3C_2 and Cr_7C_3	NiCr	Cr_2O_3
3	HVOF-sprayed NiCrBSi [C1]	H11	Ni	Cr_7C_3, Ni_3B, and CrB	—
4	AP-sprayed NiCrBSi [C2]	H11	Ni	Cr_7C_3	NiC and CrB
5	HVOF-sprayed 65%(NiCrSiFeBC)+ 35%(WC-12Co)[C4]	H11	Ni and WC	CrB	W and W_2C
6	AP-sprayed 65%(NiCrSiFeBC)+ 35%(WC-12Co)[C5]	H11	Ni and W_2C	FeSi, Cr_7C_3, NiB, and Co	—

4.6.4 SEM/EDS Examination along the Cross Section

The cross-sectional micrographs showed three regions namely substrate, coating, and epoxy. The average thickness of the coatings was measured along the cross section of the coatings. Figure 4.9 shows the cross-sectional morphologies of the HVOF spray [C1] coating on two steel specimens. The HVOF coatings showed a dense appearance with nearly uniform thickness. Few pores and voids were present in the coating. The interface between the coating and the substrate was almost free from defects with exception of some interfacial voids. The EDS analysis showed the existence of all the vital elements of the coating powder. The presence of some amount of oxygen indicated the probability of oxidation of coating elements. The cross-sectional coating microstructure for the HVOF spray [C4] coating on the H11 and H13 steel mainly comprises laminar splats (Figure 4.10). The splats were seen to be bounded with each other and with the substrate as well. The microstructures contained mainly W and C, indicating the probability of formation of the WC phase in the light regions. The darker phase in the micrographs showed the presence of mainly Ni, Cr, B, Si, and C. The WC grains remained almost solid during spraying. The grains were flattened by their own impact and the compressive forces generated by the subsequent layers. However, a small number of voids are present at the interface (Figure 4.11).

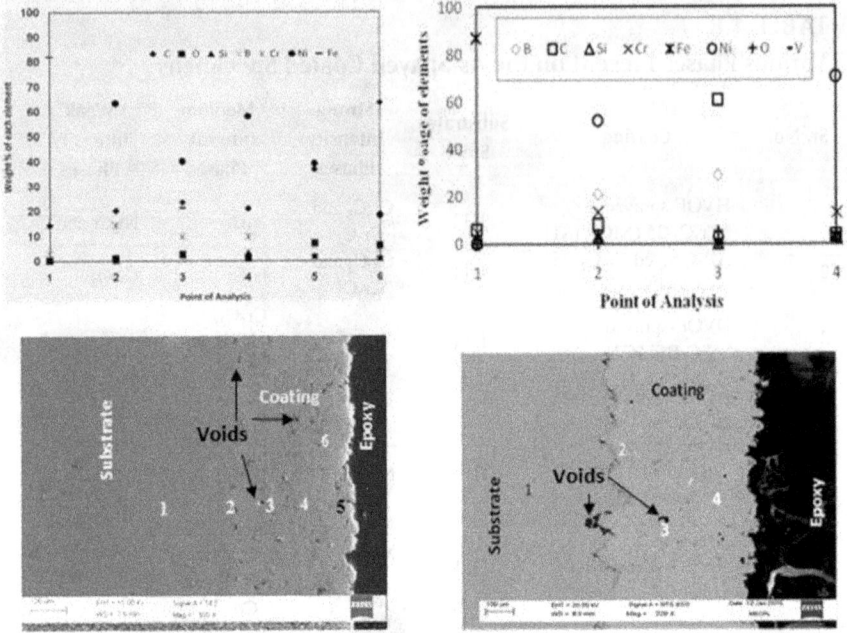

FIGURE 4.10 The SEM morphology along the cross section and the elemental composition for the NiCrBSi coating developed on (a) H11 die steel and (b) H13 die steel using HVOF technique.

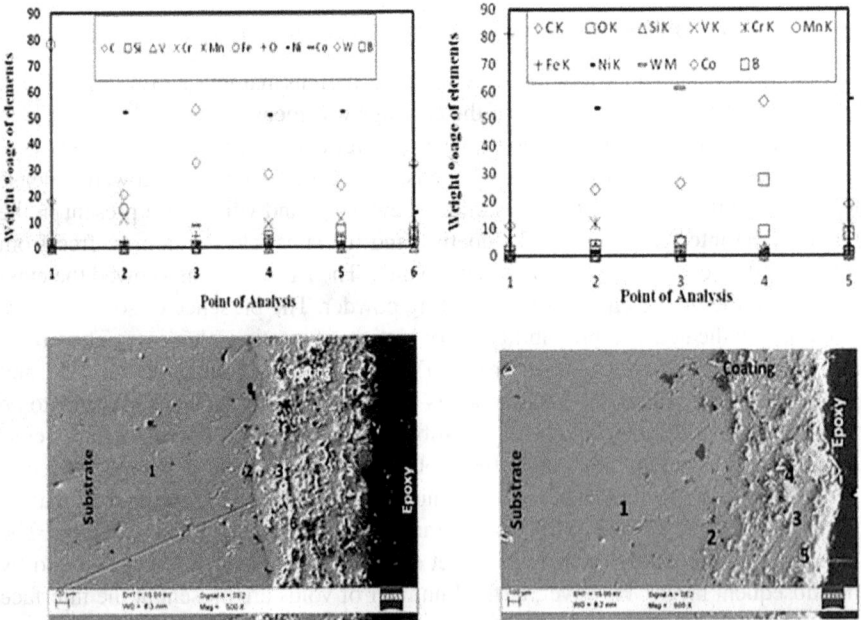

FIGURE 4.11 The SEM morphology along the cross section and the elemental composition for the [(NiCrFeBSiC)-35%(WC-12Co)] coating developed on (a) H11 die steel and (b) H13 die steel using HVOF technique.

4.7 RESULTS OF THE TRIBOLOGICAL STUDIES OF THE WORN-OUT SPECIMENS

4.7.1 Visual Examination of the Uncoated Steel Specimens

The camera photos of the worn-out uncoated specimens are shown in Figure 4.12. At room temperatures and at the two test loads, both the die steel pins showed deep grooves on the pin surface. At 400°C and 800°C, the macrographs indicate the shallow sliding marks along with some material pull-out as shown in Figure 4.12. Some black-colored patches and the glazed top layer were observed toward the end of the experimentation. Deep parallel scratches were seen on the surface.

4.7.2 Wear and Friction Behavior of the Uncoated Steel Specimens

Figures 4.13 (a) and (b) demonstrate the average values of coefficients of friction (COF) vs. temperature for the uncoated H11 and H13 specimens evaluated under two test loads. At 400°C and at 800°C, the average COF values decreased from the values obtained at the RT. The wear rates of the worn H11 steel were found as $609.91 \pm 13 \times 10^{-6}$ mm^3/Nm and $487.17 \pm 10 \times 10^{-6}$ mm^3/Nm under 25 N load and 50 N load, respectively (Figure 4.14) at RT. The wear rates for the H13 steel were found as

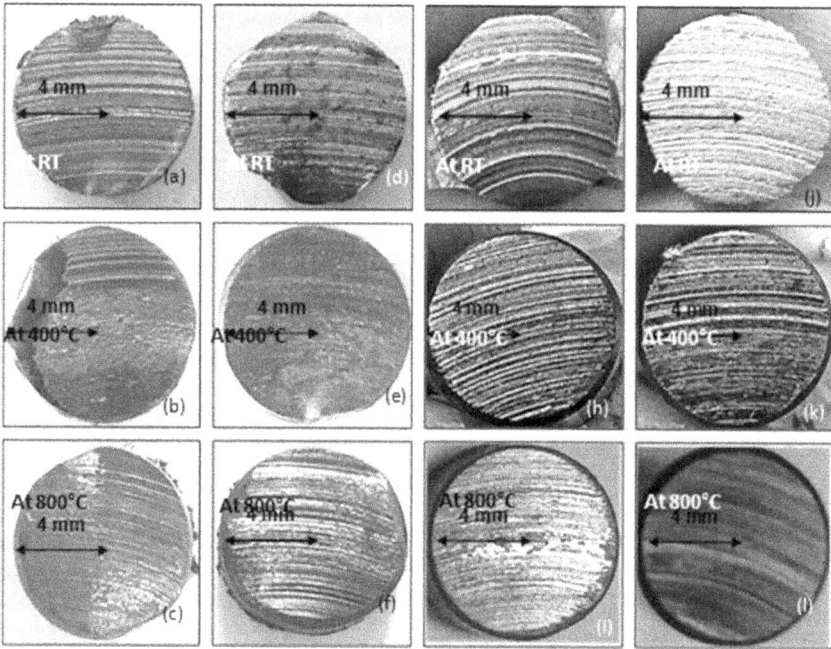

FIGURE 4.12 Camera pictures of the worn-out surface of H11 steel specimens (a), (b), and (c) at 25 N load; H11 specimens (d), (e), and (f) at 50 N load; H13 specimens (g), (h), and (i) at 25 N load; H13 specimens (j), (k), and (l) at 50 N load after sliding wear tests for 50 minutes duration.

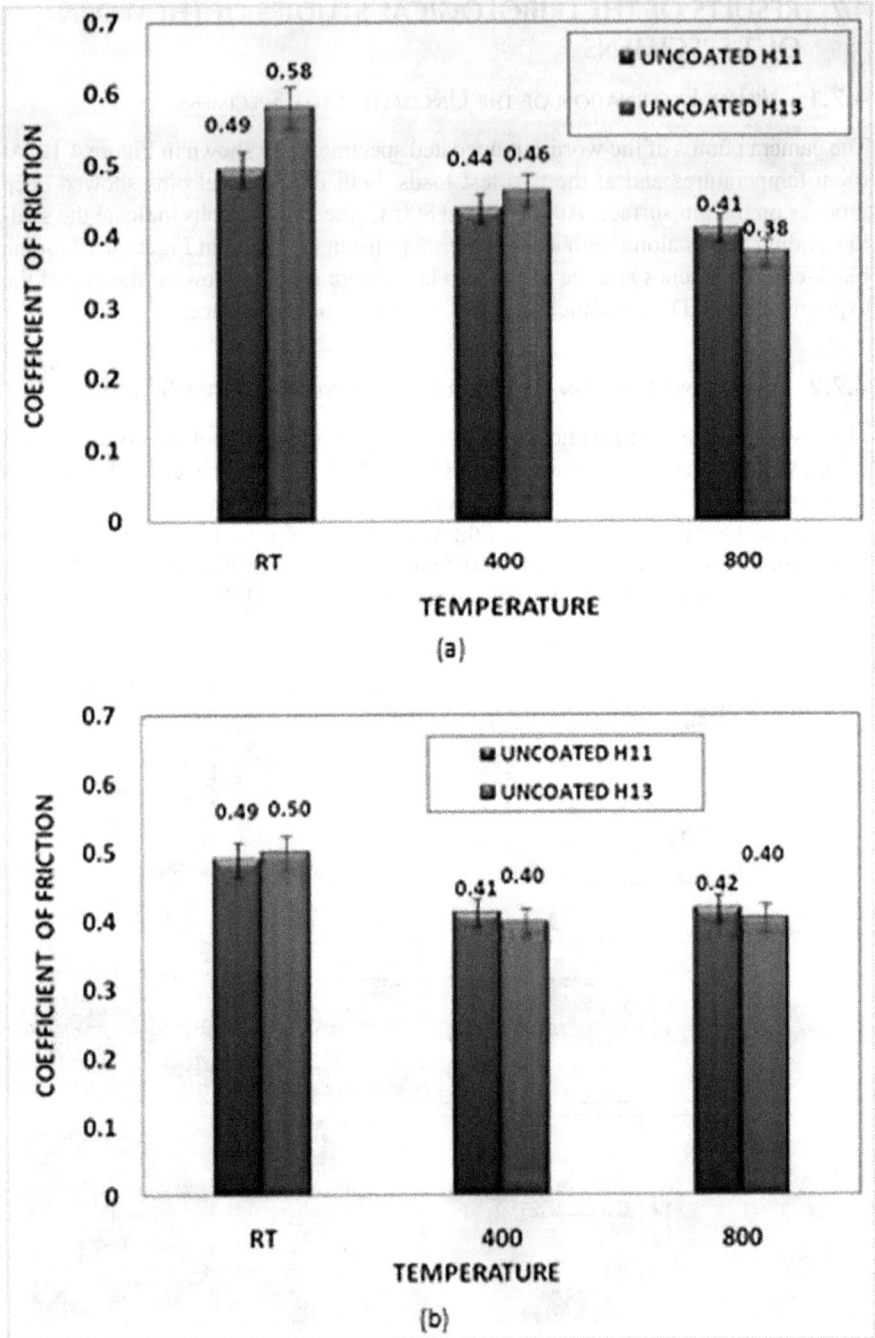

FIGURE 4.13 The average COF values of the uncoated steels after sliding wear tests under (a) 25 N loads (b) 50 N loads for 50 minutes duration.

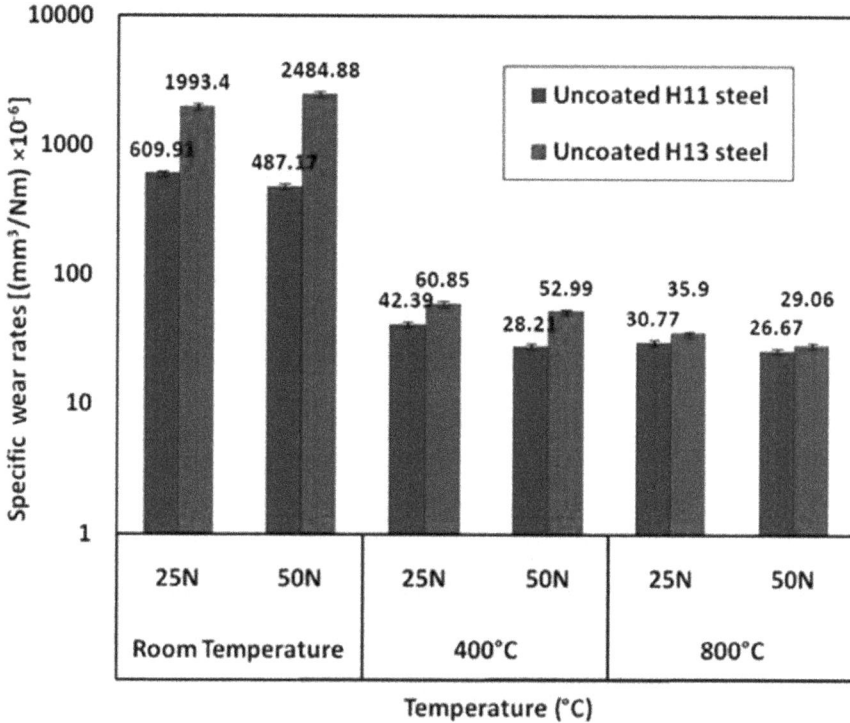

FIGURE 4.14   Specific wear rates [(mm³/Nm) ×10⁻⁶] vs. temperature plots for the uncoated steels.

1993.40 ± 88 × 10⁻⁶ mm³/Nm and 2484.88 ± 97 × 10⁻⁶ mm³/Nm at both the loads. At 400°C, the wear rates decreased for both the steels. The decrease was noticeable. The wear rates values for the H11 steels were 42.39 ± 0.9 × 10⁻⁶ and 28.21 ± 0.45 × 10⁻⁶ mm³/Nm at 25 N load and 50 N load, respectively. Similarly, for the H13 steel, the wear rate values were reported to be 60.85 ± 3.1 × 10⁻⁶ mm³/Nm and 52.99 ± 2.65 × 10⁻⁶ mm³/Nm at the two loads. On increasing the test temperature to 800°C, the wear rates again decreased. For the H11 steel, the wear rates were evaluated as 30.77 ± 0.75 × 10⁻⁶ mm³/Nm and 26.67 ± 0.55 × 10⁻⁶ mm³/Nm at the two test loads. For the H13 steel, the wear rate values were evaluated as 35.9 ± 0.96 × 10⁻⁶ mm³/Nm and 29.06 ± 0.61 × 10⁻⁶ mm³/Nm for 25 N load and 50 N load, respectively.

4.7.3   SEM-EDS EXAMINATION OF THE UNCOATED STEEL SPECIMENS

Figures 4.15 to 4.18 show the SEM images and the EDS results of the worn steel specimens at room temperature, 400°C, and 800°C at both the testing loads. At RT, the worn track exhibits a small amount of debris, material pullouts, abrasive wear marks, and some adhered material (Figure 4.15a). At higher loads, Figure 4.16a), worn surface exhibits delaminated layers with more amounts of scabs and debris. In the H13 specimen (Figure 4.17a), abrasive wear marks and delaminated regions

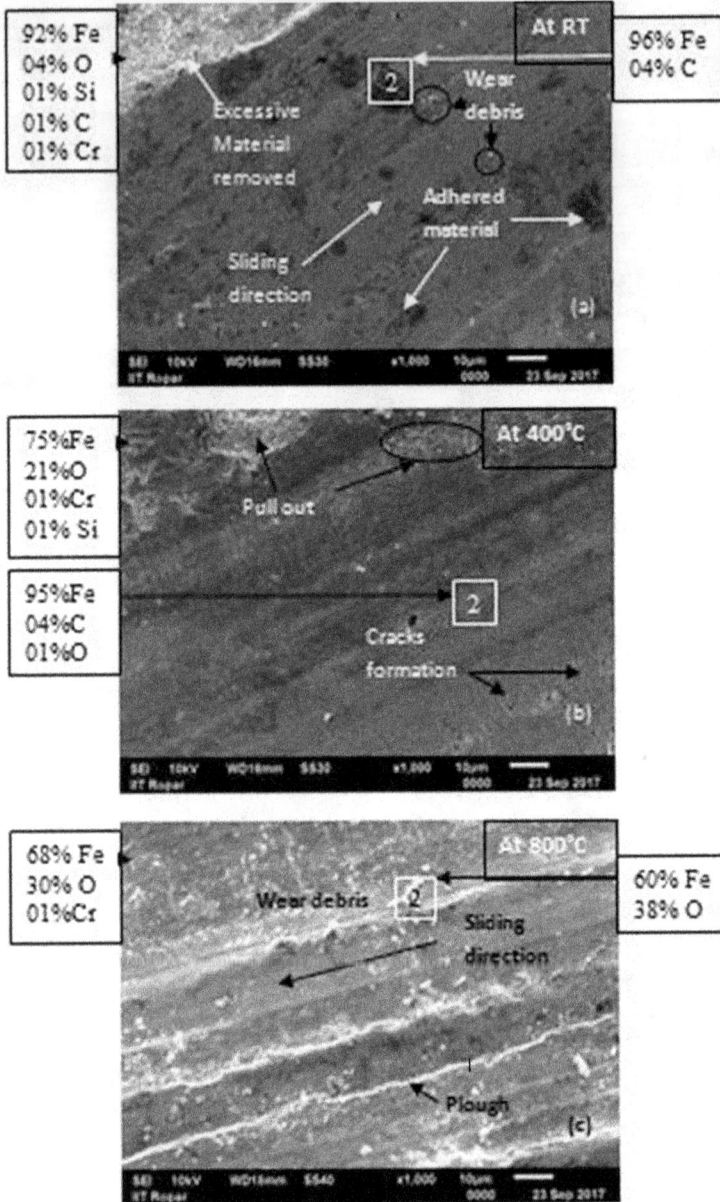

FIGURE 4.15 The SEM/EDS results after wear tests at three temperatures for the uncoated H11 steels specimens under a constant load of 25 N.

are visible. At 50 N load, (Figure 4.18a) the abrasive wear tracks appear deeper, and delaminated regions are more widespread than that at the lower loads. During the experimentation, as time progressed, the delaminated layers peeled off and there was a huge material loss. Higher specific wear rates shown in Figure 4.14

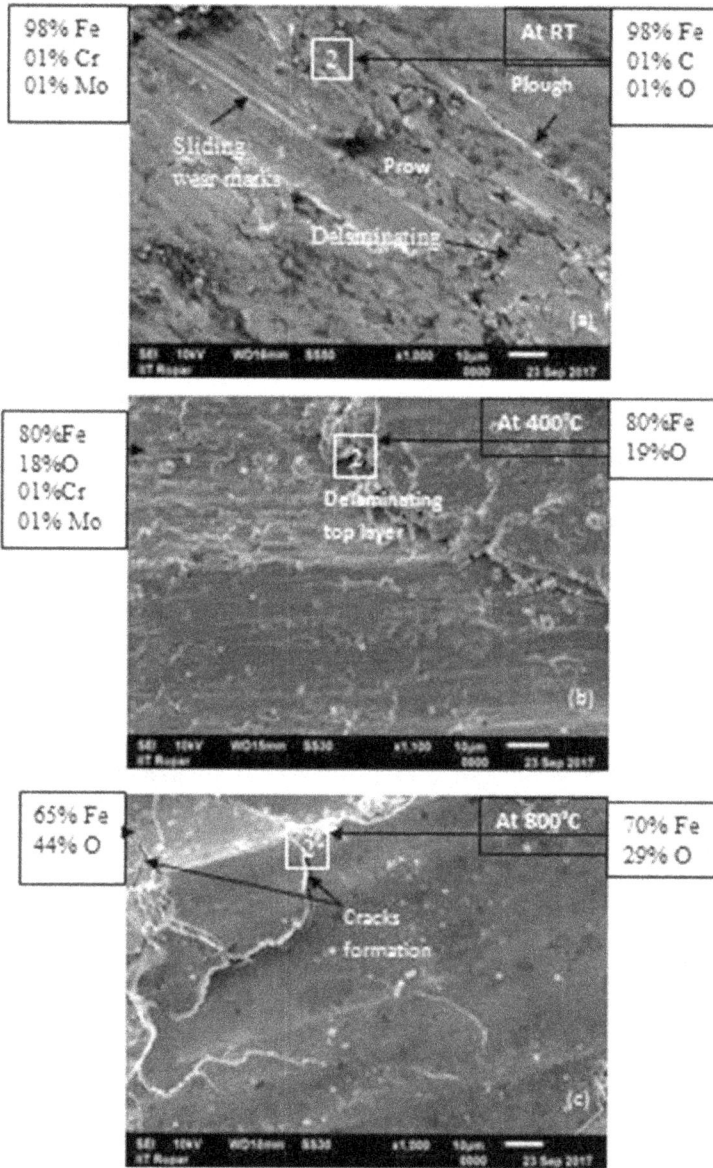

FIGURE 4.16 The SEM/EDS results after wear tests at three temperatures for the uncoated H11 steels specimens under a constant load of 50 N.

endorse these observations. The adhesive and abrasive wear mechanisms were recognized.

Figures 4.15b and 4.16b show the worn-out SEM images of the H11 steels tested at 400°C and two loads, respectively. With the increase in temperature, the surface of the pin showed the presence of an oxidized layer with delamination of material at

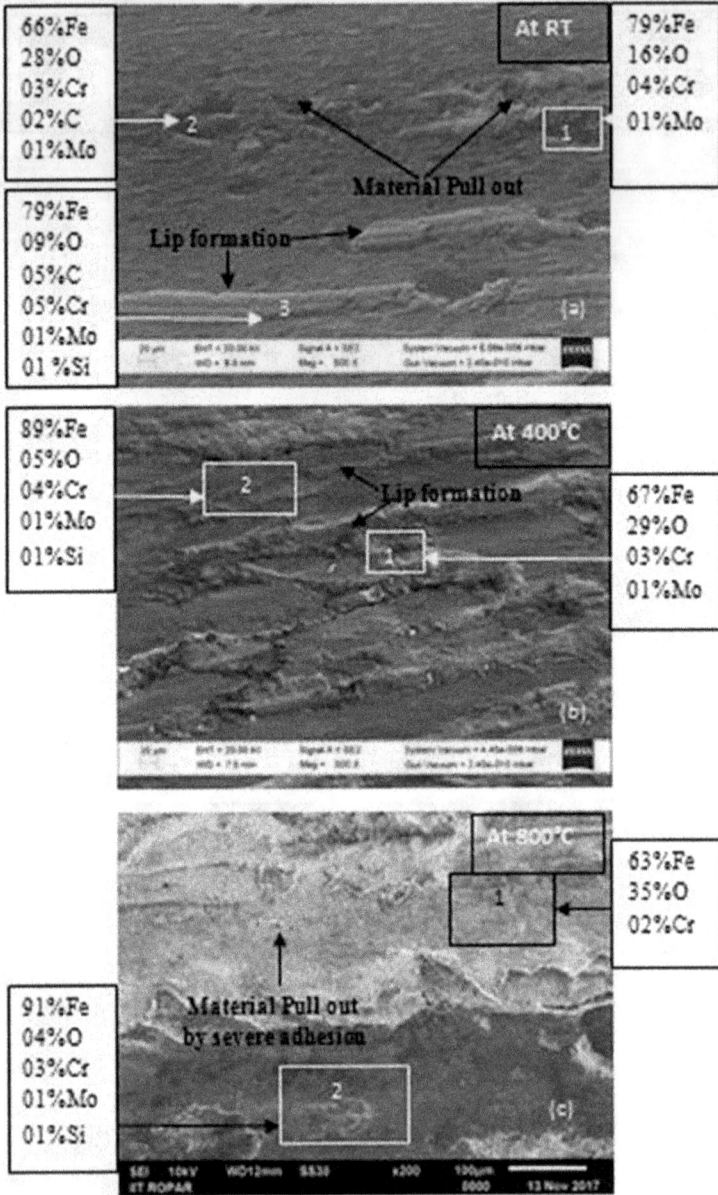

FIGURE 4.17 The SEM/EDS results after wear tests at three temperatures for the uncoated H13 steels specimens under a constant load of 25 N.

certain sites, micro-cracks, pull-out, and wear debris. The SEM morphology indicated mild oxidation and adhesion as the main wear mechanisms at 50 N load (Figure 4.16b). The morphology of worn-out H13 specimens are shown in Figures 4.17b and 4.18b. The images showed the occurrence of worn oxide layers at both loads.

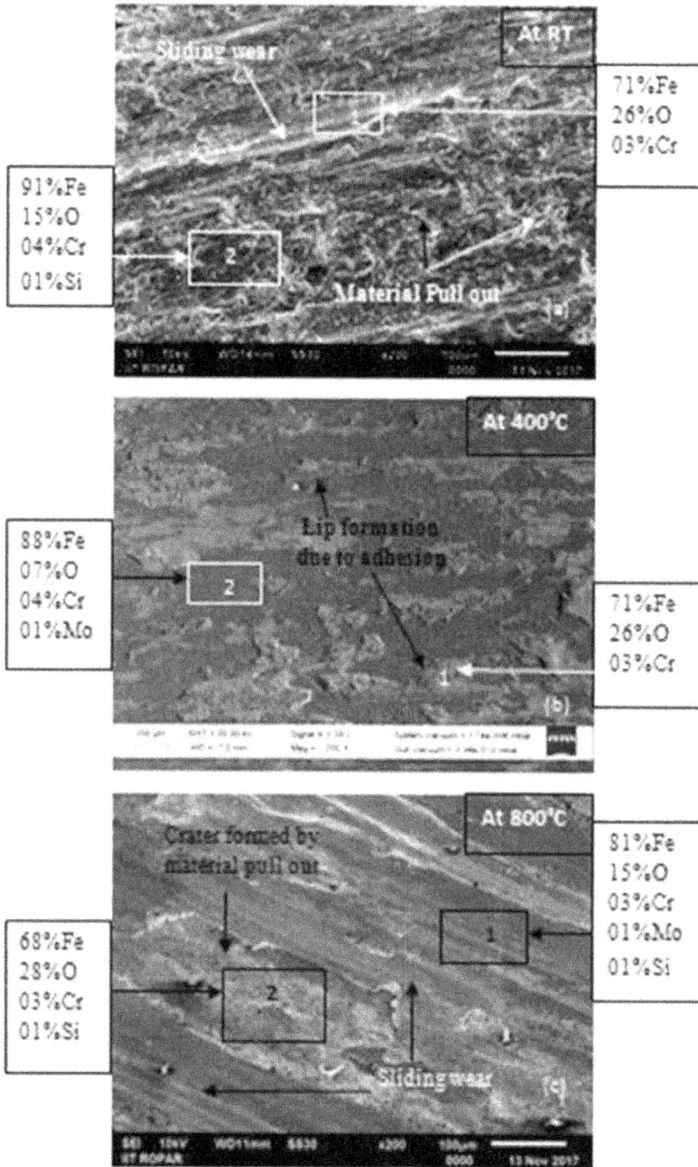

FIGURE 4.18 The SEM/EDS results after wear tests at three temperatures for the uncoated H13 steels under a constant load of 50 N.

At 800°C and 25 N (Figure 4.15c), the H11 specimen showed deep plowing marks and debris clusters on the surface. At 50 N load (Figure 4.16c), deep cracks are visible on the surface along with light scratches. The mode of wear was found to be oxidative and abrasive at 25 N load and the combination of all three oxidative, abrasive, and adhesive at higher loads. The SEM images of the worn H13 pins

(Figures 4.17c and 4.18c) also show large delaminated sites filled with wear debris. Also, abrasive wear tracks are visible on the surface of pins tested at 50 N load. The EDS analysis indicates the occurrence of Fe and O on the worn surface of the pin. The O is found in significant amounts when exposed to wear at higher temperatures. This probably indicates the formation of iron oxide on the top layer of the specimens.

4.8 TRIBOLOGICAL BEHAVIOUR OF THE HVOF-COATED H11 AND H13 STEELS

4.8.1 Visual Examination

The camera macrographs of worn surfaces of the HVOF spray Cr_3C_2-25(NiCr) [S1] coated steel specimens are shown in Figure 4.19 at all testing conditions. At room temperatures, faint burnishing marks and material pull-outs covering a part of the pin surface were observed. More material pull-out sites were seen with the increase in load at 50 N. The specimens tested at 400°C and 800°C temperatures showed clear sliding marks covering the portion of the pin surface. Also, some adhered dark patches could be seen on the pin surface. The deeper sliding marks with some shiny scratches were seen on the specimens tested at 800°C.

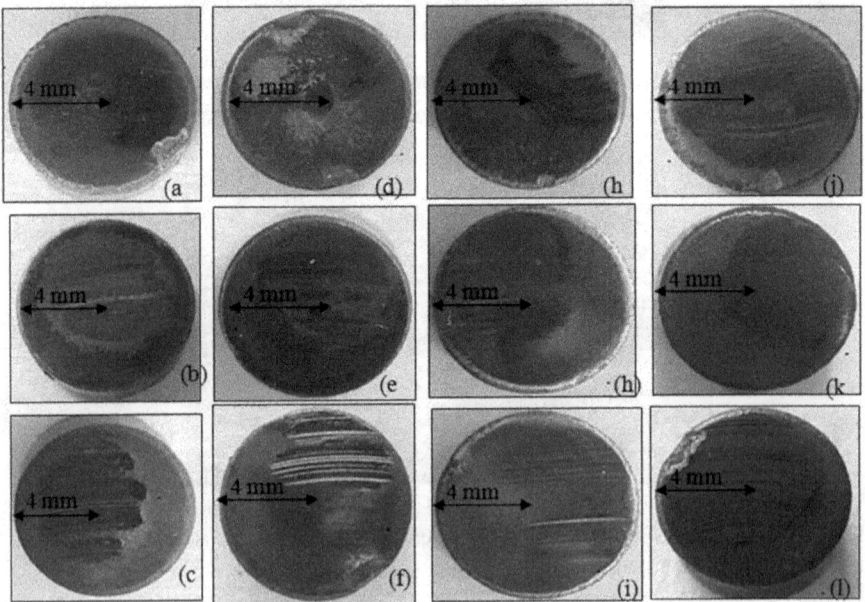

FIGURE 4.19 Macrographs of worn surfaces of HVOF spray Cr_3C_2-25(NiCr) [S1] coated H11 specimens at 25 N load (a) RT, (b) 400°C, (c) 800°C, and at 50 N load (d) RT, (e) 400°C, (f) 800°C, HVOF spray Cr_{3C2}.25(NiCr) [S1] coated H13 specimens at 25 N load (g) RT, (h) 400°C, (i) 800°C, and at 50 N load (j) RT, (k) 400°C, (l) 800°C after wear tests for 50 minutes duration.

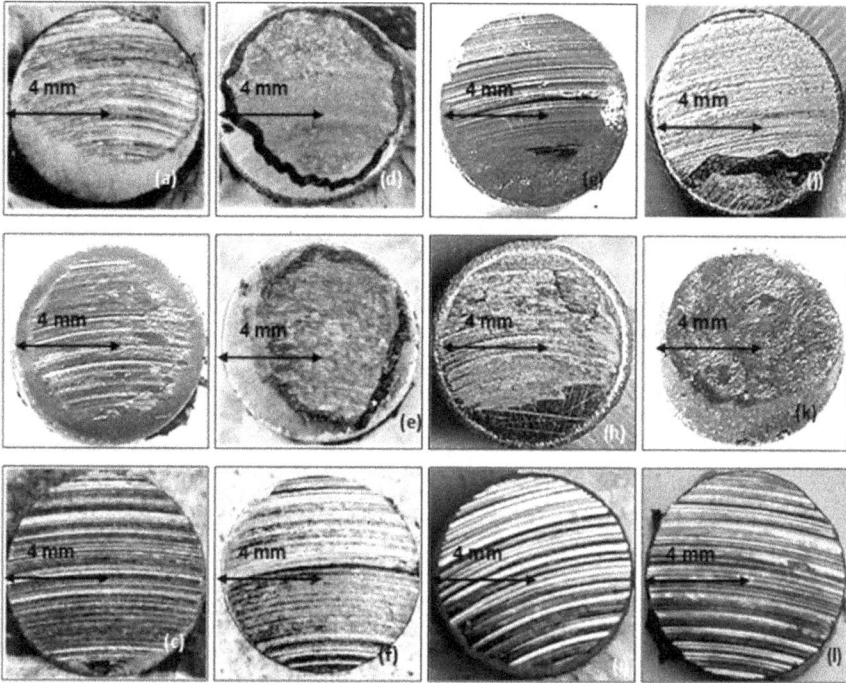

FIGURE 4.20   Macrographs of worn surfaces of HVOF spray NiCrBSi [C1] coated H11 specimens at 25 N load (a) RT, (b) 400°C, (c) 800°C, and at 50 N load (d) RT, (e) 400°C, (f) 800°C, HVOF spray NiCrBSi [C1] coated H13 specimens at 25 N load (g) RT, (h) 400°C, (i) 800°C, and at 50 N load (j) RT, (k) 400°C, (l) 800°C after wear tests for 50 minutes duration.

The worn surfaces of the HVOF spray NiCrBSi [C1] coated and the HVOF spray (65%NiCrBSi-35%(WC-Co)) [C4] coated steel specimens subjected to sling wear experimentation at three test temperatures as observed by the naked eye are shown in Figures 4.20 and 4.21. Both the coated pin specimens showed more wear resistance as compared to the uncoated ones. At elevated temperatures (400°C), the macrographs appeared to be dark with fine scratch marks. Some dark patches were present on the top surface (Figures 4.27 and 4.28). At 800°C, three distinct regions were observed. The first region has deep sliding wear marks, the second was a black colored layer, and the third, a shiny layer was observed at the end of the experimentation. Some adhered material was also observed on the pin surface.

4.8.2   TRIBOLOGICAL BEHAVIOUR

Table 4.9 shows the average values of the COF for the H11 and H13 steels coated with the HVOF spray technique obtained after testing at three temperature conditions and at two loads, respectively. It was observed that at higher loads, the average COF values for both the Cr_3C_2-25(NiCr) [S1] coated steel specimens got reduced. At 400°C test temperature, there was a notable decrease in the average

FIGURE 4.21 Macrographs of worn surfaces of HVOF spray (NiCrBSi)-35%(WC-Co) [C4] coated H11 specimens at 25 N load (a) RT, (b) 400°C, (c) 800°C, and at 50 N load (d) RT, (e) 400°C, (f) 800°C, and HVOF spray (NiCrBSi)-35%(WC-Co) coated H13 specimens at 25 N load (g) RT, (h) 400°C, (i) 800°C, and at 50 N load (j) RT, (k) 400°C, (l) 800°C after wear tests for 50 minutes duration.

COF values. The average COF values enhanced when testing temperature increased to 800°C and at 25 N loads while the average COF values remained stable at the higher loads.

Figure 4.22a and b shows the specific wear rate plots for the [S1] coating on both the tool steels at all test conditions. It was observed that the wear rates for the coated specimens decreased significantly for the coated steels at room temperature testing, in comparison to the uncoated steel specimens. At 400°C, the wear rate values again decreased for the coated ones. Further, at 800°C, the specific wear-rate values for the coated specimens were found to increase.

Figures 4.23 and 4.24 showed the average COF values obtained during experimentation for the HVOF spray [C1] and [C4] coated specimens, respectively. The values were found to be higher at RT and a further decrease in the values was observed at 400°C and 800°C for both the coated steel specimens all the testing conditions. Figure 4.25 shows the specific wear rates of the HVOF-sprayed specimens at three temperatures and two loads. The appreciable decrease in the specific wear rate values at 400°Cand at 25 N load were observed. The values increased slightly at 800°C. However, the values were found to be lower than the values of the uncoated specimens.

TABLE 4.9
Average Values of Friction Coefficients Obtained during Testing

Sr. No.	Coating	Substrate steel	Average COF value at three test temperatures					
			RT		400°C		800°C	
			25 N Loads	50 N Loads	25 N Loads	50 N Loads	25 N Loads	50 N Loads
1	HVOF-sprayed Cr_3C_2-25(NiCr) [S1]	H11	~0.71	~0.57	~0.40	~0.43	~0.43	~0.42
2	HVOF-sprayed Cr_3C_2-25(NiCr) [S1]	H13	~0.64	~0.53	~0.41	~0.42	~0.51	~0.45
3	HVOF-sprayed NiCrBSi [C1]	H11	~0.59	~0.49	~0.42	~0.43	~0.38	~0.38
4	HVOF-sprayed NiCrBSi [C1]	H13	~0.48	~0.51	~0.33	~0.32	~0.47	~0.50
5	HVOF-sprayed 65%(NiCrSiFeBC)+ 35%(WC-12Co) [C4]	H11	~0.53	~0.52	~0.30	~0.43	~0.51	~0.59
6	HVOF-sprayed 65%(NiCrSiFeBC)+ 35%(WC-12Co) [C4]	H13	~0.42	~0.52	~0.33	~0.32	~0.50	~0.56

4.8.3 SEM-EDS Examination of Worn Specimens

The FE-SEM images of worn-out Cr_3C_2-25(NiCr) [S1] coated H11 and H13 specimens after the wear and friction experimentation at room temperatures and at two loads are shown in Figures 4.26a, 4.27a, 4.28a, and 4.29a respectively. The abrasive wear marks were mild and shallow on the coated specimens. The wear mechanism was found to be mainly adhesive in nature. Mild abrasive grooving was also observed. Small craters, material pull-outs, and wear debris were observed on the surfaces. The presence of Cr, C, and Ni elements in major percentages was observed in EDS analysis reports. This proved the existence of the coating elements on the worn surface of the tested pins.

The FE-SEM images for the [S1] coated specimens at 400°C and at both the loads are shown in Figures 4.26b, 4.27b, 4.28b, and 4.29b respectively. The adhered wear debris, shallow abrasive wear marks, and the solid compact oxide layers were observed. In Figure 4.29b, material pull-out and craters could be seen. The wear mechanism was found to be a blend of adhesive, abrasive, and oxidative wear. The EDS analysis confirmed the existence of coating elements along with the noticeable fraction of oxygen. This indicated the oxidation on the coating surface. Further, the EDS analysis of adhered debris sites showed a higher percentage of Fe and O. This indicates the transfer of disc material in form of oxidized debris to the pin surface forming compact oxide layers. The development of these solid compact oxide layers resulted in the reduction in specific wear rate values.

The FE-SEM images for the [S1] coated H11 and H13 specimens at 800°C and at 25 N and 50 N loads are shown in Figures 4.26c, 4.27c, 4.28c, and 4.29c respectively.

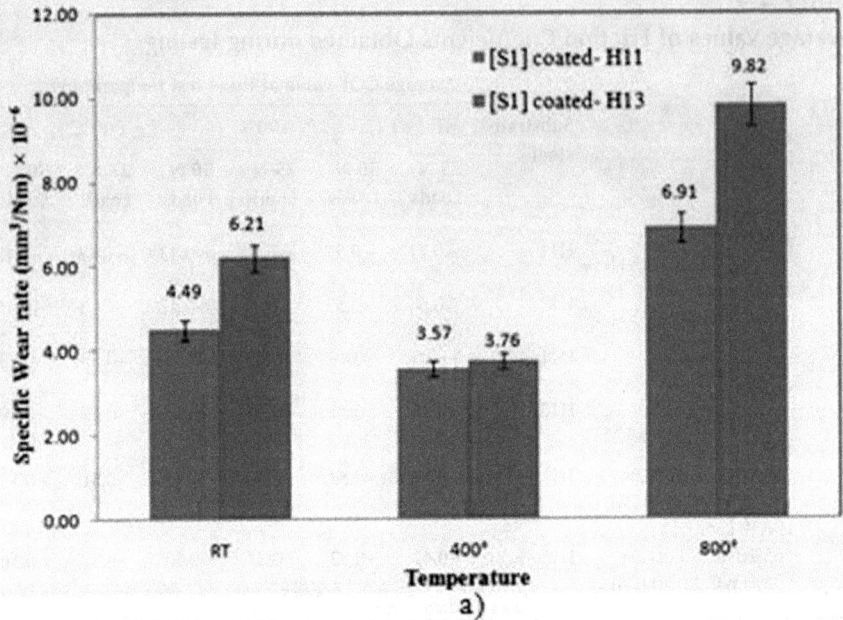

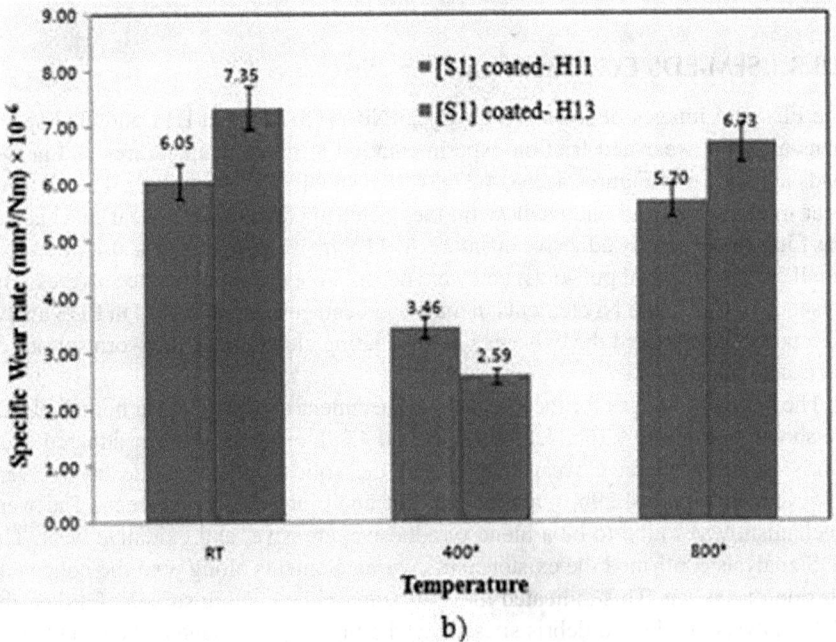

FIGURE 4.22 The specific wear rates plots of the HVOF-coated Cr_3C_2-25(NiCr) [S1] specimens obtained after wear tests under a constant load of (a) 25 N and (b) 50 N for 50 minutes duration.

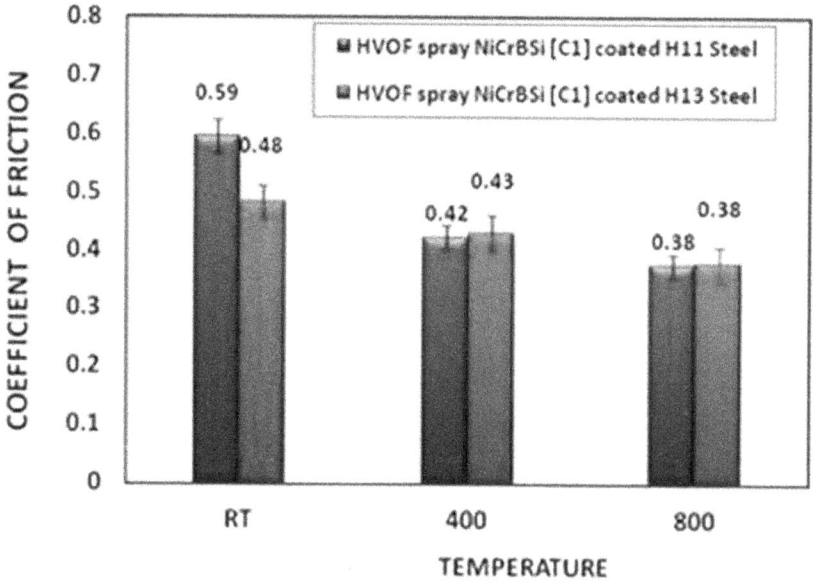

a)

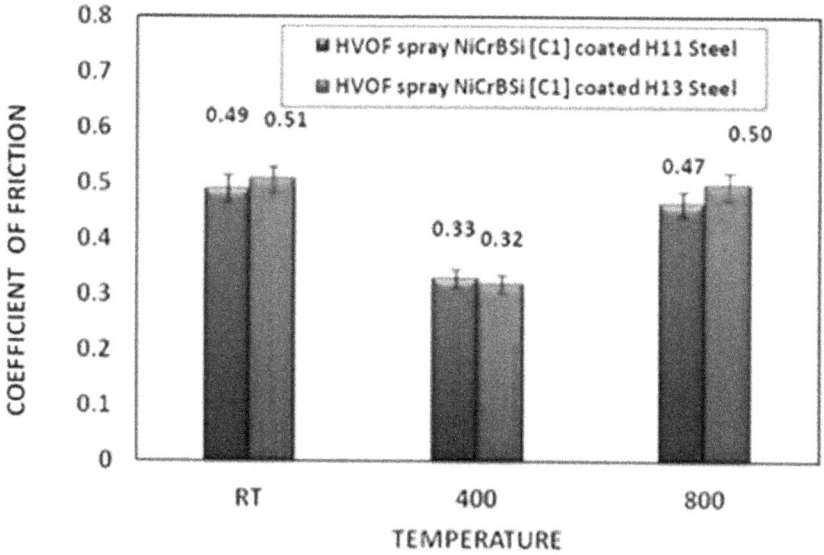

b)

FIGURE 4.23   The average friction coefficient vs. temperature plots for the HVOF spray NiCrBSi [C1] coated specimens obtained after wear test at (a) 25 N loads and (b) 50 N loads for 50 minutes duration.

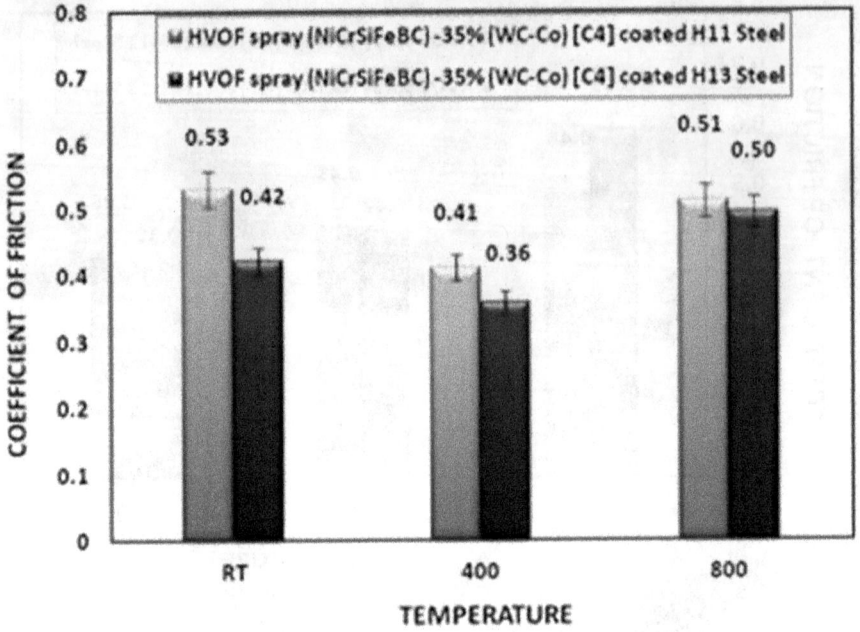

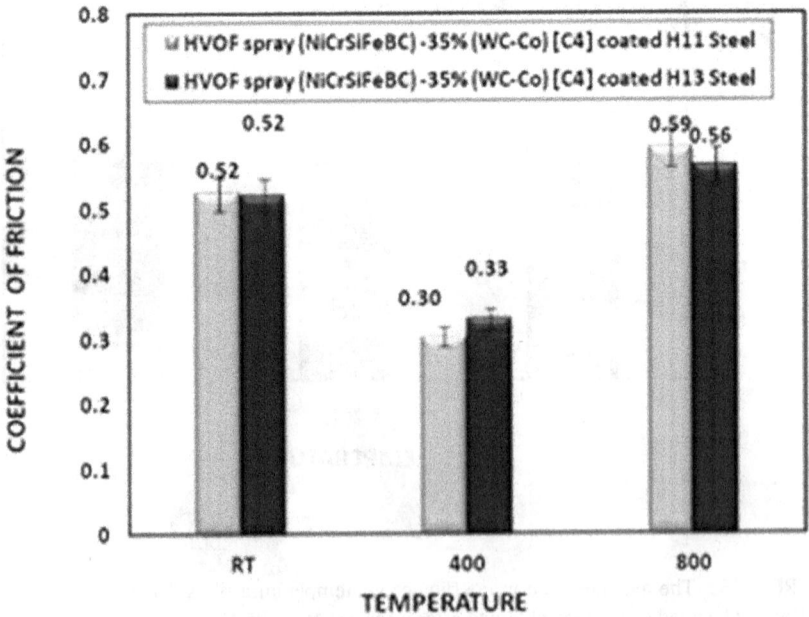

FIGURE 4.24 The average friction coefficient vs. temperature plots for the coated steel specimens obtained after wear tests at (a) 25 N loads and (b) 50 N loads for 50 minutes duration.

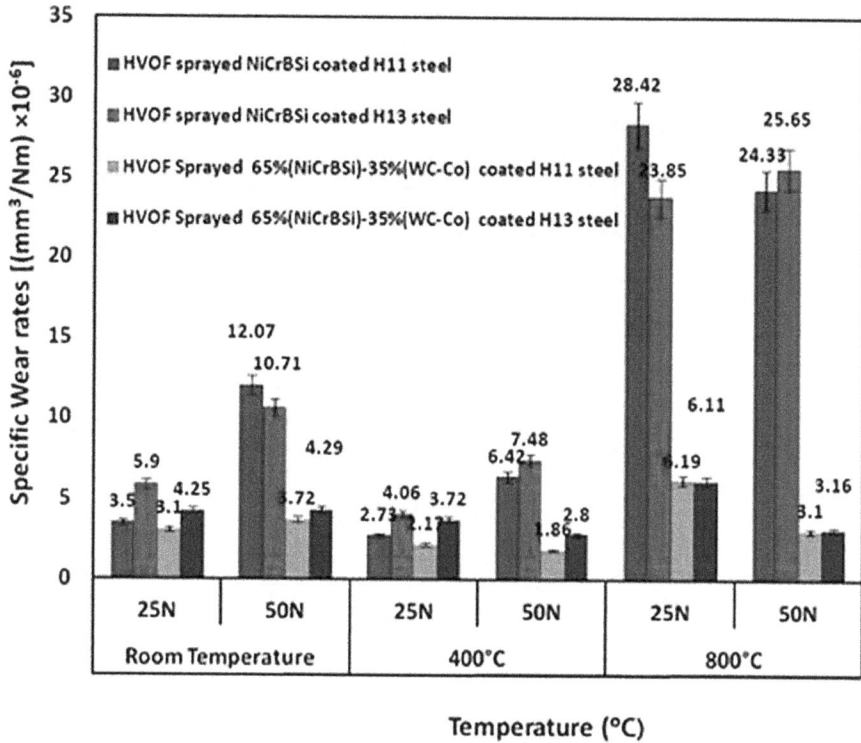

FIGURE 4.25   Specific wear rates vs. temperature plot for the coated specimens at different testing conditions.

The worn surface SEM images of the HVOF spray [C1] specimens at both the loads and at all the test temperatures are shown in Figures 4.30 and 4.31. At room temperatures, fewer pull-outs were seen as compared to the uncoated specimens. The adhesive wear mechanism was observed (Figures 4.30a and 4.31a). The disc material was found to adhere to the pin surface. The occurrence of Fe was found by the EDS analysis. A combination of fatigue and adhesive wear mechanism was observed at higher loads.

At 25 N loads and 400°C (Figures 4.30b and 4.31b), the wear mechanism in the coated specimens was found to be the combination of adhesion, abrasion, and oxidation. The occurrence of solid compact oxide layers and sliding wear marks were seen on the worn surfaces. Some patches of adhered material in the grooves could also be seen. These patches were formed from the oxides of coating elements and the disc material elements. The occurrence of the oxide layers might be one of the prominent factors in raising the wear resistance of the pin specimens. At 50 N loads (Figure 4.31b), a few delaminating sites and adhered debris were found to be present next to oxide patches. The cluster of adhered debris might be generated due to the breaking and spalling of these layers with the increase in load. However, the coating, by and large, seemed to be intact. The EDS results confirmed the existence of the main coating elements in the worn surface layer.

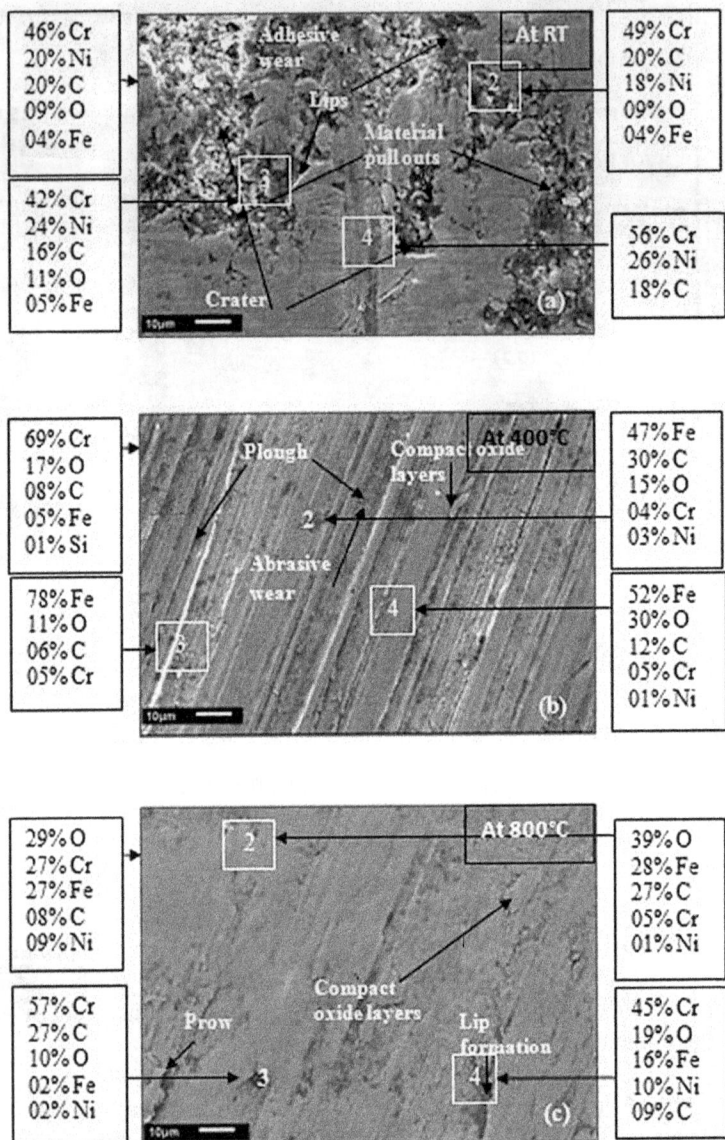

FIGURE 4.26 The SEM/EDS results after wear tests at three temperatures for the HVOF-coated Cr_3C_2-(NiCr) [S1] H11 steels under a constant load of 25 N.

The existence of abrasive wear marks, adhered material, and compact oxide layer was observed on the worn surface. All three abrasive, oxidative, and adhesive wear mechanisms were observed. The EDS analysis of the coated specimens confirmed the presence of coating elements, with significant amounts of O and Fe present in the top surface layer. The coating was found to be intact with the substrate at all test conditions.

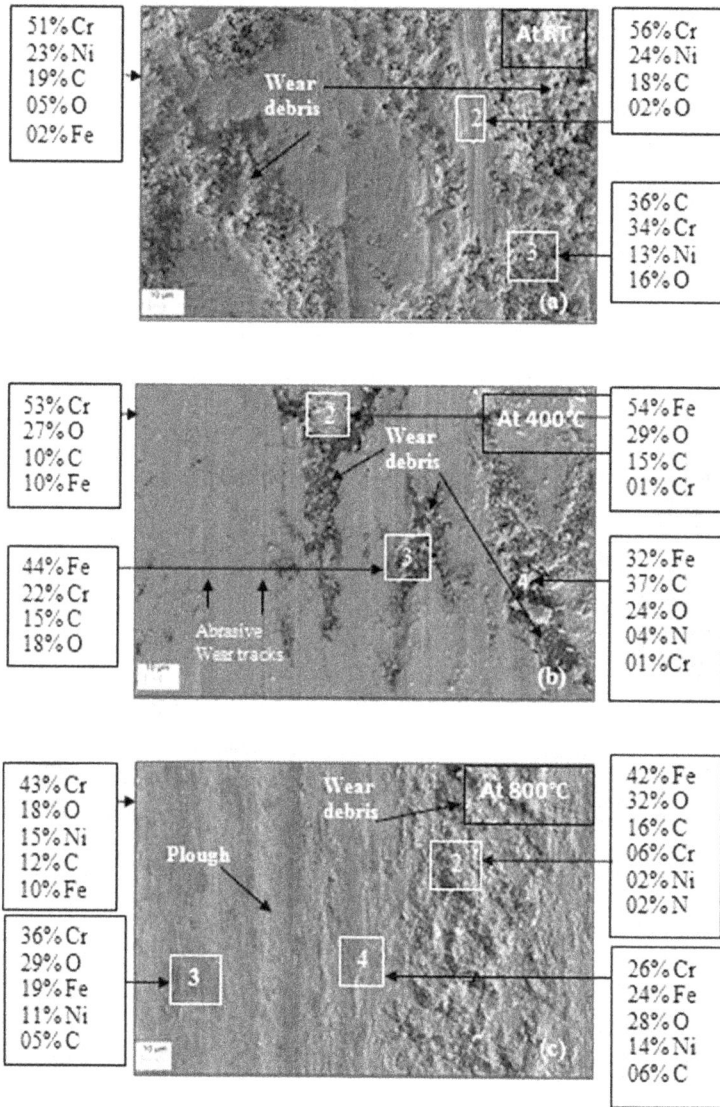

FIGURE 4.27 The SEM/EDS results after wear tests at three temperatures for the HVOF-coated Cr_3C_2-(NiCr) [S1] H11 steels under a constant load of 50 N.

At 25 N load and 800°C (Figures 4.30c and 4.31c), the morphology indicated the thick worn layer with the adherence of the material onto it. Deep abrasive wear marks were observed. The SEM morphology indicated that the wear mechanism is adhesive and abrasive. Moreover, at higher loads, the oxide layer was found to disintegrate at many sites. Large amounts of wear debris were observed. These factors might contribute to the higher wear rates at 800°C. It was notable that at both the loads, the

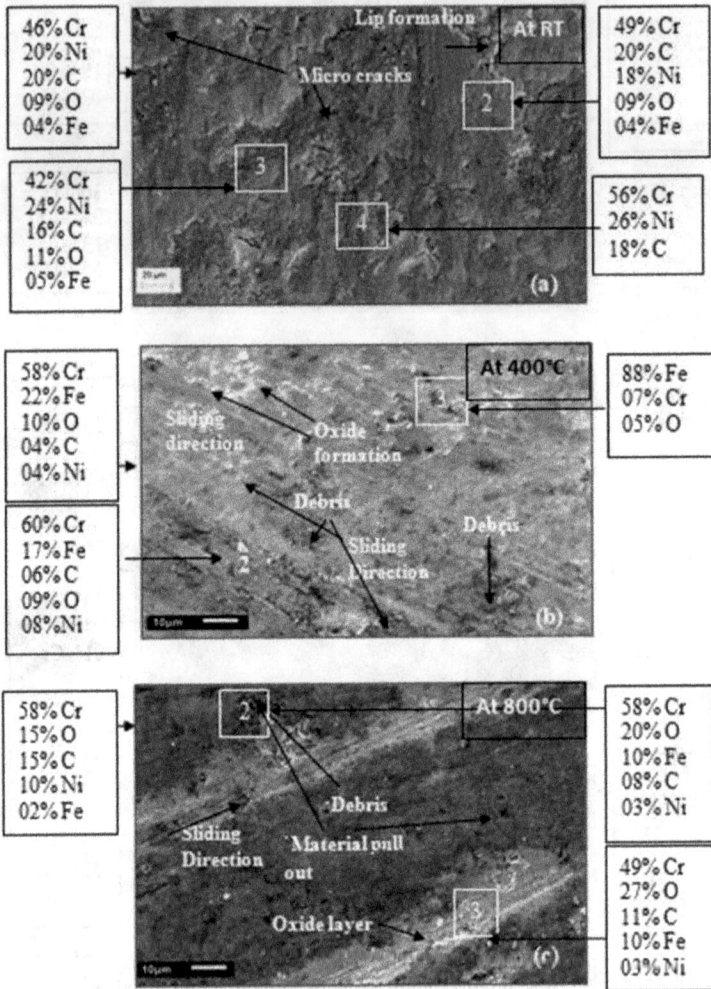

FIGURE 4.28 The SEM/EDS results after wear tests at three temperatures for the HVOF-coated Cr_3C_2-(NiCr) [S1] H13 steels under a constant load of 25 N.

coating and the substrate remained intact. The wear of the top layer could be seen, yet the worn scale seemed to be adherent to the substrate. The results were endorsed by the EDS analysis.

At room temperature, the shallow plowing marks and fewer delamination sites were observed in the SEM images of both the HVOF-sprayed [C4] worn-out specimens. The EDS results showed the existence of C and O, indicating the probability of the formation of hard carbides and oxides on the scale. The main wear mechanism was light abrasion along with adhesion. However, no major surface damage, such as cracks or WC detachment zones, were observed on the surface at both loads.

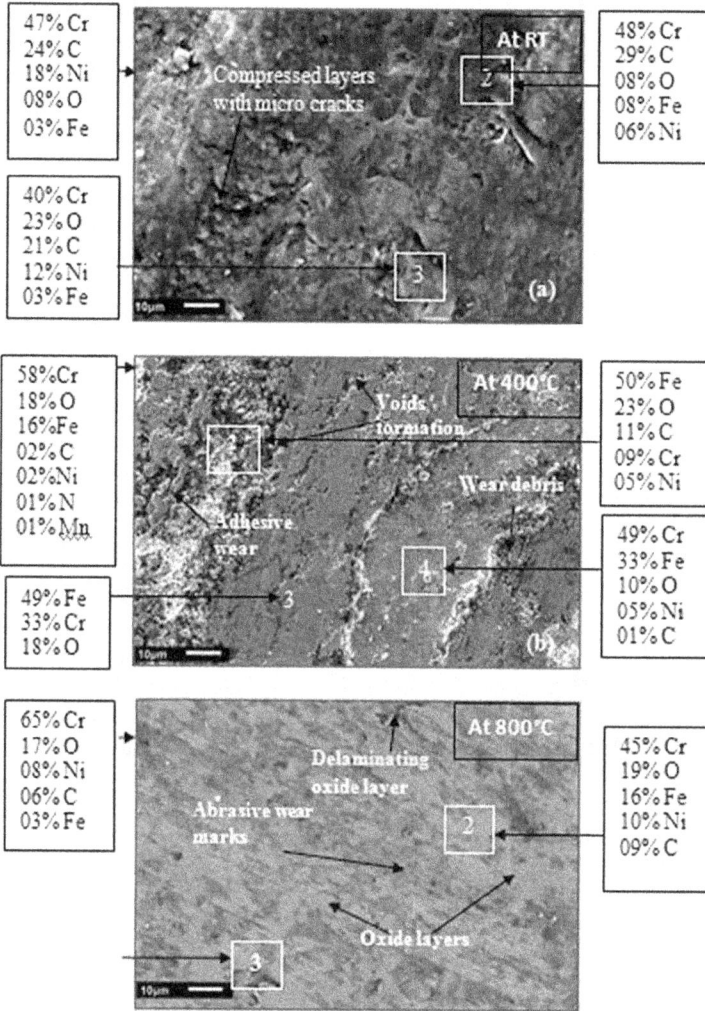

FIGURE 4.29 The SEM/EDS results after wear tests at three temperatures for the HVOF-coated Cr3C2-(NiCr) [S1] H13 steels under a constant load of 50 N.

The microstructures of the coated steels tested at 400°C showed the existence of evenly spread dark-colored oxides patches on the pin surface with no signs of delamination. The wear mode mechanism was abrasion, adhesion, and oxidation. Again at 800°C, the wear was due to a combination of adhesion, tribo-oxidation, and fatigue. The SEM images indicated the spallation of oxide layers and exposure of the fresh surface to oxidation. The phenomenon of such wear is called tribo-oxidative wear. The images showed the adhered wear debris along with disintegrated oxide layers on the pin surface. The formation of pits was observed. Moreover, the cracks were observed in WC phase at some sites, indicating the fatigue wear. Similarly, the tribological behaviour of atmospheric plasma coated H11 and H13 steels were also analyzed in a similar manner.

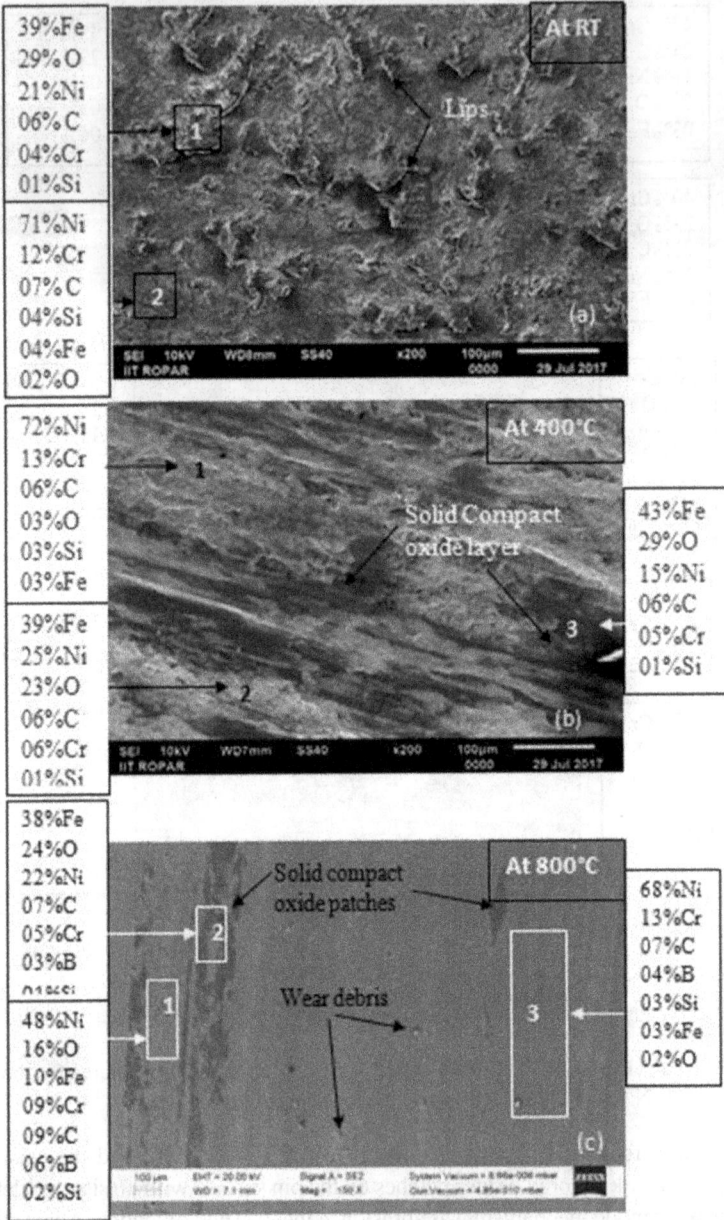

FIGURE 4.30 The SEM/EDS results after wear tests at three temperatures for the HVOF-coated NiCrBSi [C1] H11 steels under a constant load of 25 N.

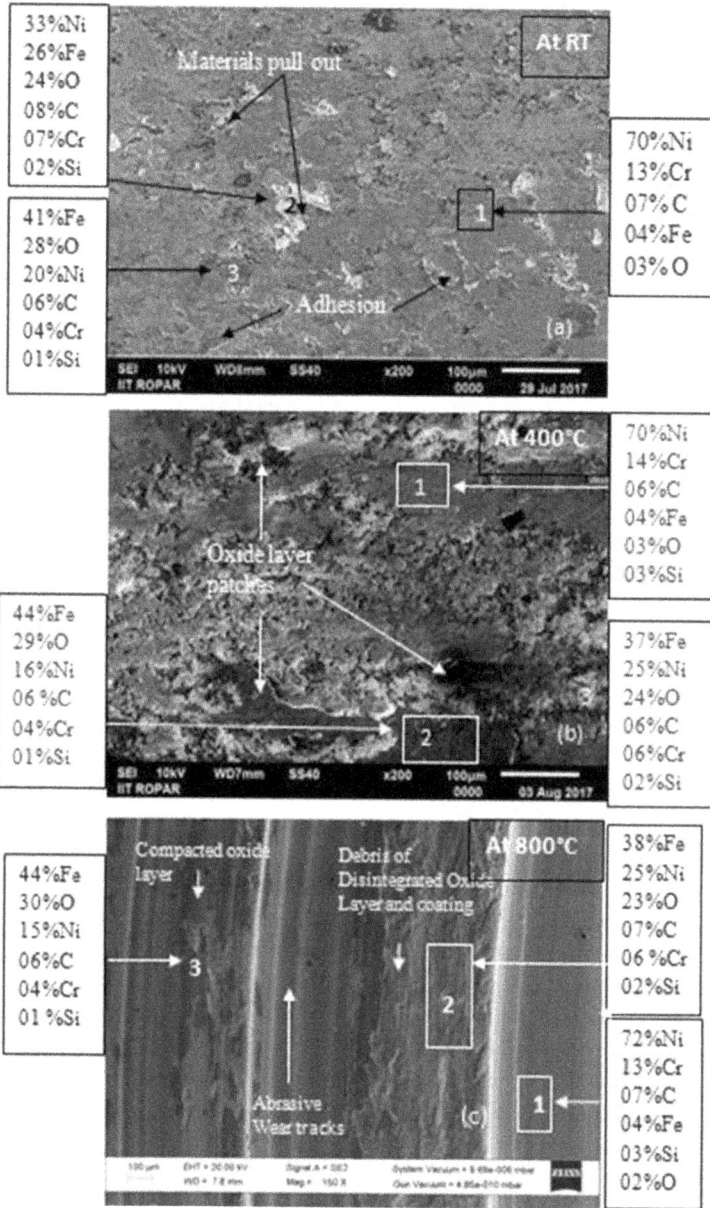

FIGURE 4.31 The SEM/EDS results after wear tests at three temperatures for the HVOF-coated NiCrBSi [C1] H11 steels under a constant load of 50 N.

4.9 DISCUSSION

The wear behaviour of uncoated H11 and H13 steels revealed that the steel underwent severe wear at room temperatures. This was also confirmed by the higher wear rate values of the uncoated steels in comparison to the coated specimens. The main wear mechanisms identified in both the uncoated steel specimens were a combination of severe adhesion and abrasion. The abrasive wear marks and material pullouts with severe plastic deformation were observed on the worn-out pin surface. The high resemblance between elements of the pin and counter-face disc led to adhesive wear mechanism and plastic deformation. Furthermore, the pin surface was abraded due to a great difference in the hardness of the hardened counter-face disc and pin surface. Also, at room temperatures, there was metal-to-metal contact between the pin and disc, and no stable oxide layers formed on the surface.

Hua et al. [23] carried out room and elevated wear testing of H13 steel using a pin-on-disc tribometer. The counter-body disc material was Cr12-MoV steel. The authors reported severe adhesive wear of H13 pins at room temperatures. It was reported that due to the lack of tribo-films on the H13 pin surface high wear rates were observed. Cui et al. [24] also reported similar reasons for the severe adhesion of the H13 pin tested at room temperatures. The SEM images also revealed the presence of an oxide layer on the pin surface and the EDS analysis confirmed a higher percentage of oxygen in the layers. Wei et al. [25] carried out high-temperature tribological testing of H13 steel at temperatures in the range of 25–400°C. D2 steel was used as a counter-body disc. The authors reported a drop in wear rates with rising testing temperatures due to change in wear modes that is from adhesive wear to oxidative wear. This phenomenon has also been reported by other researchers such as Hua et al. [23] and Cui et al. [24].

With the further rise in temperatures to 800°C, the COF and wear rates decreased marginally. The reason for the decrease could be due to the formation of glazed oxides on the surface of tool steels as suggested by Barrau et al. [26]. Also, the softening of the counter-face disc material occurred at 800°C that prevented transition from mild to severe oxidative wear. Hua et al. [23], also reported this phenomenon. The authors revealed that when the H13 pin rubbed against a softer counter-face material, the H13 pin material bore lower surface stresses and there was limited de-lamination of oxide layers. Hence, wear rates remained similar even with the rise in temperature.

The HVOF-sprayed [S1], [C1] and [C4] coated H11 and H13 specimens showed lower specific wear rates at all temperatures and load conditions in comparison to the uncoated steels. This was mainly due to the higher hardness, adhesive and cohesive strength, lower porosity, and thermal stability of the coatings. The higher wear resistance of HVOF-sprayed Cr_3C_2-NiCr [S1] coatings had been reported by many researchers [27–34]. At room temperatures, a higher coefficient of friction was observed for all the coated specimens on both the steels. During dry sliding conditions, higher local pressure was developed between the contacting carbides. This caused the plastic deformation, adhesion, and formation of local junctions, thus resulting in higher COF values. Picas et al. [29], Houdkova et al. [30], and Bolelli et al. [33] showed similar values for Cr_3C_2-NiCr coatings in a similar range.

With an increase in temperature to 400°C, the COF and wear rates decreased. This was mainly due to the transition from severe adhesion to a dominant mild oxidative and adhesive wear.

A decrease in COF value was observed in the coated specimens at higher loads. This was mainly due to frictional heating and work hardening. Huang et al. [35] studied the wear behavior of Cr_3C_2-NiCr coatings deposited on low carbon steel. The wear tests were performed at loads from 10 N to 150 N. The authors observed the decreased COF values at the higher loads. The authors explained that the magnitude of the normal load was important because it increased the area of contact between the pin and the disc surface. The depth below the surface at which the maximum shear stress occurred was also increased and they both affected the elastic or plastic deformation states. In addition, owing to the frictional heating, the increase in temperature contributed to reducing the friction coefficients. As the load was increased, the temperature of the frictional surface increased as well. It gave rise to the decrease of plastic deformation resistance of frictional pairs. The decrease of mutual hindering effect between micro-peaks in the surface also occurred and reduced the friction coefficients. Houdkova et al. [30] opined that in the case of Cr_3C_2-NiCr coating due to its lower fracture toughness, a higher amount of carbides, and bigger wear debris were pulled out and caused the fluctuation of the coefficient of the friction curve.

At 400°C, a decline in the COF and wear rates were observed for all the coated specimens. At this temperature, the oxidation of wear debris and counter-face material occurred under the compressive forces between the two mating surfaces. Hardell [6] explained in his work that the compact layers consisted of oxides of coating and the counter-face material debris. These layers acted as the solid lubricant which further lowered the value of friction and specific wear rates. The increase in compressive forces due to the increase in load made such a layer more compact, thus further decreasing the specific wear rate. The XRD investigations showed the presence of Cr_3C_2 along with oxides of Ni, Cr, and Fe. This indicated the stability of Cr_3C_2 at elevated temperatures providing higher wear resistance at elevated temperatures.

At 800°C, the COF values almost remained similar with the increase in temperature for the Cr_3C_2-NiCr coating. The wear rates increased with an increase in temperature. Poirier et al. [31] and Zhou et al. [34] also reported similar results at higher temperatures. The authors opined that the long-term isothermal holding of Cr_3C_2-NiCr coating at elevated temperatures above 650°C led to the embrittlement of the matrix and re-precipitation of carbides. The wear debris formed by the carbides in the layers promoted the three-body abrasive wear, thereby giving rise to more wear. Further, the wear mechanism was found to be a blend of abrasive, oxidative, and adhesive. Poirier et al. [31] and Zhou et al. [34] also reported similar wear mechanisms in their studies. The XRD investigations showed the existence of the Cr_3C_2 phase in the coating indicated the stability of the developed coating at 800°C. Zhou et al. [34] also found similar phases for the Cr_3C_2-NiCr coating at 650°C.

The HVOF-sprayed [C1] and [C4] coatings showed better wear resistance than the uncoated specimens at all test parameters. This could be attributed to superior mechanical properties obtained by the coatings. Also, HVOF [C4] coatings retained a large amount of hard WC phase which further augmented their wear resistance. The

WC particles bore most of the generated stresses, and therefore, prevented adhesive wear and deformation in the soft matrix.

At room temperatures, [C1] coatings revealed the adhesive wear mechanism, and the [C4] coatings showed the wear was mainly caused by mild abrasion at 25 N loads. At higher loads, the signs of adhesion, fatigue, and abrasion were seen on the worn surface. The coated specimens were able to resist cracking and severe de-lamination, owing to superior mechanical properties. Stoica et al. [36] and Yao [37] reported similar SEM images for the worn HVOF-sprayed NiCrBSi/WC(Co) specimens at different loads.

The wear behaviour of the HVOF coatings was affected by the increase in testing temperature. At 400°C, both wear rates and COF were lower in comparison to the values at room temperature. Similar observations are reported in literature studies. With the increase in temperature, the oxide layers consisting of Ni, Cr, and Fe oxides were formed on the contacting surfaces. These stable thin layers were lubricious in nature. Under this layer, a wear-resistant oxide layer was formed by the compaction of wear debris and sintering of subsurface oxides under conditions of high pressure and temperature. These layers cease metal-to-metal contact that led to the reduction in friction coefficient and the specific wear rate [38–40].

At 800°C, an increase in the wear rates were observed for [C1] and [C4] coatings on both substrates. However, the obtained values were lower in comparison to those for the uncoated specimens. The increase in specific wear rate with rising testing temperature was mainly due to softening of the Ni-based matrix material. Different researchers reported an increase in the wear rate of Ni-based coatings beyond 600°C [41–42]. Also, the high resemblance between Ni (in pin material) and Fe (in counter-face material) led to adhesive wear and plastic deformation. Further, the softening of the Ni-based matrix reduces the cohesion between the matrix and the oxide layers in the [C1] coating. In the case of the [C4] coating along with the above-said factor, the matrix could not support WC particulates as effectively as it did at a lower temperature. Due to the weakening of cohesive strength between coating layers, the formed oxide layers spalled and reformed continuously causing oxidative wear. The pullout and cracking of WC particles led to fatigue wear. Three-body abrasive wear was also produced by the hard oxides and the WC particulates that were struck between the two surfaces. However, wear rates were still much lower than the uncoated specimens and the cross-sectional images revealed that the wear phenomenon was limited to the top surface and the immediate subsurface.

4.10 CONCLUSIONS

1. All the three coatings namely [75Cr$_3$C$_2$-25(80Ni-20Cr)]; NiCrBSi; and [65%(NiCrSiFeBC)+ 35%(WC-12Co)] were successfully obtained on the given tool steels by the selected thermal spray techniques. The values of porosity for the coated specimens were observed between 1.7% and 3.00%. The maximum value of bond strength evaluated in the study is found to be 73.61 MPa for the HVOF-sprayed 75Cr$_3$C$_2$-25(80Ni-20Cr) coating.

2. The microhardness of the coatings was found to vary with the distance from the coating–substrate interface. The increase in the hardness values was

observed for the coated steels as compared to the uncoated steels. 65%(NiCrSiFeBC)-35%(WC-Co) [C4] coating on the H11 steel developed by the HVOF spray process showed the highest average microhardness value as 933±69 HV. The microhardness of the HVOF-sprayed coatings was found to increase by 46% and 32% with the addition of 35%(WC-Co) for the cases of H11 and H13 substrates, respectively. Similarly, an addition of 35%(WC-Co) in APS NiCrBSi coating enhanced the microhardness of the coating on the H11 and H13 base steel by 47% and 38%, respectively.

3. The splat-like microstructures free from cracks were observed for most of the coating compositions. In the WC-Co blended compositions, the WC-Co-rich splats were seen distributed in the NiCrBSi-rich matrix and the resulting microstructure appeared to be composite.

4. During sliding wear tests, the operating temperature was found as the main parameter that influences the wear and friction behaviour of die steels. The uncoated specimens showed severe adhesion and plastic deformation. Deep abrasive grooves were also visible on the surface of worn samples. Higher specific wear rates of the uncoated samples were mainly attributed to their lower hardness and higher similarity between the pin and disc elements.

5. At elevated temperatures (400°C), wear rates, and COF values were found to be lesser for the coated specimens in comparison to those observed at room temperatures. This might be due to the development of lubricating oxide layers on the coating surface, which reduced direct metal-to-metal contact and led to oxide–oxide contact. This phenomenon lowered the friction coefficients as well as the wear rates of the samples. The SEM microstructures of the coated worn samples revealed the presence of dark oxide patches on the worn surface. Mild oxidative wear and mild adhesion wear were found as main wear mechanisms.

6. At 800°C, the wear rates increased at both the testing loads. The wear mechanism was found to be the combination of oxidative, adhesive, and abrasive wear.

7. In general, the results showed that all coated specimens showed more wear resistance in comparison to the uncoated steels. The coatings were also successful in maintaining integrity with their respective substrates throughout the experiment at all the three temperatures of the study, without any significant damages beyond the coating subsurface.

ACKNOWLEDGEMENTS

The authors gratefully acknowledge the research grant from Department of Science and Technology, New Delhi, India, under SERB, Science and Engineering – Engineering Scheme (File No. SR/S3/MERC/0072/2012, dated February 28, 2013) titled *Development of Thermal Spray Coatings to Control Wear during High-Temperature Applications*, and (EMR/2015/000234, dated March 11, 2016) titled *Development of Thermal Spray Coatings to Control Wear during High-Temperature*

Applications-Phase II, to carry out this research and development work. Moreover, the authors would like to thank the Metallizing Equipment Company, Pvt. Ltd., Jodhpur, India, for providing the HVOF and Plasma spray coatings services for the die materials. The authors owe special thanks to Dr. Harpreet Singh for extending the necessary facilities and support in conducting the detailed analysis at the Indian Institute of Technology Ropar, Roopnagar, Punjab, India, and Dr. S. Prakash at the Indian Institute of Technology Roorkee, Roorkee, India. The authors would also like to thank IKGPTU, Jalandhar, India, for supporting this work.

REFERENCES

1. Hutchings I., Gee M., and Santner E., (2006), "Friction and Wear," In: Czichos H., Saito T., Smith L., (eds) *Springer Handbook of Materials Measurement Methods*. Springer Handbooks. Springer, Berlin, Heidelberg, pp. 685–709.
2. Jost H.P., (1966), *"Lubrication (Tribology) Education and Research - A Report on the Present Position and Industry's Needs,"* H. M. Stationary Office, London.
3. Bay N., (2002), *"Class Notes,"* Technical University of Denmark, Lyngby, Denmark.
4. Luo D., (2009), "Selection of Coatings for Tribological Applications," Thesis, Master Universite Jiaotong du Sud-Ouest (Chine).
5. Hardell J., Kassfeldt E., and Prakash B., (2008), "Friction and Wear Behavior of High Strength Boron Steel at Elevated Temperatures of up to 800°C," *Wear*, 264(9–10), pp. 788–799.
6. Hardell J., (2009), "Tribology of Hot Forming Tools and High Strength Steels," Doctoral Thesis, Lulea University of Technology, Department of Applied Physics and Mechanical Engineering, Lulea.
7. Lange K., Cser L., Geiger M., and Kals G.A.J., (1992), "Tool life and Tool Quality in Bulk Metal Forming," *J. Eng. Manuf.*, 207, pp. 223–239.
8. Kang H.J., Park W.I., Jae S.J., and S. S. Kang, (1998), "A Study on Die Wear Model of Warm and Hot Forgings," *Met. Mater.*, 4 (3), pp. 477–483.
9. Sjostrom J. and Bergstrom J., (2005), "Thermal Fatigue in Hot-Working Tools," *Scand. J. Metall.*, 34 pp. 221–231.
10. Bulletin M.R.S., (2009), *"Technology Advances,"* Cambridge University Press, Cambridge, pp. 792.
11. Roberts G., Krauss G., and Kennedy R., (1998), *"Tool Steels,"* 5th ed., ASM International, Novelty, OH.
12. Okorafor O.E., (1987), "Fracture Toughness of M2 and H13 Alloy Tool Steels," *Mater. Sci. Technol.*, 3 pp. 118–124.
13. Kheirandish S. and Noorian A., (2008), "Effect of Niobium on Microstructure of Cast AISI, H13 Hot Work Tool Steel," *J. Iron Steel Res. Int.*, 15(4) pp. 61–66.
14. Knoerr M. and Shivpuri R., (1989), "Failure in Forging Dies," Report No.ERC/NSM-B-89-15, Engineering Research Center for Net Shape Manufacturing.
15. Dahl C., Vazquez V., and Altan T., (1998), "Effect of Process Parameters on Die Life and Die Failure in Precision Forging," Report No. PF/ERC/NSM-98-R-15, Engineering Research Center for Net Shape Manufacturing.
16. Tulsyan R., Shivpuri R., and Altan T., (1993), "Investigation of Die Wear in Extrusion and Forging of Exhaust Valves," Report No. ERC/NSM-B-93-28, Engineering Research Center for Net Shape Manufacturing.

17. Gronostajski Z., Hawryluk M., Widomski P., Kaszuba M., Nowak B., Polak S., Rychlik M., Ziemba J., and Zwierzchowski M., (2019), "Selected Effective Methods of Increasing the Durability of Forging Tools in Hot Forging Processes," *Procedia Manuf.*, 27, pp. 124–129.

18. ASM Handbook, (1992), "*Vol. 18: Friction, Lubrication, and Wear Technology*," ASM International, Materials Park, OH, USA.

19. Illavsky J., Pisacka J., Chraska P., Margandant N., Siegmann, S. Wagner, W. Fiala P., and Barbezat G., (2000), "*Microstructure-Wear and Corrosion Relationships for Thermally Sprayed Metallic Deposits*," *Proc. '1st Inter. Thermal Spray Conf.'*, Montreal, Quebec, Canada, May 8–11, pp. 449–454.

20. Simunovic K., (2010), "*Thermal Spraying*," Welding Engineering and Technology, Encyclopedia of Life Support Systems.

21. Sobolev V.V., Guilemany J.M., Nutting J., and Miquel J.R., (2013), "Development of Substrate-Coating Adhesion in Thermal Spraying," *Int. Mater. Rev.*, 42(3), pp. 117–136.

22. Tucker R.C., (1994), "Thermal Spray Coatings, Surface Engineering", *ASM Handbook 5 ASM Int.*, pp. 497–509.

23. Hua H., Zhou Y., Li X., Zhang Q., Cui X., and Wang S., (2015), "Variation of Wear Behavior of H13 Steel Sliding Against Different Hardness Counterfaces," *Proc. IMechE Part J.: J. Eng. Tribol.* 229(6), pp. 763–770.

24. Cui X.H., Wang S.Q., Wei M.X., and Yang Z.R., (2011), "Wear Characteristics and Mechanisms of H13 Steel with Various Tempered Structures," *J. Mater. Eng. Perform.*, 20, pp. 1055–1062.

25. Wei M.X., Wang S.Q., Wang L., Cui X.H., and Chen K.M., (2011), "Effect of Tempering Conditions on Wear Resistance in Various Wear Mechanisms of H13 steel," *Tribol. Int.*, 44, pp. 898–905.

26. Barrau O., Boher C., Vergne C., and Rezai-Aria F., (2002), "*Investigations of Friction and Wear Mechanisms of Hot Forging Tool Steels*," *6th International Tooling Conference*, Karlstad University, Sweden, pp. 95–111.

27. Sahraoui T., Fenineche N.E., Montavon G.C., and Coddet C., (2003), "Structure and Wear Behaviour of HVOF Sprayed Cr_3C_2-NiCr and WC-Co Coatings," *Mater. Des.*, 24(5), pp. 309–313.

28. Roy M., Pauschitz A., Polak R., and Franek F., (2006), "Comparative Evaluation of Ambient Temperature Friction Behaviour of Thermal Sprayed Cr_3C_2-25(Ni20Cr) Coatings with Conventional and Nano-Crystalline Grains," *Tribol. Int.*, 39, pp. 29–38.

29. Picas J., Punset A.M., Menargues S., Campillo M., Baile T. M., and Forn A., (2009), "The Influence of Heat Treatment on Tribological and Mechanical Properties of HVOF Sprayed CrC-NiCr Coatings," *Int. J. Mater. Form.*, 2(S1), pp. 225–228.

30. Houdkova S., Kasparova M., Zahalka S., and Vyzkum F., (2010), "The Friction Properties of the HVOF Sprayed Coatings Suitable for Combustion Engines, Measured in Compliance with ASTM G-99," *WIT Trans. Eng. Sci.*, 66, pp. 129–139.

31. Poirier D., Legoux J.G., and Lima R.S., (2013), "Engineering HVOF Sprayed Cr_3C_2-NiCr coatings: the Effect of Particle Morphology and Spraying Parameters on the Mircrostructure, Properties, and High Temperature Wear Performance," *J. Therm. Spray Technol.*, 22 (2–3), pp. 280–289.

32. Shabana Sarcar M.M.M., Suman K.N.S., and Kamaluddin S., (2015), "Tribological and Corrosion Behavior of HVOF Sprayed WC-Co, NiCrBSi and Cr_3C_2-NiCr Coatings and Analysis Using Design of Experiments," *Mater. Today Proc.*, 2(4–5), pp. 2654–2665.

33. Bolelli G., Berger L.M., Borner T., Koivuluoto V., Matikainen V., Lusvarghi L., Lyphout C., Markocsan N., Nylen P., Sassatelli P., Trache R., and Vuoristo P., (2016), "Sliding and Abrasive Wear Behaviour of HVOF and HVAF-Sprayed Cr_3C_2–NiCr Hard Metal Coatings," *Wear*, 358–359, pp. 32–50.

34. Zhou W., Zhou K., Li Y., Deng C., and Zeng K., (2017), "High Temperature Wear Performance of HVOF-Sprayed Cr_3C_2-WC-NiCoCrMo and Cr_3C_2-NiCr Hardmetal Coatings," *Appl. Surf. Sci.*, 416, pp. 33–44.

35. Huang C., Du L., and Zhang W., (2009), "Microstructure and Tribological Properties of Plasma Sprayed NiCr/Cr_3C_2 and NiCr/Cr_3C_2-BaF_2-CaF_2 Composite Coatings," *Adv Tribol.*, pp. 669–675.

36. Stoica V., Ahmed R., and Itsukaichi T., (2005), "Influence of Heat-Treatment on the Sliding Wear of Thermal Spray Cermet Coatings," *Surf. Coat. Technol.*, 199 pp. 7–21.

37. Yao S.H., (2014), "Tribological Behaviour of NiCrBSi-WC (Co) Coatings," *Mater. Res. Innov.*, 18(2) pp. 332–337.

38. Stott F.H., Lin D.S., and Wood, G.C., (1973), "The Structure and Mechanism of Formation of the 'Glaze' Oxide Layers Produced on Nickel-based Alloys during Wear at High Temperatures," *Corros. Sci.*, 13, pp. 449–469.

39. Karaoglanli A.C., Oge M., and Doleker K.M., (2017), "Comparison of Tribological Properties of HVOF Sprayed Coatings with Different Composition," *Surf. Coat. Technol.*, 318, pp. 299–308.

40. Torgerson T.B., Harris M.D., Alidokht S.A., Scharf T.W., Aouadi S.M., Chromik R.R., Zabinski J.S., and Voevodin A.A., (2018), "Room And Elevated Temperature Sliding Wear Behavior Of Cold Sprayed Ni-WC Composite Coatings," doi: 10.1016/j.surfcoat.2018.05.090.

41. Zikin M., Antonov M., Hussainova I., Katona L., and Gavrilovic A., (2013), "High Temperature Wear of Cermet Particle Reinforced NiCrBSi Hardfacings", *Tribol. Int.*, 68, pp. 45–55.

42. Bolelli G., Borner T., and Milanti A., (2014), "Tribological Behavior of HVOF and HVAF-Sprayed Composite Coatings Based on Fe-Alloy + WC-12% Co", *Surf. Coat. Technol.* 248, pp. 104–112.

5 Thermal Conductivity, Failure Life Prediction of Thermal Barrier Coatings and Hot Erosion of Oxidation Resistant Coatings

Kuiying Chen and Prakash C. Patnaik

CONTENTS

DOI: 10.1201/9781003213185-5

5.1   INTRODUCTION

A typical thermal barrier coating (TBC) is comprised of four main parts: the substrate, topcoat (TC), bond coat (BC) and thermally grown oxide (TGO), usually α-Al_2O_3 generated between the surface of the bond coat and topcoat. These desired layers have been designed to provide thermal insulation and a long life for hot-section components [1]. Reducing high-temperature attack on a substrate is one of the major objectives in developing TBCs. One function of the bond coat is to optimize the bonding between the underlying superalloy and the topcoat and to protect the superalloy from hot corrosion and oxidation [2]. The ceramic coating microstructure is extremely inhomogeneous, containing defects of voids, pores and vacancies as well as cracks of various sizes and shapes. Generally, the presence of such defects [3] and spraying parameters [4] influences the thermal conductivity of TBCs.

To attain lower thermal conductivity, there must be an optimized distribution of pores and cracks in coatings [5]. Therefore, it is important to understand the basic microscopic structural properties of TBCs to fabricate an optimized coating. In this chapter, Bruggeman two-phase model was used as a reference to develop a six-phase model of thermal conductivity in heterogeneous mediums. It was known that the Bruggeman model is suited to provide an estimation of the thermal properties of porous materials as it can be used to cover a wide range of defects. The morphology of porosity was investigated in relation to the estimation of the thermal conductivity of porous TBCs. This investigation of porosity was performed by a quantitative study of the pore morphology using Image J. This software is developed by the National Institutes of Health, USA, and gained popularity due to its simple user interface.

It has been understood that failure of air plasma-sprayed TBC (APS-TBC) systems is largely attributed to the formation of TGO as large stresses could be generated while TGO thickens upon progressive oxidation of the bond coat [5–10]. Meanwhile, extensive cracks nucleate from the sites where transient mixed oxides such as spinel form, leading to a reduction of fracture toughness [11–13]. Based on the identified failure mechanisms, various life models of APS-TBCs have been explored. One early model developed by Miller [14] attempted to describe APS-TBCs life prediction, an essential role of the TGO thickness in APS-TBCs life prediction, in which the number of thermal cycles to failure was correlated to critical strain range and the role of critical TGO thickness was addressed. The life of APS-TBCs was evaluated using ratios of TGO thickness δ over the critical TGO thickness, as well as the ratio of the strain over the critical strain.

Another life model proposed by Beck et al. [15] divided the entire life of APS-TBCs into two parts associated with crack nucleation and propagation, where the trends of TGO thickness as well as crack length were used to define the boundary between these two life periods. The residual stress generated due to a difference of coefficient of thermal expansion (CTE) between the topcoat and TGO together with TGO growth stress were incorporated into the life model. The life in [15] was numerically evaluated by calculating the crack growth rate iteratively during thermal cycles up to the specific measured failure crack length.

Vaßen et al. [16] investigated the life of APS-TBC systems through examining subcritical crack growth rate

$$\int_{a_0}^{a_f} \frac{da}{a^{m/2}} = \frac{A^* Y^m}{K_{IC}^m} \int_t^t \sigma(T)^m \, dt \tag{5.1}$$

where $\sigma(T)$ is the residual stress acting on the APS-TBC topcoat, m is an exponent parameter fitted to testing data, a is the crack length in the topcoat, K_{IC} is the fracture toughness of the topcoat, T stands for temperature, t is the exposure time. In the stress model, the coating interface profile such as the amplitude of roughness, the wavelength, the TGO thickness were included in the stress evaluation. However, in the life model [16], the fitting parameter $A^* Y^m / K^m_{IC}$ in Equation (5.1) was fitted as a constant independent of testing temperatures of APS-TBCs.

In this chapter, temperature-dependent model parameters are identified and fitted to the test life data from the literature [17, 18]. A stress model was proposed and then used to describe the stress at the valley of the topcoat, where a CTE mismatch strain is shown to be the main contributor of residual stress in the vicinity of the topcoat/TGO interface. The proposed stress model illustrates an inversion characteristic versus the TGO thickness, and the stress model parameters were fitted to the 3-D FEA stress calculations [19], while the life model parameters were fitted to existing burner rig test life of APS-TBC systems from the literature [17, 18].

In a sandy environment, engine hot-section components, such as combustor liners, turbine nozzle guide vanes and turbine blades, often suffer from solid particle erosion in addition to oxidation and hot corrosion. The combination of all these damage processes will accelerate the component degradation significantly. Therefore, erosion–oxidation interaction is a factor that can affect engine performance, flight safety, maintenance costs of the fleet, and consequently, this problem must be given a serious attention. In this chapter, the erosion–oxidation degradation mechanisms were initially analyzed, and then a kinetic equation was derived for oxide scale evolution under solid particle impact to describe the erosion–oxidation interaction. Subsequently, the variation of oxide scale thickness under particle impact was calculated as a function of particle velocity, temperature and incident angles of particles for Cannon Muskegon Superalloy X-4 (CMSX-4).

Recently, multi-physics/multi-length scale modeling was conducted to study mechanical and thermal physical properties of oxidation resistant coatings and APS-TBCs/EB-PVD TBCs [20–29]. Using density functional theory (DFT), Ozfidan et al. [22] studied the interface adhesion strength β-NiAl/α-Al$_2$O$_3$ along with the effect of elements doping on the interface. By combining DFT and mechanics-based interface stress model, the effect of Platinum (Pt) on β-NiAl/α-Al$_2$O$_3$ interfacial tensile stress [20] was evaluated [20]. The result shows that Pt is capable of reducing the interfacial tensile stress, thus reducing the probability of interface crack nucleation and propagation, consequently increasing durability of coatings life. The effect of sulfur (S) diffusion in β-NiAl and the effect of Pt additive on sulfur was also explored [21]. To study possible electronics-related mechanisms that may govern the Pt effect, the electron localization function (ELF) is evaluated for S distributed in β-NiAl and Pt-doped β-NiAl. The ELF amplifies the bonding features, and allows one to analyze the electron distribution on an absolute scale. It was found that it is the repulsive interaction between Pt and S atoms that stops and/or slows S diffusion toward the β-NiAl/α-Al$_2$O$_3$ interface, consequently reducing S detrimental effect. Life prediction of APS-TBC and EB-PVD TBC using temperature-dependent model parameters were was carried out [26, 27]. By applying specific failure mode, the crack growth in TBC was calculated versus thermal cycles. Chen et al. [29] developed a physics-based model of temperature-exposure-dependent interfacial fracture toughness of TBCs. Based on the proposed model, the toughness of APS-TBC was predicted and compared with available experimental data.

5.2 ANALYTICAL MODELS FOR EVALUATING THERMAL CONDUCTIVITY

5.2.1 BRUGGEMAN MODEL

One approach to minimize the thermal conductivity of TBCs is to use porous microscopic structures [30]. To evaluate the overall thermal conductivity of ceramic coatings deposited by different deposition methods, a substantial number of models have been developed [31]. Bruggeman developed a model to estimate the thermal conductivity of porous materials that have a random distribution of spherical particles of various sizes. It was found that if large spherical particles are dropped into dispersions that contain smaller particles, the disturbance of the field around large spheres will most likely be negligible as a result of the small spheres. Using this model, Bruggeman demonstrated that the limitation on the volume fraction of dilute dispersion could be removed, and the model is then written as [32]

$$\frac{k - k_m}{k + 2k_m} = f \frac{k_d - k_m}{k_d + 2k_m} \tag{5.2}$$

where f is the volume fraction of the t_{ih} phase, κ is the composite thermal conductivity, κ_d is the dispersed phase thermal conductivity and κ_m is the matrix thermal conductivity.

In the Bruggeman model, many spheres are introduced into a dilute dispersion consisting of many different sizes. Furthermore, each sphere is a part of the continuous matrix. This can eliminate the constraint on a volumetric fraction of the dilute dispersion, which was impossible with respect to other models ($0 \leq f < 1$). This model was designed for a porous TBC system. However, the model considered only one type of defect that has a single orientation, i.e., spherical particles. For further simplification of the Bruggeman two-phase model, we assumed that thermal conductivities of defects/pores are negligible; i.e., κ_d equals to 0. In the condition of non-radiating pores, Equation (5.2) was minimized to

$$(1 - f)^{\frac{3}{2}} = \frac{k}{k_m} \tag{5.3}$$

The equation is useful, especially when the pores have a spherical shape. More generally, the Bruggeman model is an alteration of the Equation (5.3) for the dispersion of an ellipsoid, and can be written as [33]

$$\psi\left(f_1\right)=\left(1-f\right)^X=\frac{k}{k_m} \tag{5.4}$$

where the value of X depends on factors such as the ellipsoid shape factor F. α is an angle between the axis of revolution and the heat flux following equation [33]:

$$X=\frac{1-\cos^2\alpha}{1-F}+\frac{\cos^2\alpha}{2F} \tag{5.5}$$

When there are only spherical pores in a continuous matrix, X is assigned to 1.5. Up to a specific temperature limit, this assumption of non-conducting defects/pores is valid. However, its primary role is to simplify the model. Actually, coatings are composed of various types of defects [34]. Thus, a superposition of various types of defects on the overall thermal conductivity is necessary for coatings to be realistically modeled. Bruggeman's two-phase model is comprised of spherical pores and a binary mixture of dense material. The two-phase model can be used iteratively to generate a three-phase model for composites. This can be achieved as follows. Assume f_1 and f_2 represent the final percentages of type 1 and type 2 porosities, respectively, then the coating overall porosity can be described by f_{tot}, where f_{tot} equals to a sum of f_1 and f_2. In a continuous matrix, two ways are used to add these defects. First, consider that the first type of porosity f_1 is embedded into a continuous matrix. Next, consider the second type of porosity f_2 is embedded into a continuous matrix. This will produce an expression

$$\left\{\phi\left[\frac{f_2}{1-f_1}\right]\psi\left(f_1\right)\right\} \tag{5.6}$$

However, if the type 2 porosity is embedded first in the continuous matrix, then the type 1 is embedded later, it will produce the following formula:

$$\left\{\psi\left[\frac{f_1}{1-f_2}\right]\phi\left(f_2\right)\right\} \tag{5.7}$$

The proposed model is an average of selected values of the constituents making up the composite materials. Using an average of the two possible cases normally produces the thermal conductivity of three-phase mixture as [33]

$$k=\frac{k_0}{2}\left\{\psi\left[\frac{f_1}{1-f_1}\right]\phi\left(f_2\right)+\phi\left[\frac{f_2}{1-f_1}\right]\psi\left(f_1\right)\right\} \tag{5.8}$$

where κ_o is the matrix thermal conductivity and the functions $\psi(f)$ and $\phi(f)$ describe the impact of defects on coating thermal conductivity. This equation is symmetrical about the product of $\psi(f)$ and $\phi(f)$ to comply with the dilution process commutative properties.

5.2.2 SIX-PHASE MODEL FOR POROUS TBCs

Porosities distributed in coatings can be analytically modeled. Nevertheless, one has to consider various scenarios to ensure that the microstructural details are enough to represent the material properties [35]. To simulate different kinds of defects, the spheroids (ellipsoids that have a revolution axis corresponding to the ellipsoid axis 'a', whose two other axes are equal, i.e., b equals to c were selected in this study. Spheroids can be used to account for a wide range of actual circumstances through defining the revolution angle (α) between dispersed particle and heat flux as well as by adjusting the axes ratio a/c. Figure 5.1 represents a spheroidal shape, where the ratio between the axes of the spheroid is varied. This shape can be used to model spheres, lamellas, spheroids and cylinders. In Equation (5.5), the shape factor F has different values for various shapes considered. The ratio of axes a over c affects F. The image J is used to evaluate values of aspect ratios. Figure 5.2 illustrates a graphical representation of the shape factor F as a function of the aspect ratio a/c. For the sphere-shaped defect, its minor and major axis are equal, a equals to c, then the shape factor F is equal to 0.33333. When the axes b and c are bigger than 'a' axis, i.e., c is larger than a, the spheroid is horizontally stretched to form an oblate spheroid. In the case of oblate pores (lamellas), F falls within 0 and 0.3. While the pores are stretched vertically, a prolate-shaped spheroid is obtained and the shape factor falls between 0.3 and 0.5. A representation of a graph of factor X as a function of the aspect ratio a/c is presented in Figure 5.3. The curve labeled as (1) is the factor X for randomly oriented spheroid pores. Curve (2) shows values of factor X for a spheroid oriented horizontally to heat flux and with the revolution axis 'a', while curve (3) stands for the factor X of spheroids oriented vertically to heat flux and with the revolution axis 'a'. The points where all these three curves meet are the factor X for a sphere-shaped spheroid. At this point, the factor X is equal to 3/2, and this corresponds to an aspect ratio a/c =1.

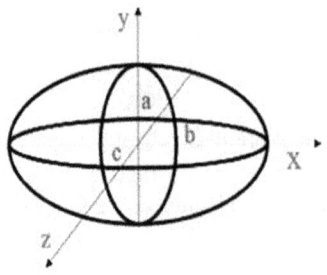

FIGURE 5.1 Representation of spheroidal shape used in defect modeling [28].

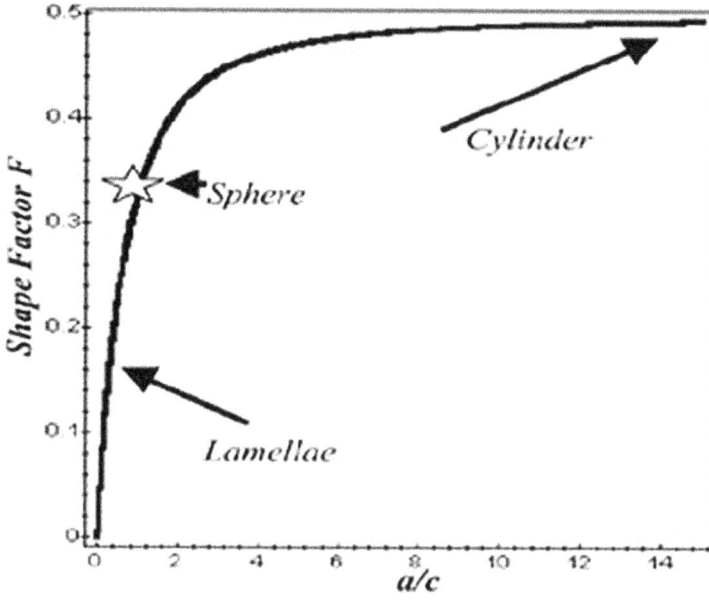

FIGURE 5.2   Shape factor F versus the aspect ratio a/c [33].

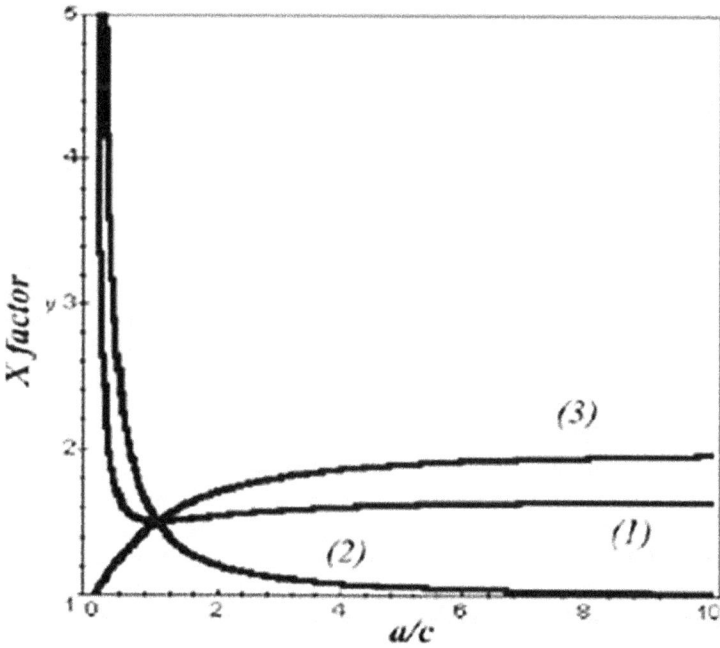

FIGURE 5.3   Relationship between factor X and aspect ratio for various shapes [33].

The six-phase model was implemented by using an iteration of the averaging method. This study aims to improve the capability of the proposed analytical model for a better prediction of overall thermal conductivity. For the proposed model, we assumed that five types of defects were added into a continuous matrix. We also assumed that there is no heat transfer within these pores. Thus, k_d equaling to 0 for pore thermal conductivity still holds. The total porosity f_{tot} is the sum of different types of porosities, hence, f_{tot} equals to a sum of f_1, f_2, f_3, f_4 and f_5. The values of different functions $\psi(f_1)$, $\phi(f_2)$, $\theta(f_3)$, $\beta(f_4)$ and $\chi(f_5)$ are provided in Equation (5.5). The values of these f_1, f_2, f_3 and f_4 are acquired using Image J. The iterative approach allows for the simultaneous addition of five various types of defects into a continuous matrix performed by the averaging technique. This leads to an equation of the six-phase porous composite materials that have up to 120 equations. Therefore, the equation of the six-phase model is separated into five sections, and all five sections were simply summed to produce the equation. The six-phase model equation is described by the following equation:

$$k = \frac{k_0}{120}\left(A + B + C + D + E\right) \tag{5.9}$$

In Equation (5.9), A, B, C, D and E were the five sections of equations [28]. The formula takes an average of all possible scenarios in which the five types of defects can be summed up in various sequences.

5.2.3 RESULTS AND DISCUSSION

SEM and optical micrographs were collected from literatures [36–38]. The following coating materials were used in this study: strontium zirconate $SrZrO_3$, yttria partially stabilized zirconia (YPSZ) $8Y_2O_3$-ZrO_2, magnesia-stabilized zirconia (MSZ) $22MgO$-ZrO_2, ceria-stabilized zirconia (CSZ) $25CeO_2$-$2.5Y_2O_3$-ZrO_2, including three different powder coatings of YSZ, namely hollow sphere (HOSP), fused and crushed (F&C) and agglomerated and sintered (A&S). The powder coating materials such as A&S were evaluated at different annealing temperatures for different time period and the results yield different thermal conductivity. Generally, five different regions of both optical and SEM micrographs for each as-sprayed and after heat treatment of selected coating materials were analyzed using Image J. The analyses of these five images were then averaged to obtain a representative value of porosity. The procedure of these analyses is explained in [28].

As an example, Figure 5.4 presents SEM micrographs of $SrZrO_3$ annealed at 1400°C for selected time periods. As a demonstrated example to show image analysis procedure, Figure 5.5 represents an SEM micrograph of $SrZrO_3$ annealed at 1400°C for 230 hrs. It is obvious that the image on the left top corner shows the original grayscale while the top right image illustrates the binary image with the total porosity. The image located on the left middle shows non-flat porosity, while the image located in the right middle region displays microcrack content. The images positioned in both the left and right bottom depict vertical (penny-shaped) and horizontal (interlamellar) porosity, respectively. The total porosity present in $SrZrO_3$

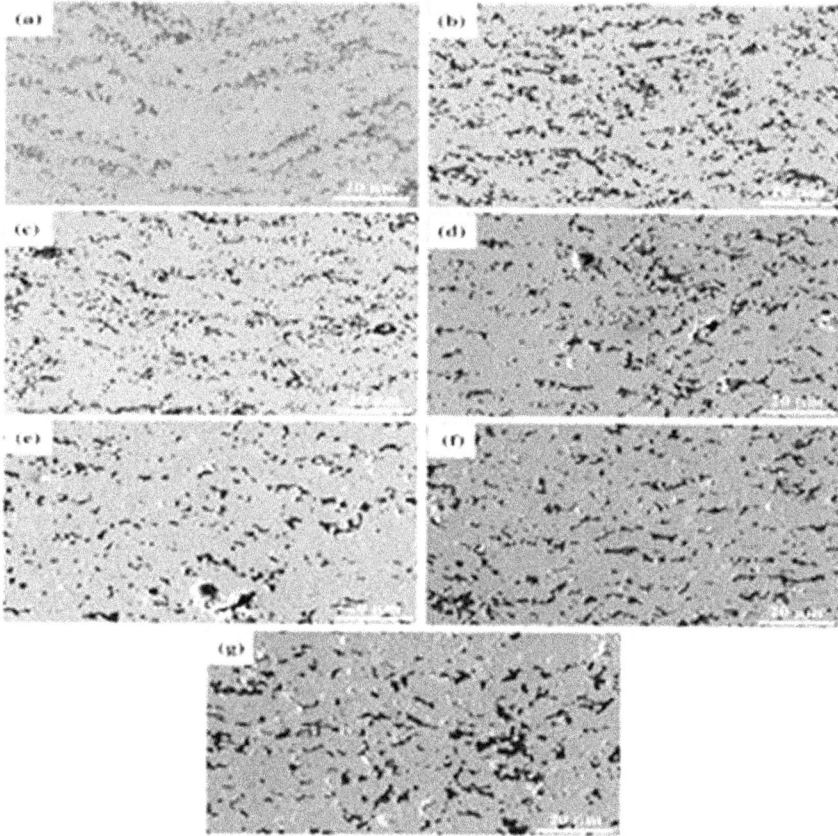

FIGURE 5.4 The SEM micrographs of SrZrO₃ coating after annealing at 1400°C for (a) 0 hrs., (b) 5 hrs., (c) 10 hrs., (d) 20 hrs., (e) 100 hrs., (f) 230 hrs. and (g) 360 hrs. [36]. The white magnification scale bars in (a) to (g) are 20 μm.

coating selected 0, 5, 10, 20 and 100, 230 and 360 hrs. of annealing is plotted in Figure 5.6. In general, it can be clearly visualized from Figure 5.6 that the total porosity was initially reduced and then increased for a long period (i.e., 230 and 360 hrs.) of annealing. Furthermore, the total porosity of the as-sprayed coating was approximately 17.89% and was then raised moderately to approximated 18.63% for 5 hrs. annealing. Additionally, there was a gradual drop in overall porosity to 100 hrs. annealing period. In contrast, the observed increase in the total porosity for the annealed coating for 230 and 360 hrs. may be ascribed to grain pull-out [36].

Table 5.1 shows the total porosity and its components present in SrZrO₃ for the selected annealing period. The image showed that the non-flat porosity was dominated in all annealing period with the highest value of 14.07%, whereas the content of the open randomly oriented porosity seemed to have the minimum value for all analyses with the exception of the as-sprayed coating which was higher (2%) in comparison to the penny-shaped porosity (1.71%). Also, the amount of microcracks was

Microstructure of $SrZrO_3$ for 230h → Thresholded image with total porosity

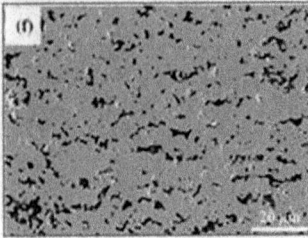

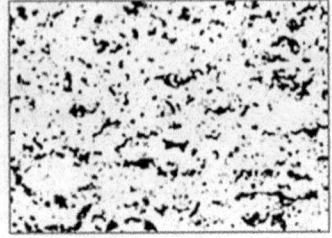

Image with non-flat porosity↓crocracksImage with mi

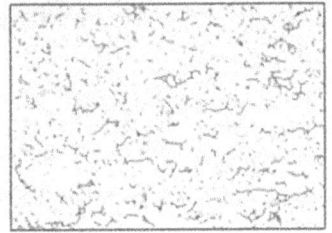

Image with vertical porosity↓ Image with horizontal porosity

FIGURE 5.5 Sample of image analysis process for segmenting microstructure of defects [28]. The white magnification scale bars in (a) to (g) are 20 μm.

slightly higher than that of penny-shaped and interlamellar porosities over the annealing period. Similar procedure was followed for optical micrographs of various assprayed coatings (i.e., 8YPSZ, 25CSZ and 22MSZ).

It is necessary to investigate the microstructure and thermal conductivity of porous TBCs before and after the heat annealing process. Therefore, extensive microstructure analyses and thermal conductivity evaluations of several powder ceramic materials of YSZ (i.e., F&C, A&S and HOSP) were conducted. SEM images of morphology of the as-sprayed and after annealing treatment at 1200°C for 225 hrs. are shown in Figure 5.7 and presented for a reference. Figure 5.8 depicts the process employed by image analysis for segmenting the different classes of defects of F&C powder. By comparing the overall porosity present in the as-sprayed and heat-treated powder ceramic samples HOSP, F&C and A&S, it is found that the as-sprayed coating HOSP

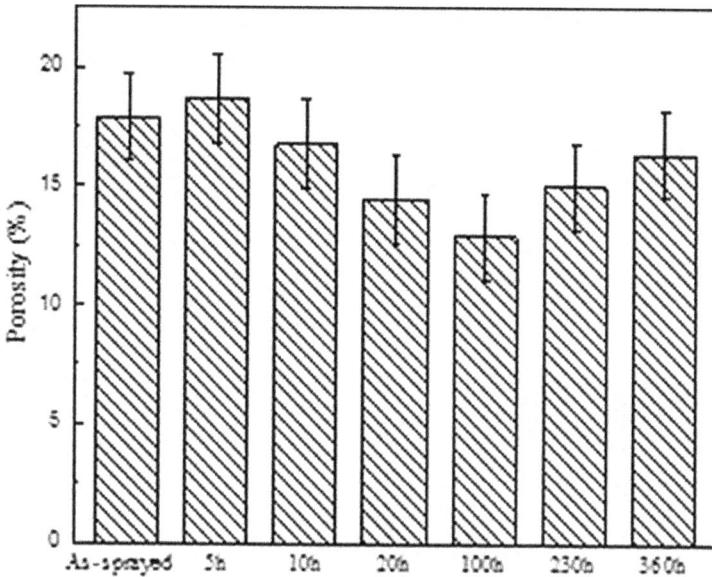

FIGURE 5.6 Porosity of SrZrO$_3$ at different times after heat treatment at 1400°C performed by image analysis [28].

TABLE 5.1
Porosity Content of SrZrO$_3$Coating Samples Annealed at 1400°C for Different Times Performed by Image Analysis (IA)

Time of Heat Treatment (h) for SrZrO$_3$Coating Material	As-sprayed	5	10	20	100	230	360
Total porosity (%)	17.89	18.63	16.75	14.49	12.87	14.98	16.37
Open randomly oriented porosity (%)	2	0.17	0.70	0.76	0.46	0.39	0.43
Microcracks (%)	2.78	1.91	2.05	1.95	1.87	1.86	2.13
Non-flat porosity (%)	9.4	14.07	11.18	9.02	8.64	10.44	11.05
Penny-shaped porosity (%)	1.71	1.09	1.12	1.15	0.83	0.93	1.23
Interlamellar porosity (%)	2	1.73	1.70	1.61	1.07	1.36	1.53

yields the greatest content of the total porosity in correlation to other two powder samples. Also, all of the annealing samples illustrated a lower value of porosity in comparison to the as-sprayed samples that affects the thermal properties of the coating materials. Variation in microstructure occurred due to the sintering effect taking place during the annealing process [40]. It is useful to compare the results of total porosity with annealing to the as-sprayed materials which were found to range between 4% and 23%.

Annealed coatings

↓   ↓   ↓

F&C   A&C   HOSP

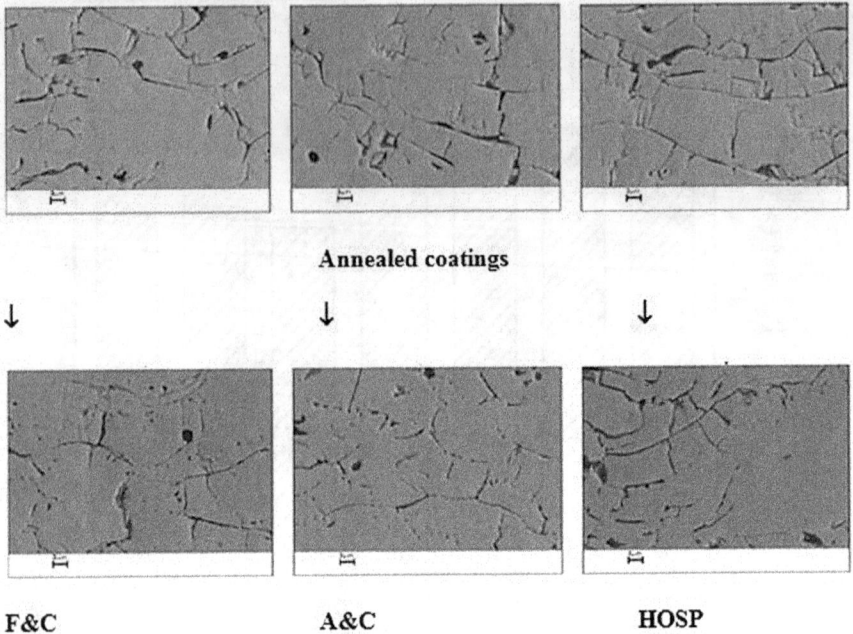

FIGURE 5.7 The SEM images of various as-sprayed powder coating materials, annealed at 1200°C for 225 hrs. [39]. The black scale bars in the above figures are 5 μm.

A significant drop in the total porosity of A&S coating was observed after heat-treating the sample for 5–15 hrs. This was followed by a gradual decreasing in the overall porosity of up to 200 hrs. It is worth mentioning that as the annealing time prolonged, the total porosity of A&S coating was slightly reduced. The results for the overall porosity and its constituents of A&S powder coating are given in Table 5.2 and Figure 5.9. The classification of porosities present in A&S powder coating materials annealed at 1250°C for different periods performed by image analysis is given in Figure 5.9. As expected, during the annealing period, the overall porosity diminished to 15.28%. This variation was mainly associated with the decrease in the content of non-flat porosity with a lesser impact on the other types of porosities which was mainly caused by the sintering process. As an exception, the volumetric fraction of the open randomly oriented porosity was greatly influenced for 200 hrs. annealing period. It is concluded that no remarkable variations were observed in the case of microcracks, interlamellar and penny-shaped porosities.

As stated above, the thermal conductivity of porous TBCs is largely affected by the presence of defects in coatings. Thus, the six-phase model was used to predict the thermal conductivity of porous TBCs based on the microstructure of coatings. Moreover, the defects present in coatings plays a major role in optimizing the thermal resistance of coatings. Different shapes and orientations of porosities have various impacts on reducing the thermal conductivity of TBCs [41].

Microstructure of as-sprayed F&C coating ↓ Thresholded image with total porosity

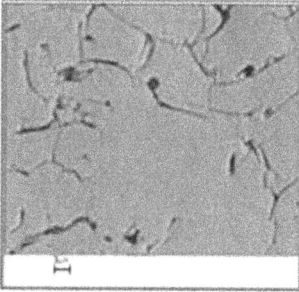

Image with non-flat porosity↓Image with microcracks

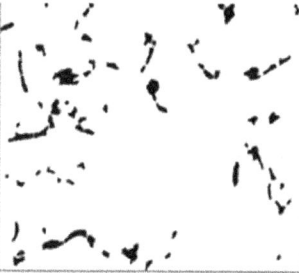

image with horizontal porosityↆImage with vertical porosity

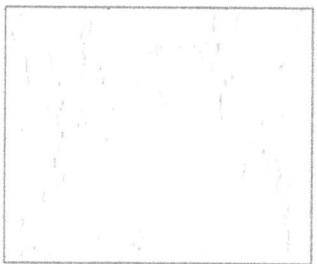

FIGURE 5.8 Samples of image analysis process for segmenting microstructure of defects [28].

The heat treatment considered in this model is another factor that can affect the thermal conductivity of TBCs. Furthermore, this section also compares the predicted thermal conductivity with earlier published measured values and those by FEA calculations in an attempt to validate the results by using the six-phase model. To clarify, the measured and the thermal conductivity calculated by FEA were acquired from the literature [36, 39]. The predicted and measured overall thermal conductivity of the as-sprayed and heat-treated $SrZrO_3$ coating material at 1400°C for 5, 10, 20, 100, 230 and 360 hrs. are displayed in Table 5.3. The bulk thermal conductivity value of $SrZrO_3$ the sample is 2.08 W/m K at 1000°C plotted in Figure 5.10. The white bars

TABLE 5.2
Porosity Measurements of Agglomerated and Sintered Powder Coating Materials Annealed at 1250°C, Performed by Image Analysis

Time of Heat Treatment (h) for Agglomerated and Sintered Coating Material	As-Sprayed	5	15	50	200
Total porosity (%)	21.47	19.71	16.31	15.75	15.28
Open randomly oriented porosity (%)	3.03	4.21	3.62	3.64	1.35
Microcracks (%)	2.52	2.79	2.67	2.64	2.12
Non-flat porosity (%)	11.57	8.25	6.40	6.03	8.54
Penny-shaped porosity (%)	1.53	1.58	1.18	1.11	1.17
Interlamellar porosity (%)	2.82	2.88	2.44	2.33	2.10

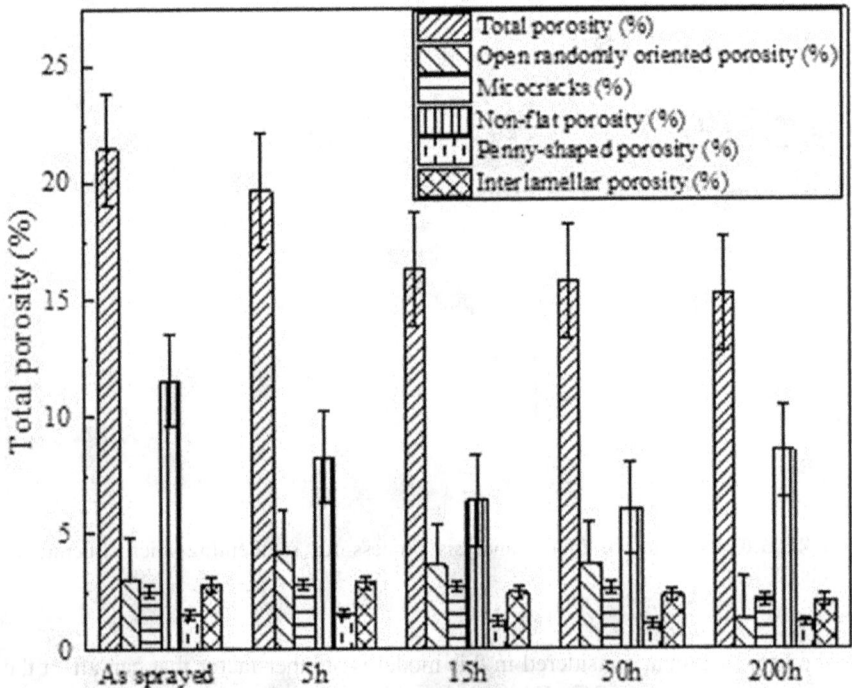

FIGURE 5.9 The classification of porosities presents in agglomerated and sintered powder coating materials annealed at 1250°C for different periods performed by image analysis.

represent the predicted values while the black bars represent the measured data. Both the predicted and measured thermal conductivities increased with further increasing heat treatment time. In one hand, the predicted and measured thermal conductivities were increased at the first stages of the heat treatment (i.e. from 5 to 100 hrs.). This

TABLE 5.3

A Comparison between Experimental and Predicted Thermal Conductivities of SrZrO$_3$ (Bulk Value = 2.08 W/m K)

SrZrO$_3$Coating Material for Different Times of Heat Treatment (hrs.)	Experimental Thermal Conductivity (W/m K)	Predicted Thermal Conductivity (W/m K)
As-sprayed	1.25	1.2
5	1.2	1.235
10	1.28	1.287
20	1.29	1.36
100	1.31	1.43
230	1.2	1.36
360	1.27	1.29

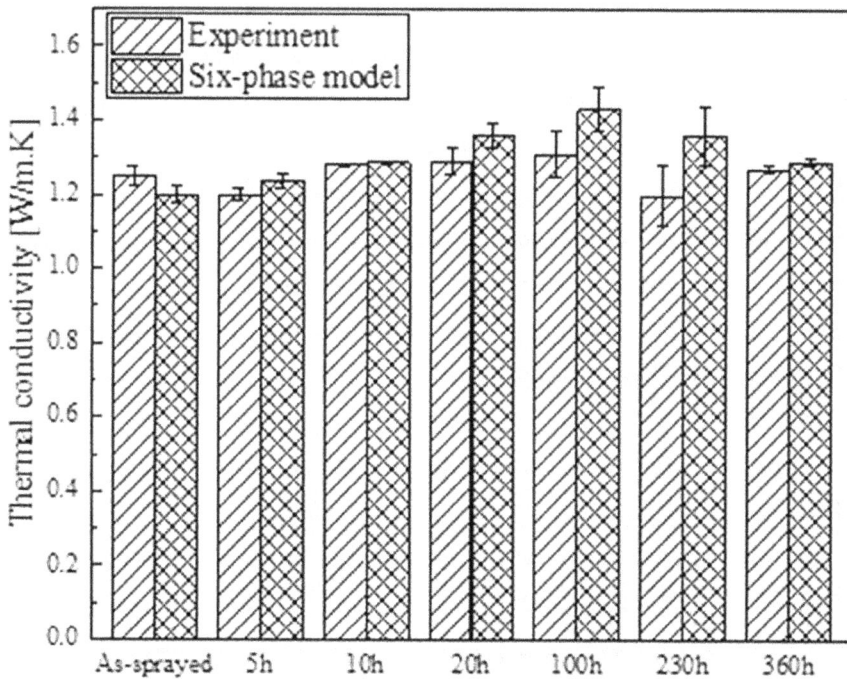

FIGURE 5.10 A comparison of predicted and experimental thermal conductivity of SrZrO$_3$ coating.

increment in thermal conductivities was due to the reduction in interlamellar porosity which was caused by pore healing. On the other hand, although the heat treatment was increased to 230 and 360 hrs., both the predicted and measured thermal conductivities were reduced.

TABLE 5.4

Comparison between the Experimental and Calculated Thermal Conductivity of Three Different As-Sprayed Coating Samples 8YPSZ, 22MSZ and 25CSZ (Bulk Value of 8YPSZ and 25CSZ=2.8 W/m K and Bulk Value of 22MSZ=2.2W/m K)

Material Name	Experimental Thermal Conductivity (W/m K)	Calculated Thermal Conductivity (W/m K)
8YPSZ	1.2	1.42
22MSZ	1.1	1.19
25CSZ	1.19	1.3

Another comparison was made for the calculated results and measured thermal conductivity of three coatings: 8YPSZ, 22MSZ and 25CSZ. The comparison is presented in Table 5.4 and Figure 5.11. The bulk thermal conductivity of 8YPSZ and 25CSZ is 2.8 W/m K, while the bulk thermal conductivity of 22MSZ is 2.2 W/m K.

The calculated thermal conductivity for three different coatings was relatively higher than those experimentally measured. Specifically, the calculated values for coatings 22MSZ and 25CSZ were closer to the measured ones in comparison to the 8YPSZ coatings. Even though 8YPSZ coating sample yields a higher content of the overall porosity than the other two coating samples, the calculated thermal conductivity was higher. This higher trend may be attributed to a lower content of microcracks as previously discussed. The variation between the calculated and measured thermal conductivities may be caused by imprecise evaluation of microstructural properties due to Image *J* limitation.

A further comparison of thermal conductivities of three as-sprayed and heat-treated coatings of YSZ, namely HOSP, A&S and F&C, was made between the computed results using the six-phase model and those of the measured and FEA results. This comparison is shown in Table 5.5 and Figure 5.12 with a bulk value of 2.5 W/m K. The thermal conductivity was raised after the heat-treatment process. As previously stated, this increase in thermal conductivity results from a change in microstructures due to the sintering effect. It is evident that the overall porosity decreases, while the thermal conductivity increases. The thermal conductivity of both as-sprayed and heat-treated HOSP coatings was lower than the other two powder coatings. This result could be ascribed to the hollow spherical structure of powder coatings [42]. The prediction using the six-phase model yields thermal conductivity closer than that using FEA modeling to the measured values except for the case of heat-treated HOSP and F&C coatings which exhibited higher values than the FEA modeling.

The computed thermal conductivities reveal a gradual increase with further increasing annealing time as expected because of the decrease in the overall porosity due to the sintering effect. Similar trends were observed for the measured data. The significant overestimation in the case of the as-sprayed coating may be

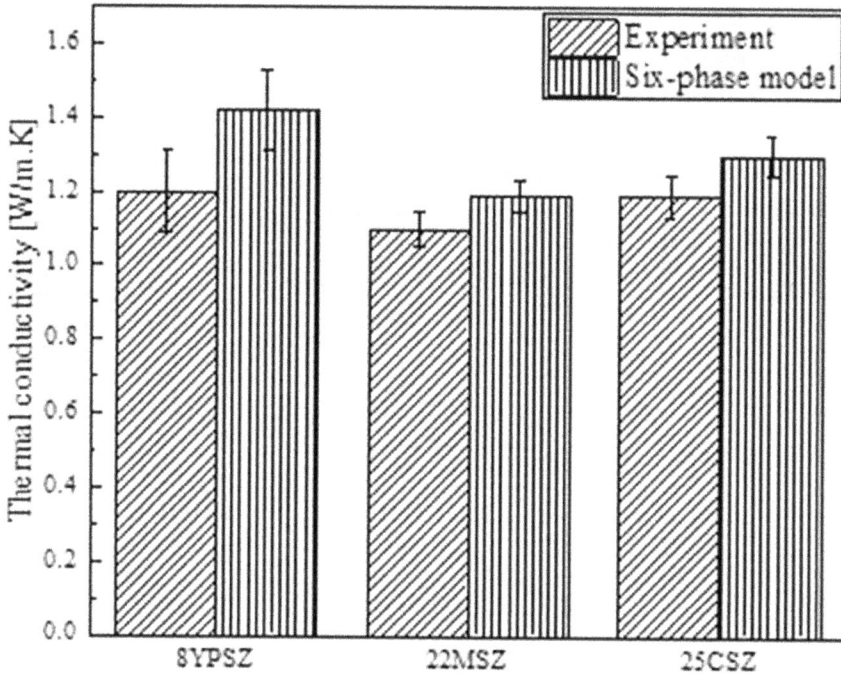

FIGURE 5.11 Comparison of calculated six-phase model and measured thermal conductivity of different coating materials.

TABLE 5.5

Comparison between Experimental, Calculated Six-Phase and FEA Model Thermal Conductivity Values (Bulk Value = 2.5 W/m K)

Sample Name	Experimental Thermal Conductivity (W/m K)	Six-Phase Thermal Conductivity (W/m K)	FEA Thermal Conductivity (W/m K)
HOSP as-sprayed	1.1	1.28	1.32
HOSP annealed	1.3	1.54	1.44
F&C as-sprayed	1.23	1.53	1.56
F&C annealed	1.85	1.73	1.78
A&S as-sprayed	1.12	1.59	1.75
A&S annealed	1.43	1.73	1.77

from following factors: (1) the connected porosity was not considered in this model; (2) the 2D microstructure image only provides an approximation of the real 3D microstructure of a coating; and (3) a software limitation in evaluating the porosity.

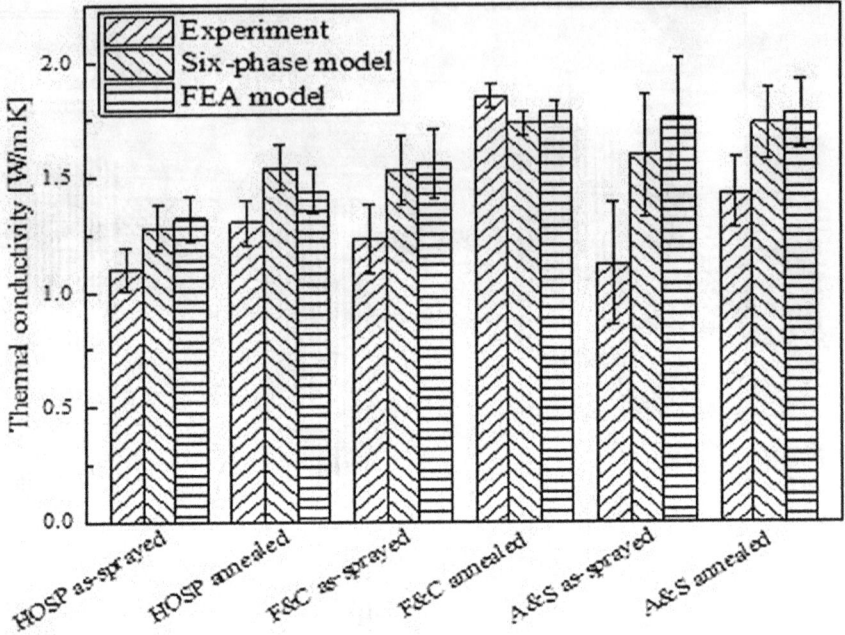

FIGURE 5.12 Comparison of simulated six-phase, FEA models and experimental thermal conductivity of various powder coating materials.

5.3 FAILURE ANALYSIS OF APS-TBCS

5.3.1 FAILURE ANALYSIS

Life prediction of APS-TBCs relies primarily on roughness analysis of the coating interface between the topcoat and TGO, as shown in Figure 5.13. To study residual stress upon cooling and its effect on crack nucleation and propagation, it is assumed that the coating is at a stress-free state during high-temperature dwell period. Upon reaching the cooling stage, large tensile stress normal to the coating interface develops at the peak location where the crack nucleation occurs and subsequent propagation proceeds along the interface until inhibited at the valley of the topcoat due to compressive stress. As the thermal cycle continues, the TGO thickens due to progressive oxidation of the bond coat. The compressive stress normal to the interface at the valley attenuates, and at a certain thickness of TGO, the topcoat at the valley location is under tensile stress.

This tensile stress will promote fatigue crack nucleation and propagation. A spallation of the topcoat occurs when these neighboring cracks link and collapse, which indicates a failure of APS-TBCs.

5.3.2 STRESS MODEL

Based on Eshelby protocol and elasticity theory, Mumm and Evans [43] derived analytically, the residual stress model $\sigma_{rr}(T)$ at the topcoat of TBC in a tri-material system (topcoat, TGO and substrate) with spherical symmetry

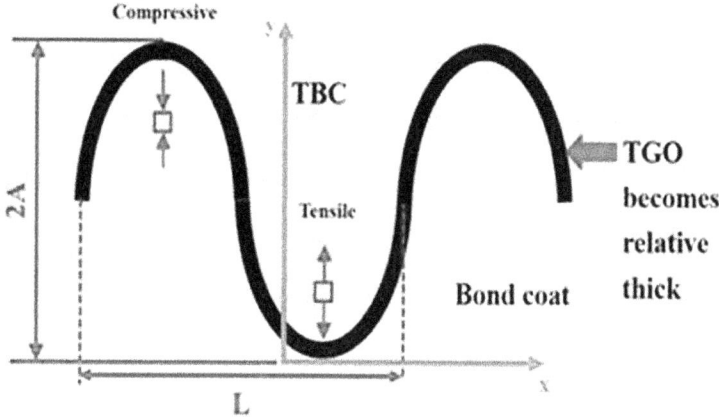

FIGURE 5.13 Failure mechanism of APS-TBCs. The crack nucleates at the valley of topcoat as TGO thickens where a stress inversion was shown as thermal cycle proceeds [16].

$$\sigma_{rr}(r) = -\Lambda \left[\alpha_{TGO} - \alpha_{TBC} + (\alpha_{BC} - \alpha_{TGO})(1 - h/R)^3 \right] \left(R/r \right)^3 \qquad (5.10)$$

where r is the radial coordinate of tri-materials, R is the curvature of the sphere, h is the TGO thickness, where

$$\Lambda = 4\kappa\mu_{TBC}\Delta T / (\kappa + 4\mu_{TBC}/3),$$
$$\Lambda = 4\kappa\mu\Delta T\kappa + 4\mu 3\mu = \text{ETC21} + \upsilon,$$
$$\kappa = \text{ETC31} - 2\upsilon \ \ \kappa = E_{TBC}/3(1 - 2\upsilon_{TBC})$$

ν_{TBC} is the Poisson ratio of the topcoat, ΔT is the temperature drop causing the CTE-induced residual stress, α_{TGO}, α_{BC} and α_{TBC} represent coefficients of thermal expansion of TGO, bond coat and topcoat, respectively. This high spherical symmetry stress model, however, may not be suitable for a coating interface described using a sinusoidal function characterized by various ratios of its wavelength over amplitude. To calculate the residual stress of topcoat with an undulated TGO/topcoat interface such as the wave-like TGO/topcoat interface, the stress model of Equation (5.10) needs modification accordingly. We proposed a stress model for the topcoat at the valley location for the wave-like TGO/topcoat interface

$$\sigma_{valley}(y) = -\alpha\Lambda \left\{ (\alpha_{TGO} - \alpha_{TBC}) + (\alpha_{BC} - \alpha_{TGO})\left(1 - \beta\frac{d_{TGO}}{A}\frac{R}{y}\right)^3 \right\} \left(\frac{R}{y}\right)^3 \qquad (5.11)$$

where two model parameters α and β will be fitted to the 3-D FEA calculated stresses [19]. A is the amplitude of interfacial roughness, d_{TGO} is the TGO thickness and y is the valley location of the TBC topcoat, where the residual stresses will be evaluated for crack propagation.

5.3.3 Topcoat Sintering and TGO Growth

The sintering effect of Young's modulus E_{TBC} of the topcoat was also taken into account upon evaluating the crack growth rate using Equation (5.1). The Young's modulus sintering effect can be described by [16]

$$E_{TBC}(t) = \frac{\beta_s E_{TBC}^0 E_{TBC}^\infty}{\beta_s E_{TBC}^0 + E_{TBC}^\infty - E_{TBC}^0} \qquad (5.12)$$

$$\beta_s = 1 + A_{sint} \exp\left(-\frac{E_{sint}}{\kappa_B T}\right) t^{p_s}$$

where k_B is the Boltzmann constant, A_{sint}, E_{sint} and p_s are the sintering coefficient, sintering activation energy and sintering exponent of the topcoat in APS-TBC, respectively. T is the temperature. Although the model Equation (5.11) describes the residual stress located at the topcoat, it involves the effect of TGO, bond coat and topcoat through combined parameters such as the difference of CTEs between the TGO and topcoat, the difference of CTE the bond coat and TGO, TGO thickness, as well as the elastic moduli of bond coat and topcoat. In this chapter, it is assumed that the TGO growth follows a parabolic law given by [16]

$$d_{TGO} = A_{TGO} \exp\left(-\frac{E_{TGO}}{\kappa_B T}\right) t^{p_{TGO}} \qquad (5.13)$$

where A_{TGO}, E_{TGO} and p_{TGO} are the TGO growth rate coefficient TGO growth activation energy and TGO growth exponent, respectively. t is an exposure dwell period. It is assumed that the interfacial profile is approximately described by a sinusoidal curve, the curvature radius R of imperfection is given by $L^2/4\pi^2 A$, with L representing the mean value of the wavelength. The valley location at $y = 8$ μm was selected according to the Scanning Electron Microscope (SEM) measurement of the topcoat [16].

As previously explained, the radial stress at the valley of topcoat develops initially under compressive state for a thinner TGO. With the TGO thickening, this stress develops into a tensile state at a certain TGO thickness, i.e., the sign change of residual stress occurs from an initially negative compressive stress to a positive tensile stress. The corresponding critical TGO thickness can be derived at the vanishing point of the residual stress, where the critical time can be predicted for such stress conversion.

5.3.4 Life Prediction Procedure

By combining the stress model and subcritical crack growth formula Equation (5.1), the life was evaluated numerically. On the right-hand side of Equation (5.1), the integral of the residual stress on the topcoat is conducted from an initial time t_0 to t_f, the failure lifetime to be estimated. The left-hand side of Equation (5.1) is an integral of the crack length initially starting from a_0, an assumed half wavelength, to the a_f of the entire wavelength, indicating the spallation of the topcoat from the bond coat completely. Equation (5.1) can be explained as follows: the right-hand side of Equation (5.1) represents the driving force of the crack growth within a cyclic process, while the left-hand side of Equation (5.1) is a consequence of fatigue crack growth driven by the right side during thermal cycles, leading to crack propagation. Upon the crack length a reaching the critical value a_f, the cracks collapse, resulting in a spallation of the topcoat, where the failure time t_f can consequently be obtained for the coating. This is the procedure that was used in estimating the life of APS-TBC systems in the present research.

5.3.5 MODEL PARAMETERS

The burner rig test results of APS-TBC system [18] were used to fit model parameters of Equation (5.1). The thermal radial stress versus the TGO thickness at the valley of the topcoat calculated using FEA [19] is shown in Figure 5.14 together with the plotted curve using the proposed stress model of Equation (5.11). It can be seen that a stress conversion takes place from a compressive state to the tensile state at the TGO thickness of $h = 4.4\ \mu m$. The amplitude $A = 7\ \mu m$ and wavelength $\lambda = 65\ \mu m$ in describing the interface roughness were used to fit the residual stress model parameters α and β in Equation (5.11) (Figure 5.14).

The fitted α and β are listed in Table 5.6, and the parameters in describing Young's modulus sintering and TGO growth in the model are listed in Table 5.7. It is assumed that two residual stress model parameters α and β are temperature-independent, while the temperature-dependent characteristics of the life model is described by the life model fitting parameters A^*Y^m/K^m_{IC} A^*YmKI, cm in the subcritical crack growth formula in Equation (5.1). In the fitting process, the model parameter A^*Y^m/K^m_{IC} was

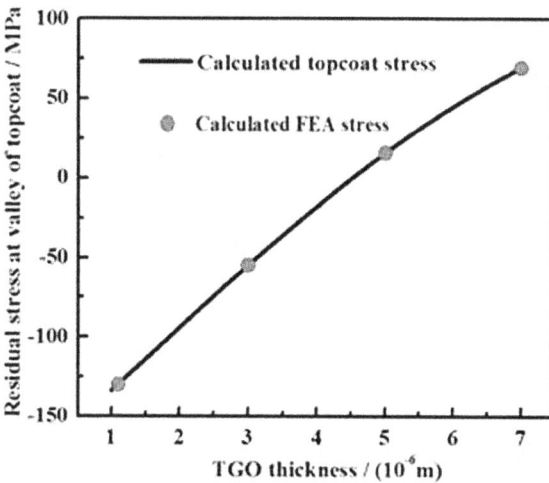

FIGURE 5.14 The calculated residual stresses using Equation (5.3) versus TGO thickness compared to the stresses calculated by FEA [19].

TABLE 5.6
Stress Model Parameters in Equation (5.11) and Their Related Variables

Parameters	A	R	y	[16]	[16]
Abbreviation	Amplitude	Curvature radius	Valley location	CTE of TGO	CTE of BC
Value	7 μm	15.3μm	8μm	8×10^{-6} K^{-1}	1.6×10^{-5} K^{-1}
Parameters	α_{TBC}[44]	A	β	\	\
Abbreviation	CTE of TBC	Residual stress-related coefficient		\	\
Value	1×10^{-5}K^{-1}	0.174816	0.29585	\	\

TABLE 5.7
Model Parameters in TGO Growth Formula Equation (5.13) and Topcoat Young's Modulus Sintering Equation (5.12)

Parameters	d_{TGO}	A_{TGO}	E_{TGO}	p_{TGO}	T
Abbreviation	TGO thickness	TGO growth rate coefficient	TGO growth activation energy	TGO growth exponent	Temperature during a hold period
Value	\	7.48×10^{-4}m/ s^p	0.907 eV	0.25	1273.15 K
Parameters	t	E_{TBC}	N	E_{TBC}^0	E_{TBC}^∞
Abbreviation	exposure holding time	Young's modulus of TBC	Poisson ratio of TBC	Initial modulus of TBC	Bulk modulus of TBC
Value	\	\	0.33	20GPa	136GPa
Parameters	K_B	A_{sint}	E_{sint}	P_s	d_{TGO}^C
Abbreviation	Boltzmann constant	Sintering coefficient	Sintering activation energy	Sintering exponential coefficient	Critical TGO thickness
Value	1.38×10^{-23} J/K	2×10^{10} s^{-P}	3 eV	0.25	4.4 μm

fitted to the test life data of coating system at specific temperatures. At different temperatures, the test life data vary, and as a result, A^*Y^m/K^m_{IC} shows different values. It was observed that the fitted A^*Y^m/K^m_{IC} parameter can be described using an exponential formula

$$\frac{A^*Y^m}{K^m_{IC}} = a_{fit}\exp\left(b_{fit}T\right) \qquad (5.14)$$

The fitting parameters are $a_{fit} = 7.55\times10^{-17}$ and $b_{fit} = -0.16$. As the parameter A^*Y^m/K^m_{IC} involves a combination of parameters including the topcoat fracture toughness K_{IC}, it demonstrates that the TBC topcoat toughness K_{IC} is a temperature-dependent property rather that a constant.

5.3.6 LIFE PREDICTION AND DISCUSSIONS

The bond coat used for the present life prediction is the NiCoCrAlY. The selected wave amplitude $A = 7$ μm was close to the measured average roughness R_a. For the selected coating system, the fitted life model parameter $A * Y^m / K_{IC}^m$ in Equation (5.1) reflects these specific coating materials as well as the test conditions. The stress model and its parameters reflect the stress characteristics that show its dependence on a TGO thickness, and more importantly, it also shows a conversion behavior of the stress sign, i.e., from an initial compressive state to the tensile state at a certain TGO thickness. Table 5.8 lists related life prediction model parameters. Using the temperature-dependent model parameter, the predicted life between 1273 and 1348 K is shown in Figure 5.15. Four tested life data (black solid circles) were used for model

TABLE 5.8
Life Prediction Model Related Parameters and Variables

Parameters	a	a_f	a_0	$\sigma(T)$	t_0
Abbreviation	Crack length	Full crack length (entire wavelength)	Initial crack length (half wavelength)	Temperature-process-dependent stress	Initial time that tensile stress developed
Value Parameters	\ t_f	65μm m	32.5μm $\dfrac{A * Y^m}{K_{IC}^m}$	\	\
Abbreviation	Estimated lifetime	Exponential fitting coefficient	Temperature-dependent fitting parameter		
Value	\	18	\		

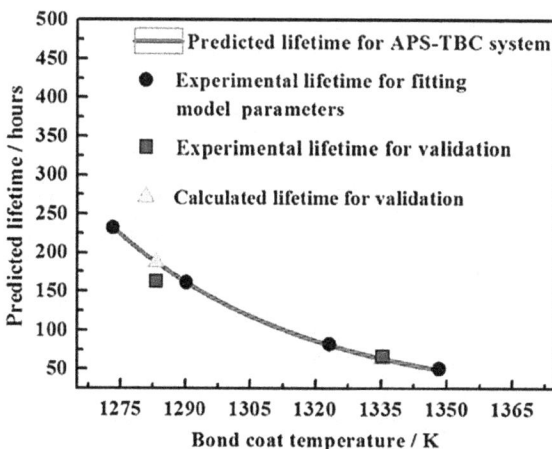

FIGURE 5.15 The predicted life versus the bond coat temperature. Four tested life data (black solid circles) were used for model parameter fitting [16, 18] and two tested life data (blue squares) were used to validate the life model.

parameter fitting [44, 45], and two tested life data (blue squares) were used to validate the life model. An exponential form was obtained to describe this life-temperature correlation

$$t_f = 8.78 \times 10^{13} \exp(-0.02T) \tag{5.15}$$

The average life of the APS-TBCs decreases versus the bond coat temperatures. Four tested life data (black solid circles) were used for model parameter $A^* Y^m / K^m{}_{IC}$ fitting, while two tested life data (blue squares) were employed to validate the proposed life model. According to the predicted life data (green triangles) compared to the test life data (black squares) at the specific temperatures illustrated in Figure 5.15, it shows that using the temperature-dependent $A^* Y^m / K^m{}_{IC}$ parameter gives more accurate prediction results.

5.4 SOLID PARTICLE EROSION AND EVOLUTION OF OXIDE SCALE THICKNESS UNDER EROSION ATTACK

5.4.1 SOLID PARTICLE EROSION

Solid particle erosion results in removal of materials from component surfaces by successive impacts of hard particles at high velocities. Materials removal by solid particle erosion is the overall result of a series of essentially independent but similar impact events. Despite the impact period being very short, solid particle erosion is completely different from other material removal processes such as sliding wear, abrasion, grinding and machining, where the contact between the abrasive tool and the work-piece target is continuous.

5.4.2 EVOLUTION OF OXIDE SCALE THICKNESS UNDER EROSION ATTACK

To describe the evolution of oxide scale under solid particle attack, Kang et al. [46] proposed the following kinetic equation:

$$\frac{dX}{dt} = \frac{K_{ce}}{X} - K_{oe} \tag{5.16}$$

where X is the instantaneous oxide scale thickness, K_{ce} is the rate constant of oxide growth under combined attack of erosion and oxidation, K_{oe} is the erosion rate depending on the impact particle velocity, incident angle as well as the oxide properties.

Oxidation contributes to the scale development. In many cases, as reported in literature, oxidation was often expressed as the weight gain with the oxide growth constant rate K_p (kg^2 m^{-4} s^{-1}). To convert K_p to the oxide scale growth rate constant, K_{ce} (m^2 s^{-1}), the following relationship was applied [47]:

$$K_p = K_{ce} \left(\frac{bM_o}{aV_m} \right)^2 \tag{5.17}$$

where, for an oxide such as M_aO_b, a and b are the stoichiometric numbers in the oxide compound, M_o is the atomic weight of oxygen (Kg/mol) and V_m is the molar volume of the metal (m³/mol). Rearranging Equation (5.17) gives

$$K_{ce} = K_p \left(\frac{aV_m}{bM_o} \right)^2$$

(5.18)

For CMSX-4 alloys, the molar volume V_m of CMSX-4 can be calculated based on the weight percent of molar volume of each specifies in the alloys, $V_m = 1.242 \times 10^{-5}$ m³/mol. Similarly, the calculated molar weight $M_0 = 0.016$ Kg/mol. For Al_2O_3, $a = 2$, $b = 3$. The parabolic growth rate constant Kp of Al_2O_3 normally obeys the Arrhenius law [44]

$$K_P = K_0 \exp\left(-\frac{Q}{RT} \right)$$

(5.19)

where K_0 is the pre-exponential factor, Q stands for the apparent energy barrier of oxide formation, R the universal gas constant and T the absolute temperature. Both Q and K_0 can be obtained from a fitting to the testing data of $K_p \sim T$ using a least square method. As an example, the experimentally measured K_p values of CMSX-4 at specific temperatures are given in Table 5.9 [45] and $K_0 = 8.97$ (kg²m⁻⁴s⁻¹) and $Q/R = 5128.0$ were obtained based on Table 5.1 and Equation (5.19).

It is noted that the oxide growth rate constant K_P of Al_2O_3 is much smaller if compared to other oxides such as Cr_2O_3 and Fe_2O_3 [48]. This smaller value K_p of Al_2O_3 will affect the erosion–oxidation interaction described by Equation (5.16), and consequently will affect the erosion behavior of CMSX-4 at elevated temperatures.

5.4.3 EROSION MODEL AND MODEL PARAMETERS

Both oxide scale and substrate can be removed by particle impacts. To study the erosion rates of oxide scale and substrate, K_{oe}, the related erosive mechanisms should be clearly identified. Bitter's work [49] illustrated a method to quantitatively describe erosion degradation based on energy conservation. It has been recognized that the total amount of erosion rate ΔW_t, measured in terms of weight loss per particle gram (mg/g) or volume loss per particle gram (mm³/g), is basically contributed from two types of erosion processes, i.e., the deformation erosion (ΔW_D) and the cutting

TABLE 5.9

Al_2O_3 Growth Rate at Different Temperatures for CMSX4 Alloys [45]

Temperature (°C)	K_p (mg²cm⁻⁴·h⁻¹)
850	9.76E-5
900	2.04E-4
950	2.61E-4

erosion (ΔW_{C1} or ΔW_{C2} depending on the impacting angle of a particle). The total erosion rate ΔW_t can be expressed as

$$\Delta W_t = \Delta W_D + \Delta W_{C1}, \text{if } \alpha \leq \alpha_0 \tag{5.20}$$

$$\Delta W_t = \Delta W_D + \Delta W_{C2}, \text{if } \alpha \geq \alpha_0 \tag{5.21}$$

where α is the incident angle of impact particles. α_0 is the impact angle at which the horizontal velocity of the impact particle equals to zero when it leaves the surface. Originally, Bitter used rather complicated forms to express each erosion component. Later, Neilson and Gilchrist [50] simplified Bitter's model for erosion degradation. The simplified model takes the following form [50]:

$$\Delta W_t = \frac{\frac{1}{2}\Delta M V^2 \cos^2 \alpha \sin n\alpha}{\phi} + \frac{\frac{1}{2}\Delta M (V \sin \alpha - K)^2}{\varepsilon}, \alpha \leq \tag{5.22}$$

and

$$\Delta W_t = \frac{\frac{1}{2}\Delta M V^2 \cos^2 \alpha}{\phi} + \frac{\frac{1}{2}\Delta M (V \sin \alpha - K)^2}{\varepsilon}, \alpha \geq \alpha_0 \tag{5.23}$$

where ΔM stands for the accumulated mass of particles attacking the target material (oxide or substrate), V is the velocity of particles, α is the impact angle of particles, n is a parameter related to the target materials and α_0 is the impact angle at which the horizontal velocity equals zero when the particle leaves the surface. ϕ and ε represent the energies required to remove a unit mass from the target materials through the cutting and deformation processes, respectively. These two model parameters can be obtained by fitting to the erosion-induced weight changes (ΔW) as a function of particle impact velocity and angle for the alloy. As two examples, the erosion rate model is plotted in Figures 5.16 and 5.17 in comparison with the test data for Al plate and FXS 414 alloys [48, 51]. It can be inferred that the ΔW_C, erosion rate due to the cutting process, achieves the maximum value at acute angles (20°~35°), while ΔW_D, erosion rate due to the deformation process, shows the maximum value at the normal angle.

To study the evolution of oxide scale, the erosion rate from Equations (5.22) and (5.23) needs to be converted into the unit (m/s) as specified in Equations (5.16) and (5.17). By infinitesimal analysis, the erosion rate ΔW_t in Equations (5.22) and (5.23) can be expressed as

$$\Delta W = \rho_T \Delta V = \rho_T A \Delta X \tag{5.24}$$

where ρ_T is the density of the target material (oxide or substrate), A is the attacking area by the flux of particles and ΔX the depth of the material removed by solid particle impact. Dividing ΔW by $A\rho_T$ in Equations (5.22), (5.23) and (5.24), we have

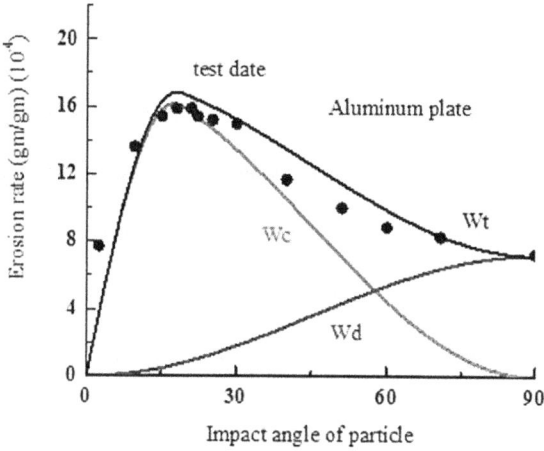

FIGURE 5.16   The erosion model fitted to Aluminum [50]. ΔW_t represents the total erosion rate, while ΔW_C and ΔW_d stand for the erosion rates due to the cutting and deformation processes, respectively [50].

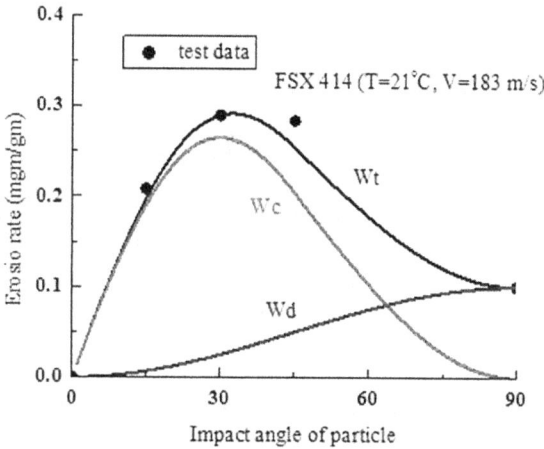

FIGURE 5.17   The erosion model fitted to FSX 414 alloys [51].

$$\Delta X = \frac{1}{2}\frac{\Delta M}{A}\frac{1}{\rho_T}\frac{V^2 \cos^2 \alpha \sin n\alpha}{\phi} + \frac{1}{2}\frac{\Delta M}{A}\frac{1}{\rho_T}\frac{\left(V \sin\alpha - K\right)^2}{\varepsilon}, \quad \alpha < \alpha_0 \quad (5.25)$$

and further

$$\frac{\Delta X}{\Delta t} = \frac{1}{2}\frac{\Delta M}{A\Delta t}\frac{1}{\rho_T}\frac{V^2 \cos^2 \alpha \sin n\alpha}{\phi} + \frac{1}{2}\frac{\Delta M}{A\Delta t}\frac{1}{\rho_T}\frac{\left(V \sin\alpha - K\right)^2}{\varepsilon}, \quad \alpha < \alpha_0 \quad (5.26)$$

Define F as

$$F = \frac{\Delta M}{A \Delta t} \tag{5.27}$$

Equation (5.27) represents the flux of solid particles (the erodent) (kg m^{-2} s^{-1}). The typical value of flux rate $F = 10.0$ (kg m^{-2} s^{-1}) in the test condition was used in the present research [44]. Equation (5.26) can be re-written as

$$\frac{dX}{dt} = \frac{1}{2}\frac{F}{\rho_T}\frac{V^2 \cos^2 \alpha \sin n\alpha}{\phi} + \frac{1}{2}\frac{F}{\rho_T}\frac{(V \sin \alpha - K)^2}{\varepsilon}, \quad \alpha < \alpha_0 \tag{5.28}$$

This yields the scale thickness variation rate V_{ero} (m/s) by erosion in terms of the erodent flux F:

$$V_{ero} = \frac{\frac{1}{2}FV^2 \cos^2 \alpha \sin n\alpha}{\phi \rho_T} + \frac{\frac{1}{2}F(V \sin \alpha - K)^2}{\varepsilon \rho_T}, \quad \alpha \leq \alpha_0 \tag{5.29}$$

And, similarly

$$V_{ero} = \frac{\frac{1}{2}FV^2 \cos^2 \alpha}{\phi \rho_T} + \frac{\frac{1}{2}F(V \sin \alpha - K)^2}{\varepsilon \rho_T}, \quad \alpha > \alpha_0 \tag{5.30}$$

Equations (5.29) and (5.30) will be applied to evaluate the change in oxide scale thickness under solid particle impact conditions when substituted into Equation (5.16).

5.4.4 SCALE THICKNESS EVOLUTION

In this section, we will combine the oxidation model of Equation (5.19) and the erosion model of Equations (5.29) and (5.30) into Equation (5.16) to simulate oxide scale evolution under solid particle impact. Substituting Equations (5.19), (5.29) and (5.30) into Equation (5.16), the instantaneous scale thickness change rate is expressed as

$$\frac{dX}{dt} = \frac{K_p \left(\frac{aV_m}{bM_o}\right)^2}{X} - \left[\frac{\frac{1}{2}FV^2 \cos^2 \alpha \sin n\alpha}{\phi \rho_T} + \frac{\frac{1}{2}F(V \sin \alpha - K)^2}{\varepsilon \rho_T}\right], \quad \alpha \leq \alpha_0 \tag{5.31}$$

$$\frac{dX}{dt} = \frac{K_p \left(\frac{aV_m}{bM_o}\right)^2}{X} - \left[\frac{\frac{1}{2}FV^2 \cos^2 \alpha}{\phi \rho_T} + \frac{\frac{1}{2}F(V \sin \alpha - K)^2}{\varepsilon \rho_T}\right], \quad \alpha > \alpha_0 \tag{5.32}$$

In Equations (5.31) and (5.32), K_p can be determined from the Arrhenius Equation (5.19) into Equations (5.31) and (5.32), we have

$$\frac{dX}{dt} = \frac{K_o \exp\left(-\dfrac{Q}{RT}\right)\left(\dfrac{aV_m}{bM_o}\right)^2}{X} - \left[\frac{\dfrac{1}{2}FV^2\cos^2\alpha\sin n\alpha}{\phi\rho_T} + \frac{\dfrac{1}{2}F\left(V\sin\alpha - K\right)^2}{\varepsilon\rho_T}\right], \alpha \le \alpha_0$$

(5.33)

$$\frac{dX}{dt} = \frac{K_o \exp\left(-\dfrac{Q}{RT}\right)\left(\dfrac{aV_m}{bM_o}\right)^2}{X} - \left[\frac{\dfrac{1}{2}FV^2\cos^2\alpha}{\phi\rho_T} + \frac{\dfrac{1}{2}F\left(V\sin\alpha - K\right)^2}{\varepsilon\rho_T}\right], \alpha > \alpha_0$$

(5.34)

The Runge–Kutta fourth-order method was used to solve these ordinary differential equations (ODE) Equations (5.33) and (5.34).

5.4.5 RESULTS AND DISCUSSIONS

To apply Equations (5.33) and (5.34) in the simulation of erosion–oxidation interaction, the model parameters ϕ and ε need to be determined for both substrate and oxide. According to the definition, ϕ stands for the energy required for the cutting wear per volume/per particle, while ε represents the energy needed for the deformation wear per volume/per particle on eroded materials. For high-strength materials such as steel, to remove materials by either cutting wear or deformation wear, large values of ϕ and ε are needed. It essentially indicates that the parameters ϕ and ε may exhibit some intrinsic correlation with the strength of materials. Following the procedures of reference [50], we used the erosion models Equations (5.22) and (5.23) to evaluate parameters ε, ϕ, n and α_0 for the selected alloys; these model parameters are given in Table 5.10. Figures 5.18 and 5.19 show the correlations of ϕ and ε with the yield strength σ_Y for the alloys tested and fitted.

As the first approximation, linear relationships are obtained to describe such correlations as

$$\phi = -1.43287 + 0.00526\sigma_Y \left(m^2/s^2 \times 10^7\right)$$

(5.35)

TABLE 5.10
Model Parameters Used in the Oxide Scale and Substrate Evolution Simulations

	ε (m²/s²)	ϕ (m²/s²)	n	α_0 (°)	ρ_T (g/cm³)
CMSX-4	1.77×10^9	3.195×10^7	2.47	36.4	8.62
Al₂O₃	4.68×10^5	∞			3.95

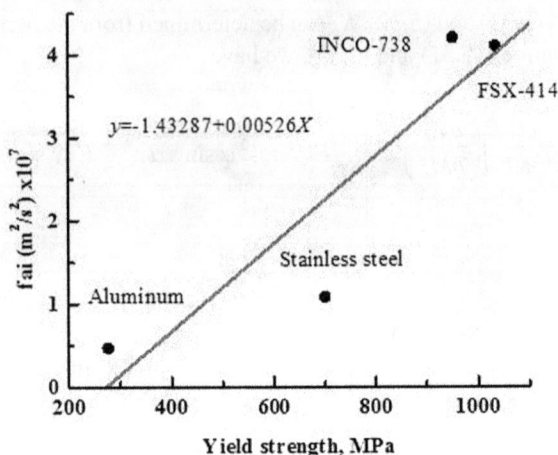

FIGURE 5.18 The correlation between erosion model parameter ϕ fitted to the test erosion data and alloy yield strength σ_y.

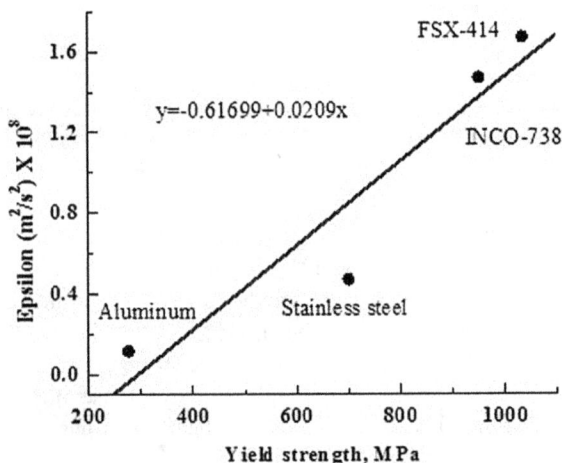

FIGURE 5.19 The correlation between the erosion model parameter ε fitted to the test erosion data and alloy yield strength σ_y.

$$\varepsilon = -0.61699 + 0.20\sigma_Y \left(m^2/s^2 \times 10^8\right) \tag{5.36}$$

Based on such linear correlations of Equations (5.35) and (5.36), for instance, the erosion model parameters ϕ and ε for CMSX-4 can be calculated using the alloy's yield strength $\sigma_Y = 880$ MPa [52]. Table 5.10 lists the model parameters used to calculate the change rate of the oxide scale formed on CMSX-4. Table 5.2 also lists the erosion model parameters for Al_2O_3 fitted based on the erosion test results [53].

In this section, we attempted to simulate erosion–oxidation interaction and Al_2O_3 scale evolution in CMSX-4 as this alloy is pertinent to turbine blade applications. Based on Equations 5.33 and 5.34, the oxide scale thickness evolves as a function of particle velocity, impact time, incident angle and temperature. The simulated results for CMSX-4 are plotted in Figures 5.20 and 5.21.

Figure 5.20 shows the evolution of Al_2O_3 scale at $T = 850°C$ under particle attack at a velocity of $V = 100$ m/s and at different impact angles. The initial thickness of oxide scale was set as 10 μm. Under particle attack, the oxide scale was completely removed in a few seconds, followed by an erosion of the substrate. It can be observed that the oxide scale was removed faster at the attack angle of 75° than at both 15° and 25°. At 75° of attack, the oxide scale was completely removed within 1 second, while

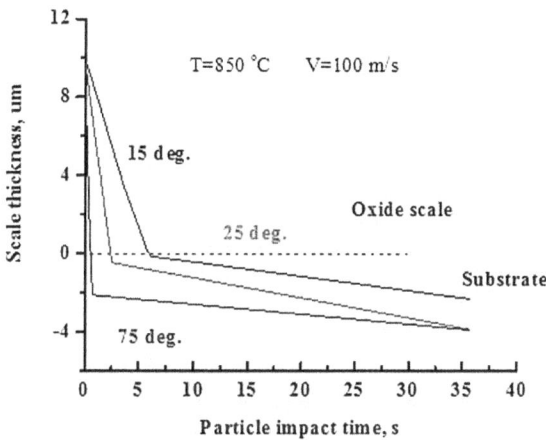

FIGURE 5.20 Evolution of Al_2O_3 scale and CMSX-4 substrate under particle impact at a speed of $V = 100$ m/s, at $T = 850°C$ under different impact angles.

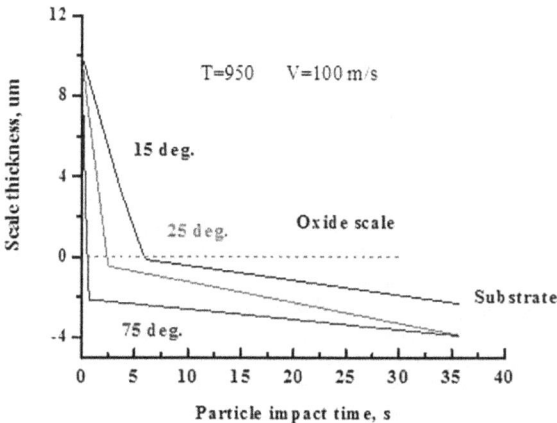

FIGURE 5.21 Evolution of Al_2O_3 scale and CMSX-4 substrate under particle impact at a speed of $V=100$ m/s, at $T = 950°C$ under different impact angles.

it took 6 seconds to completely remove the scale at 15°. This occurrence can be explained based on Al_2O_3 properties. For brittle materials such as ceramic Al_2O_3, the maximum erosion rate is usually observed at the normal angle of impact, while for ductile materials, the maximum erosion rate normally occurs at an acute angle (15°~30°). The simulation results suggest that for CMSX-4 alloys, the Al_2O_3 scale does not play a role in protecting the substrate from erosion attack. Figure 5.21 shows the oxide scale thickness evolution under attack at the particle velocity of 100 m/s, and at temperature $T = 950$°C as well as at three different impact angles. The figure shows a similar result as in Figure 5.6, indicating that at a particle velocity of 100 m/s, the temperature difference of $\Delta T = 100$°C shows little effect on the erosion behavior compared to the results at $T = 850$°C.

5.5 CONCLUDING REMARKS

This chapter has extended Bruggeman two-phase model to the six-phase model using an iterative technique to identify different types of defects, namely open randomly oriented, non-flat, penny-shaped, interlamellar porosity and microcracks present within a TBC. Microstructure characterization for selected TBCs was conducted using image analysis via Image J. The influence of various types of porosities on the thermal conductivities of TBCs was identified. It has been shown that the presence of microcracks and interlamellar porosity are prominent factors in optimizing the thermal resistance for all TBCs examined in this study.

The drastic increase in thermal conductivity after heating TBCs to high temperatures is due to the sintering effect. Small defects are greatly influenced by the heat treatment process and tend to vanish, particularly interlamellar, penny-shaped and non-flat porosity and microcracks. The predicted results using the six-phase model were compared to the measured and that by using FEA with a good agreement was obtained. However, some of the modeling values exhibit a difference in comparison to the measured values. This difference is likely due to (a) a software limitation or (b) lack of information about the exact microstructure.

The modeling approach used in this study can reasonably predict TBCs thermal conductivity. Consequently, the proposed six-phase model could be utilized as an effective tool to assess the thermal conductivity of TBCs. However, it must be noted that the radiation effect and the effect of connected pores on the thermal conductivity of TBCs were not included in this six-phase model. Furthermore, to improve the accuracy and quality of this model, additional investigations of TBCs microstructure are needed. Such investigations can be achieved by combining the image analysis technique with other microstructural characterization approaches.

Life prediction of APS-TBCs was conducted by using a proposed residual stress model and existing burner rig test life data. The proposed stress model involves using elastic and thermal physical properties of the TGO, bond coat and topcoat, the model also uses geometric characteristics of the TGO/bond coat/topcoat interfaces. Stress model parameters were fitted to the stress data evaluated using FEA calculations. Temperature-dependent model parameters in evaluating crack propagation were obtained by fitting to the burner rig test data of APS-TBC systems. The life prediction was conducted by using temperature-dependent model parameters, and this capability was significantly improved by combining stress integrals with a critical onset for stress sign inversion.

Solid particle erosion and high-temperature oxidation, especially their interplay, significantly affect the performance of hot-section components of gas turbine engines. This interplay will accelerate the component degradation, and therefore needs to be addressed from flight safety and maintenance cost perspectives. To simulate the erosion–oxidation interaction and the oxide scale evolution, an erosion–oxidation equation was derived by combing solid particle erosion and oxide growth models. The erosion model parameters were fitted to alloys ranging from lower yield strength of aluminum to higher yield strength of FSX-414 steels. An approximate linear relationship of model parameters versus the yield strength of alloys was established. Based on this linear relationship, the erosion model parameters of CMSX-4 were estimated. The evolution of Al_2O_3 scale was simulated numerically at the particle velocity of $V = 100$ m/s, at temperatures of 850°C and 950°C and at impact angles of 15°, 25° and 75°. The Al_2O_3 scale was fast removed under solid particle attack at elevated temperatures due to a low oxide growth rate compared to the erosion rate. The Al_2O_3 scale was removed fast at a higher impact angle of 75° than at an acute angle of 15°, while the opposite trend of erosion rate was found when attacking on the substrate CMSX-4 alloy. These simulation results indicate that the ceramic Al_2O_3 scale is not capable of providing protection for the CMSX-4 from high-temperature solid particle erosion due to its much slow oxide scale growth rate.

ACKNOWLEDGMENTS

This research was sponsored by the DTS program of National Research Council Canada.

REFERENCES

1. L. Wang, Y. Wang, X. G. Sun, J. Q. He, Z. Y. Pan, and C. H. Wang, "A novel structure design towards extremely low thermal conductivity for thermal barrier coatings: Experimental and mathematical study," *Mater. Des.*, vol. 35, pp. 505–517, 2012, doi: 10.1016/j.matdes.2011.09.031.
2. H. M. Tawancy, A. I. Mohammad, L. M. Al-Hadhrami, H. Dafalla, and F. K. Alyousf, "On the performance and failure mechanism of thermal barrier coating systems used in gas turbine blade applications: Influence of bond coat/superalloy combination," *Eng. Fail. Anal.*, vol. 57, pp. 1–20, 2015, doi:10.1016/j.engfailanal.2015.07.023.
3. J. Zhang and V. Desai, "Evaluation of thickness, porosity and pore shape of plasma sprayed TBC by electrochemical impedance spectroscopy," *Surf. Coatings Technol.*, vol. 190, no. 1, pp. 98–109, 2005, doi:10.1016/j.surfcoat.2004.06.019.
4. J. Wang et al., "Effect of spraying power on microstructure and property of nanostructured YSZ thermal barrier coatings," *J. Alloys Compd.*, vol. 730, pp. 471–482, 2018, doi:10.1016/j.jallcom.2017.09.323.
5. J. Moon, H. Choi, H. Kim, and C. Lee, "The effects of heat treatment on the phase transformation behavior of plasma-sprayed stabilized $ZrOr_2$ coatings," *Surf. Coatings Technol.*, vol. 155, no. 1, pp. 1–10, 2002, doi:10.1016/S0257-8972(01)01661-9.

6. M. Jinnestrand and H. Brodin, "Crack initiation and propagation in air plasma sprayed thermal barrier coatings, testing and mathematical modelling of low cycle fatigue behaviour," *Mater. Sci. Eng. A*, vol. 379, no. 1–2, pp. 45–57, 2004, doi:10.1016/j.msea.2003.12.063.

7. J. R. Van Valzah and H. E. Eaton, "Cooling rate effects on the tetragonal to monoclinic phase transformation in aged plasma-sprayed yttria partially stabilized zirconia," *Surf. Coatings Technol.*, vol. 46, no. 3, pp. 289–300, 1991, doi:10.1016/0257-8972(91)90171-R.

8. F. H. Yuan, Z. X. Chen, Z. W. Huang, Z. G. Wang, and S. J. Zhu, "Oxidation behavior of thermal barrier coatings with HVOF and detonation-sprayed NiCrAlY bondcoats," *Corros. Sci.*, vol. 50, no. 6, pp. 1608–1617, 2008, doi:10.1016/j.corsci.2008.02.002.

9. M. Ranjbar-Far, J. Absi, G. Mariaux, and F. Dubois, "Simulation of the effect of material properties and interface roughness on the stress distribution in thermal barrier coatings using finite element method," *Mater. Des.*, vol. 31, no. 2, pp. 772–781, 2010, doi:10.1016/j.matdes.2009.08.005.

10. A. C. Karaoglanli, K. M. Doleker, B. Demirel, A. Turk, and R. Varol, "Effect of shot peening on the oxidation behavior of thermal barrier coatings," *Appl. Surf. Sci.*, vol. 354, pp. 314–322, 2015, doi:10.1016/j.apsusc.2015.06.113.

11. J. Kimmel, Z. Mutasim, and W. Brentnall, "Effects of alloy composition on the performance of yttria stabilized zirconia-thermal barrier coatings," *Proc. ASME Turbo Expo.*, vol. 4, no. July, pp. 393–400, 1999, doi:10.1115/99-GT-350.

12. W. Nowak et al., "Effect of processing parameters on MCrAlY bondcoat roughness and lifetime of APS-TBC systems," *Surf. Coatings Technol.*, vol. 260, pp. 82–89, 2014, doi:10.1016/j.surfcoat.2014.06.075.

13. Y. Bai et al., "Isothermal oxidation behavior of supersonic atmospheric plasma-sprayed thermal barrier coating system," *J. Therm. Spray Technol.*, vol. 22, no. 7, pp. 1201–1209, 2013, doi:10.1007/s11666-013-9979-7.

14. R. A. Miller, "Oxidation-based model for thermal barrier coating life," *J. Am. Ceram. Soc.*, vol. 67, no. 8, pp. 517–521, 1984, doi:10.1111/j.1151-2916.1984.tb19162.x.

15. T. Beck, R. Herzog, O. Trunova, M. Offermann, R. W. Steinbrech, and L. Singheiser, "Damage mechanisms and lifetime behavior of plasma-sprayed thermal barrier coating systems for gas turbines - Part II: Modeling," *Surf. Coatings Technol.*, vol. 202, no. 24, pp. 5901–5908, 2008, doi:10.1016/j.surfcoat.2008.06.132.

16. R. Vaßen, S. Giesen, and D. Stöver, "Lifetime of plasma-sprayed thermal barrier coatings: Comparison of numerical and experimental results," *J. Therm. Spray Technol.*, vol. 18, no. 5–6, pp. 835–845, Dec. 2009, doi:10.1007/s11666-009-9389-z.

17. F. Traeger, M. Ahrens, R. Vaßen, and D. Stöver, "A life time model for ceramic thermal barrier coatings," *Mater. Sci. Eng. A*, vol. 358, no. 1–2, pp. 255–265, 2003, doi:10.1016/S0921-5093(03)00300-9.

18. D. Renusch, M. Schorr, and M. Schütze, "The role that bond coat depletion of aluminum has on the lifetime of APS-TBC under oxidizing conditions," *Mater. Corros.*, vol. 59, no. 7, pp. 547–555, 2008, doi:10.1002/maco.200804137.

19. M. Ahrens, R. Vaßen, and D. Stöver, "Stress distributions in plasma-sprayed thermal barrier coatings as a function of interface roughness and oxide scale thickness," *Surf. Coatings Technol.*, vol. 161, no. 1, pp. 26–35, 2002, doi:10.1016/S0257-8972(02)00359-6.

20. K. Chen, L. Zhao, M. Fu, and P. C. Patnaik, "The effect of platinum on β-NiAl/α-Al$_2$O$_3$ interfacial tensile stress: A combined Ab initio DFT and mechanics-based model study," *Oxid. Met.*, vol. 90, no. 1–2, pp. 65–82, 2018, doi:10.1007/s11085-017-9823-4.

21. K. Chen, "Sulphur diffusion in β-NiAl and effect of Pt additive: An ab initio study," *J. Phys. D. Appl. Phys.*, vol. 49, no. 5, p. 55306, 2016, doi:10.1088/0022-3727/49/5/055306.

22. I. Ozfidan, K. Chen, and M. Fu, "Effects of additives and impurity on the adhesive behavior of the NiAl (110)/Al$_2$O$_3$ (0001) interface: An Ab Initio study," *Metall. Mater. Trans. A Phys. Metall. Mater. Sci.*, vol. 42, no. 13, pp. 4126–4136, 2011, doi:10.1007/s11661-011-0813-x.

23. A. Kumar, P. C. Patnaik, and K. Chen, "Damage assessment and fracture resistance of functionally graded advanced thermal barrier coating systems: Experimental and analytical modeling approach," *Coatings*, vol. 10, no. 5, 2020, doi:10.3390/COATINGS10050474.

24. R. S. Ghai, K. Chen, and N. Baddour, "Modelling thermal conductivity of porous thermal barrier coatings," *Coatings*, vol. 9, no. 2, 2019, doi:10.3390/COATINGS9020101.

25. S. K. Essa, K. Chen, R. Liu, X. Wu, and M. X. Yao, "Failure mechanisms of APS-YSZ-CoNiCrAlY thermal barrier coating under isothermal oxidation and solid particle erosion," *J. Therm. Spray Technol.*, 2020, doi:10.1007/s11666-020-01124-4.

26. B. Zhang, K. Chen, N. Baddour, and P. C. Patnaik, "Life prediction of atmospheric plasma-sprayed thermal barrier coatings using temperature-dependent model parameters," *J. Therm. Spray Technol.*, vol. 26, no. 5, pp. 902–912, 2017, doi:10.1007/s11666-017-0558-1.

27. B. C. Zhang, K. Chen, N. Baddour, and P. C. Patnaik, "Failure and life evaluation of EB-PVD thermal barrier coatings using temperature-process-dependent model parameters," *Corros. Sci.*, vol. 156, no. December 2018, pp. 1–9, 2019, doi:10.1016/j.corsci.2019.04.020.

28. A. Moteb and K. Chen, "Modelling and evaluating thermal conductivity of porous thermal barrier coatings at elevated temperatures," *Ceram. Int.*, vol. 46, no. 14, pp. 21939–21957, 2020, doi:10.1016/j.ceramint.2020.04.228.

29. K. Chen, D. Seo, S. H. Lee, and E. Byon, "A physics-based model of temperature-exposure-dependent interfacial fracture toughness of thermal barrier coatings," *J. Phys. Commun.*, vol. 2, no. 12, p. 125005, 2018, doi:10.1088/2399-6528/aaf602.

30. B. K. Jang, "Thermal conductivity of nanoporous ZrO2-4 mol% Y$_2$O$_3$ multilayer coatings fabricated by EB-PVD," *Surf. Coatings Technol.*, vol. 202, no. 8, pp. 1568–1573, 2008, doi:10.1016/j.surfcoat.2007.07.017.

31. Z. Xu, X. Zhou, R. Mu, and L. He, "Structure, phase stability and thermophysical properties of (Yb 0.1La0.9)2(Zr0.7Ce 0.3)2O7 ceramics," *Mater. Lett.*, vol. 135, pp. 162–164, 2014, doi:10.1016/j.matlet.2014.07.162.

32. D. L. Buchanan and M. H. A. Davis, "Minerals engineering," *Met. Energy Financ.*, vol. 6, no. 4, pp. 189–221, 2018, doi:10.1142/9781786345882_0009.

33. F. Cernuschi, S. Ahmaniemi, P. Vuoristo, and T. Mäntylä, "Modelling of thermal conductivity of porous materials: Application to thick thermal barrier coatings," *J. Eur. Ceram. Soc.*, vol. 24, no. 9, pp. 2657–2667, 2004, doi:10.1016/j.jeurceramsoc.2003.09.012.

34. S. Wei, F. Fu-Chi, F. Qun-Bo, and M. Zhuang, "Effects of defects on the effective thermal conductivity of thermal barrier coatings," *Appl. Math. Model.*, vol. 36, no. 5, pp. 1995–2002, 2012, doi:10.1016/j.apm.2011.08.018.

35. M. Kachanov and I. Sevostianov, "On quantitative characterization of microstructures and effective properties," *Int. J. Solids Struct.*, vol. 42, no. 2, pp. 309–336, 2005, doi:10.1016/j.ijsolstr.2004.06.016.

36. W. Ma et al., "Microstructure and thermophysical properties of SrZrO$_3$ thermal barrier coating prepared by solution precursor plasma spray," *J. Therm. Spray Technol.*, vol. 27, no. 7, pp. 1056–1063, 2018, doi:10.1007/s11666-018-0744-9.

37. S. Ahmaniemi, M. Vippola, P. Vuoristo, T. Mäntylä, F. Cernuschi, and L. Lutterotti, "Modified thick thermal barrier coatings: Microstructural characterization," *J. Eur. Ceram. Soc.*, vol. 24, no. 8, pp. 2247–2258, 2004, doi:10.1016/S0955-2219(03)00639-3.

38. F. Cernuschi, P. G. Bison, S. Marinetti, and P. Scardi, "Thermophysical, mechanical and microstructural characterization of aged free-standing plasma-sprayed zirconia coatings," *Acta Mater.*, vol. 56, no. 16, pp. 4477–4488, 2008, doi:10.1016/j.actamat.2008.04.067.

39. Y. Tan, J. P. Longtin, and S. Sampath, "Modeling thermal conductivity of thermal spray coatings: Comparing predictions to experiments," *Proc. Int. Therm. Spray Conf.*, vol. 15, no. December, pp. 545–552, 2006, doi:10.1361/105996306X147216.

40. H. J. Rätzer-Scheibe and U. Schulz, "The effects of heat treatment and gas atmosphere on the thermal conductivity of APS and EB-PVD PYSZ thermal barrier coatings," *Surf. Coatings Technol.*, vol. 201, no. 18, pp. 7880–7888, 2007, doi:10.1016/j.surfcoat.2007.03.028.

41. Y. Huang et al., "Effect of different types of pores on thermal conductivity of YSZ thermal barrier coatings," *Coatings*, vol. 9, no. 2, p. 138, 2019, doi:10.3390/coatings9020138.

42. A. Kulkarni et al., "Comprehensive microstructural characterization and predictive property modeling of plasma-sprayed zirconia coatings," *Acta Mater.*, vol. 51, no. 9, pp. 2457–2475, 2003, doi:10.1016/S1359-6454(03)00030-2.

43. D. R. Mumm and G. A. Evans, "Mechanisms controlling the performance and durability of thermal barrier coatings," *Key Eng. Mater.*, vol. 197, pp. 199–230, 2001, doi:10.4028/www.scientific.net/kem.197.199.

44. G. Sundararajan, "A comprehensive model for the solid particle erosion of ductile materials," *Wear*, vol. 149, no. 1–2, pp. 111–127, 1991, doi:10.1016/0043-1648(91)90368-5.

45. L. Nalin, "*Degradation of Environmental Pretection Coatings for Gas Turbine Materials*," Cranfield University, Bedford, UK, 2008.

46. C. T. Kang, F. S. Pettit, and N. Birks, "Mechanisms in the simultaneous erosion-oxidation attack of nickel and cobalt at high temperature," *Metall. Trans. A*, vol. 18, no. 10, pp. 1785–1803, October, 1987.

47. P. Carter, B. Gleeson, and D. J. Young, "Calculation of precipitate dissolution zone kinetics in oxidising binary two-phase alloys," *Acta Mater.*, vol. 44, no. 10, pp. 4033–4038, 1996, doi:10.1016/S1359-6454(96)00054-7.

48. N. Birks, G. H. Meier, and F. S. Pettit, "Introduction to the high temperature oxidation of metals," *Engineering*, p. 338, 2006, Online. Available: http://www.cambridge.org/gb/knowledge/isbn/item1110694/?site_locale=en_GB.

49. J. G. A. Bitter, "A study of erosion phenomena Part I," *Wear*, vol. 6, pp. 169–190, 1963.

50. J. H. Neilson and A. Gilchrist, "Erosion by a stream of solid particles," *Wear*, vol. 11, no. 2, pp. 111–122, 1968, doi:10.1016/0043-1648(68)90591-7.

51. W. Tabakoff, A. Hamed, M. Metwally, and M. Pasin, "High-temperature erosion resistance of coastings for gas turbine," *J. Eng. Gas Turbines Power*, vol. 114, no. 2, pp. 242–249, 1992, doi:10.1115/1.2906579.

52. A. Sengupta et al., "Tensile behavior of a new single crystal nickel-based superalloy (CMSX-4) at room and elevated temperatures," *J. Mater. Eng. Perform.*, vol. 3, no. 5, pp. 664–672, 1994, doi:10.1007/BF02645265.

53. I. Finnie, "Some reflections on the past and future of erosion," *Wear*, vol. 186–187, no. PART 1, pp. 1–10, 1995, doi:10.1016/0043-1648(95)07188-1.

6 Investigation of Microstructural and Tribological Behavior of Metco 41C+WC-12Co Composite Coatings Sprayed via HVOF Process

C. Durga Prasad, Mahantayya Mathapati,
M. R. Ramesh, and Sharnappa Joladarashi

CONTENTS

6.1 INTRODUCTION

In the fields of marine, oil, and chemical industries, components such as boiler tubes and heat exchangers work under challenging conditions such as high temperature, vibrations, and speed, and with chemical agents flowing continuously through the valves [1–3]. Due to such conditions, the components are subjected to severe

DOI: 10.1201/9781003213185-6

material degradation in the form of sliding wear and a combination of wear and corrosion (tribo-corrosion) [4]. However, replacing the damaged components in such applications would be very expensive. The damaged part could be repaired through several ways such as thermal spray coating, laser- and electron-beam treatment, electroplating, and welding technologies [5, 6]. Among these methods, thermal spray coating process has wide applications and is very cost effective in solving industrial issues [7–9]. Thermal spray method exhibits hard thick film coatings; it is also versatile with respect to other coating and substrate materials [10, 11]. Thermal coatings utilize advanced material systems that can be sprayed on to metallic or composite surfaces [7, 8]. Once in place, the coatings form an integral part of a critical component's exhaust heat management system and environmental protection shield [9]. Critical for design engineers, coatings allow equipment to operate at higher operating temperatures [1–3]. For example, advanced thermal coatings can allow temperatures higher than the melting point of the metal tubes in boiler applications. This is an important consideration as demand increases for efficient supply of steam/water through tubes that can operate at high temperatures while still delivering high durability and long lifetimes [1, 12–14]. To be most effective, a thermal spray coating should offer the following characteristics: higher deposition rate, thick layers, good mechanical bonding, no thermal distortion, and no effects on substrate bulk properties [1–3]. In the view of boiler tube applications, high-velocity oxy fuel (HVOF) type of thermal spray coating offers better solution to enhance life of components [2–4, 12, 13]. Many authors have performed research on cobalt- and nickel-based coatings to impart wear and corrosion properties, but as of now, the cost of cobalt has increased drastically due to the material being employed in the preparation of battery applications [5, 15–17]. In order to reduce the cost of secondary process, iron (Fe)-based powders can be sprayed on the surface of boiler tubes, petrochemical valves, and heat exchangers by HVOF method [18]. Fe-based powders are commercially available, which is very economical and has similar advantages compared to other metallic and carbide-based powders [8, 19–21]. The iron-based metallic powders exhibit excellent physical, chemical, and mechanical properties, such as high strength, high hardness, and excellent wear resistance and corrosion resistance. Fe-based exhibits various feedstock's depending on the applications suitable powders need to select. However, Metco 41C is a type of Fe-based powder consisting of Fe-12; Ni17; Cr2.5; Mo2.3; Si0.03 composition, which offers better wear- and corrosion resistance at elevated temperatures. Reinforcing of carbides such as WC-12Co, WC-CrC-Ni, and SiC into Metco 41C feedstock could improve the metallurgical and mechanical properties of coating [7, 12, 22–24]. WC-Co is widely used for wear- and corrosion-resistance applications. Tungsten carbide (WC) is a cermet compound having high hardness and strength. WC is readily wetted by the molten nickel and cobalt. So, the performance of the WC can be improved by the addition of cobalt which acts as binder [8, 25–28]. WC-12Co is used because of its good sliding wear, impact, abrasion, and fretting resistance. Application of WC-12Co includes exhaust fans, pump housing, jet engine compressor blades, conveyor screws, and steel rolls. It is feasible to develop coatings with better hardness, tribological resistance, and chemical inertness by employing cermet-based WC-Co as the coating material [29–31]. The wear resistance of a coating can be improved by increasing the hardness, reducing roughness, changing the friction coefficient and chemical composition, and adding an appropriate lubricant [32–34].

In this chapter, 70% wt. Metco 41C and 30% wt. WC-12Co feedstock were mixed mechanically using ball milling and further deposited on ASTM-SA213-T11 boiler tube substrate by HVOF process. The coating was subjected to metallurgical and mechanical characterizations, and then high temperature as well as room condition sliding wear behavior was investigated using pin on disc apparatus.

6.1.1 MATERIALS AND EXPERIMENTAL STUDIES

Commercially available iron-based (Metco 41C) and WC-12Co were selected as feedstock. The chemical composition of Metco 41C is Fe 12Ni17Cr2.5Mo2.3Si0.03C. Metco 41C was selected as the base powder and WC-12Co as the reinforcement. Metco 41C powder was produced by water-atomized technique with particle size having −106+45 μm, whereas WC-12Co exhibits −45 to +15 μm size of particle. The feedstock with a combination of (70%) Metco 41C and (30%) WC-12Co was prepared. The two powders were mixed mechanically using ball milling process. The milling speed was kept constant at300 rpm and milled for a duration of 20 minutes. Further, the feedstock was sprayed on ASTM-SA213-T11 used as a substrate having 30 mm × 30 mm × 5 mm size for the coatings. The HVOF spray coating was developed by employing diamond jet 2100 gun, carried out at Aum Techno Spray Bengaluru, India. Spray parameters are listed in Table 6.1. The surface of the substrate was prepared for process by making the surface rough using aluminum oxide grit blasting of 24 size mesh grit and the unwanted or foreign particles (debris) were removed by projecting high-pressure air on the specimen surface to avoid contamination. During spraying, high gas fluxes and gas velocities were achieved. After exiting the nozzle, the gas flow was mainly directed toward the substrate. Near the substrate, this gas flow was strongly deflected. The strength of this impingement flow increases with higher gas velocities. However, the high gas velocity is advantageous for the application of dense coatings.

6.1.2 CHARACTERIZATION OF COATING

The morphology of powders was subjected to microstructure analysis using scanning electron microscope (SEM). The deposited coating was trimmed into a size of 12 mm × 12 mm × 5 mm using wire electro-discharge machining to accommodate samples for various characterization techniques. Prior to testing, the specimens were

TABLE 6.1
HVOF Spray Parameters

Parameter	Value (lpm)
Flow rate N_2	50
Flow rate O_2	250
Flow rate H_2	700

prepared by performing polishing on both surface and cross-section side of the coating using emery papers. Then, the coating sample was taken for surface- and cross-sectional microstructural analysis using SEM and its chemical composition was evaluated using energy dispersive spectroscopy (EDS). X-ray diffraction (XRD) technique employed to detect various phases (elemental phase/inter-metallic phase/crystalline phase/non-crystalline phase) present in the coating. However, the identification of phases in coating was performed using XRD with Ni filter (Cu Kα radiation with 20 mA and 30 kV) method. As per ASTM B487 standard, coating thickness measurement was performed using image analyzer (Biovis software). Ten readings were recorded on each specimen and the mean value was reported as the average coating thickness. With the help of surface roughness (Ra) measurement device (SJ-201P, Mitutoyo, Japan), coating surface roughness was evaluated. Test was performed on 5 different locations in order to report the mean value. Vickers microhardness of coating was estimated using OMNITECH microhardness tester. On the direction of cross section of coating and substrate, microhardness was measured under a load of 300 g with duration of 10 sec. Ten various indentation spots were selected to record the mean value. As per ASTM (B276-05) standard, the cross-sectional micrographs of coating was employed to evaluate the porosity of coating through image analyzer software. During testing, 20 fields of view were utilized to estimate the mean value of porosity.

6.1.3 DRY SLIDING WEAR

According to ASTM G99-05, wear test of coating and substrate were performed without supply of lubrication using pin on disc apparatus (TR-20LE-PHM 400-CHM 600, Ducom Instruments, Pvt. Ltd., Bangalore, India). Samples were tested by choosing parameters of 10 and 20 N at 200°C, 300°C, and also under room temperature conditions. The sliding velocity and sliding distance were kept constant at 1.3 m/sec and 3000 m, respectively. The test was conducted for a duration of 33 minutes, and a EN24 steel disc was used as the counter body for wear tests. The system having inbuilt heating option, which transfers heat to the sample attached in the holder and thermocouple, was inserted to measure the temperature.

With the help of a computer, specimen height loss was estimated. Further, the system generates data to calculate wear volume loss, wear rate, and friction coefficient. Characterization of tested samples like microstructural analysis was done through SEM to understand the wear mechanism. Similar steps were adopted in wear studies and reported [10, 12, 18, 22].

6.2 RESULTS AND DISCUSSION

6.2.1 COATING STRUCTURE ANALYSIS

The powder distribution of Metco 41 C and WC-12Co were observed to be uniform from the SEM images as shown in Figure 6.1a and b. The Metco 41C feedstock structure was noticed to be dendritic (see Figure 6.1a), whereas the particles of WC-12Co are mostly spherical and partially distorted from the spheroidal and elliptical shape

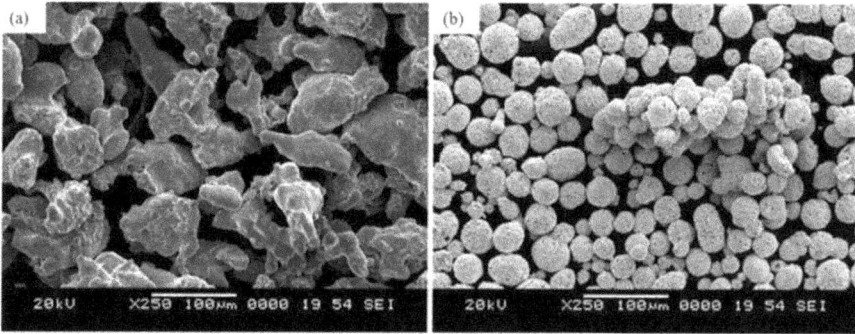

FIGURE 6.1 Morphology of powders (a) Metco 41C and (b) WC-12Co.

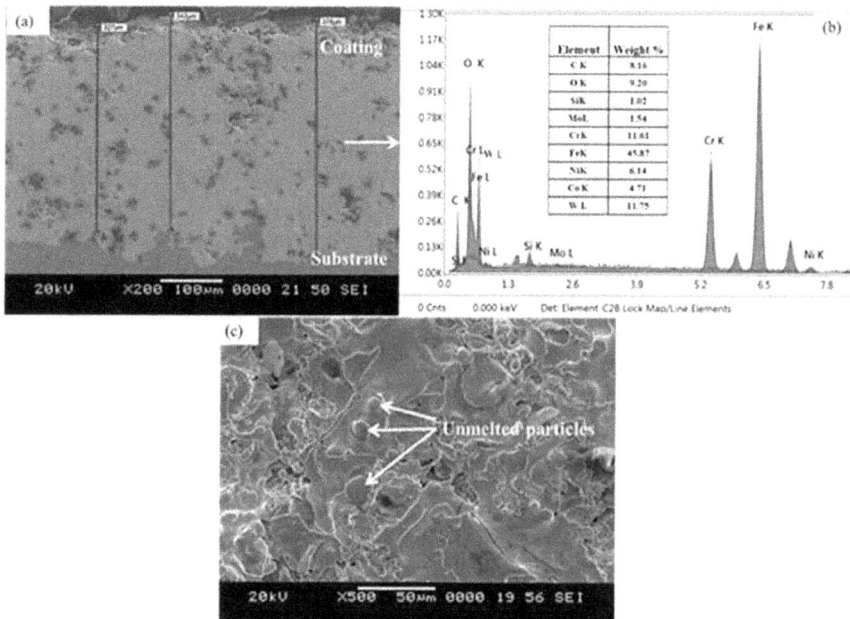

FIGURE 6.2 (a) Metco 41 C+WC-12Co coating cross section, (b) compositional EDS of the coating, and (c) Metco 41 C+WC-12Co coating surface.

depicted in Figure 6.1b. The micrograph of Metco 41C+WC-12Co coating is shown in Figure 6.2. The mean coating thickness was 330–360 µm. Regardless of the fine visible distinction between coating and substrate, gripping of deposition layers with substrate was good as seen from the micrograph shown in Figure 6.2a. There are no cracks observed along the interface, indicating strong bonding. Compact structure was clear in the micrograph, as it can be noticed that the splat units forming the layers were well mixed up and could lead to an increase in coating hardness. Coating constituents were distributed uniformly with complete melting of splats and there were minor signs of partial-melted particles strapped in the pores and voids in the cross section.

EDS measurements taken for a selected area of the cross-sectional surface of the sprayed coating indicated a uniform composition, with Cr, W, and Fe, as major elements and small quantities of Si, Ni, Co, and Mo, as shown in Figure 6.2b. In addition to these, the existence of carbon substance (8.16%) and oxygen substance (9.20%) was observed in the coating, which led to the formation of some localized oxides and carbides in the coating [12, 23]. The distribution of coating elements was uniform over the surface and the related chemical composition of the feedstock was observed as shown in Figure 6.2b.

The surface micrograph of the coating is shown in Figure 6.2c. The surface roughness of the coating was estimated to be Ra 7.274 µm. The formation of irregular-sized splats and fine fragmented particles were identified in the microstructure depicted in Figure 6.2c. Residual stress by tensile action, in which a few thin line cracks were caused, was observed during the cooling of the coating. Also, some of the unmelted particles were also evident in the coating surface microstructure. This is due to rate of acceleration of spray system being high as it minimizes the duration of feedstock passing through flame.

6.2.2 XRD ANALYSIS

Figure 6.3 presents the XRD plot of Metco 41C+WC-12Co coating. As observed in the structure of XRD, formation of some additional phases such as W_2C and Co_3W_3C were spotted. These phases were generated due to oxidation and decomposition of WC. The chemical reaction (6.1) mentioned below was obtained due to the oxidation

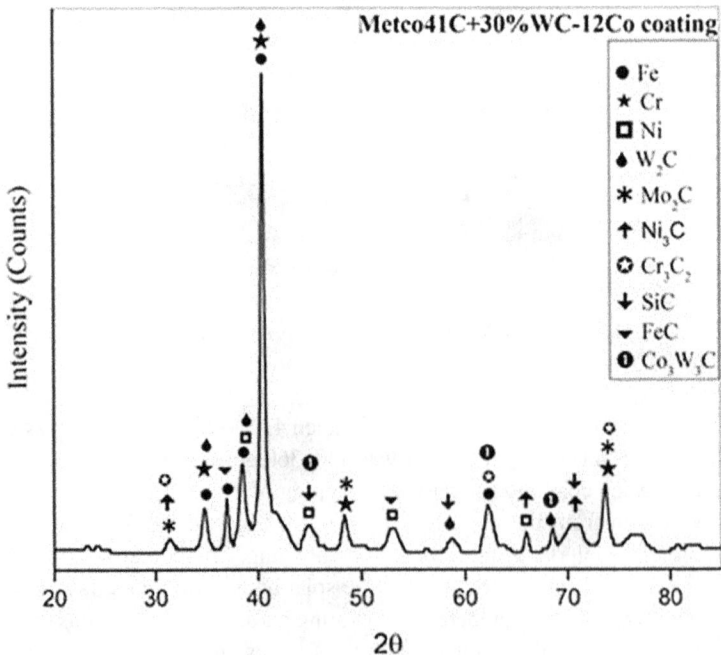

FIGURE 6.3 XRD pattern of Metco 41C+WC-12Co coating.

of WC to W_2C [24]. Creation of W_2C was considered as the product of decarburization in the thermal spray process [24, 25]. The basic carbides emerging from FeNiCr system coatings like Cr_3C_2, Ni_3C, and FeC were also observed. Minor percentages of SiC and Mo_2C phases were revealed. However, the absence of Cr_7C_3 has been confirmed by this analysis, which ensures that the decarburization of Cr_3C_2 has not taken place.

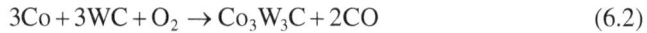

$$4WC + O_2 \rightarrow 2W_2C + 2CO \tag{6.1}$$

$$3Co + 3WC + O_2 \rightarrow Co_3W_3C + 2CO \tag{6.2}$$

6.2.3 MICROHARDNESS ANALYSIS

The microhardness plot of Metco 41C+WC-12Co coating is shown in Figure 2.4. The average microhardness value of coating and substrate was observed to be 965 ± 25 Hv and 185 ± 15 Hv, respectively. Varying the distance of the substrate–coating interface region results in the variation of coating microhardness due to the creation of compressive residual stress [10, 12, 18, 22]. The microhardness tends to fluctuate along the thickness of the WC-12Co reinforced coating may be due to the distribution of the WC hard phase in FeNiCr-based alloy matrix. The optical micrographs of indentation marks at different spots such as near interface and surface of coating are presented in Figure 6.4. The size of the indentation mark near the interface was

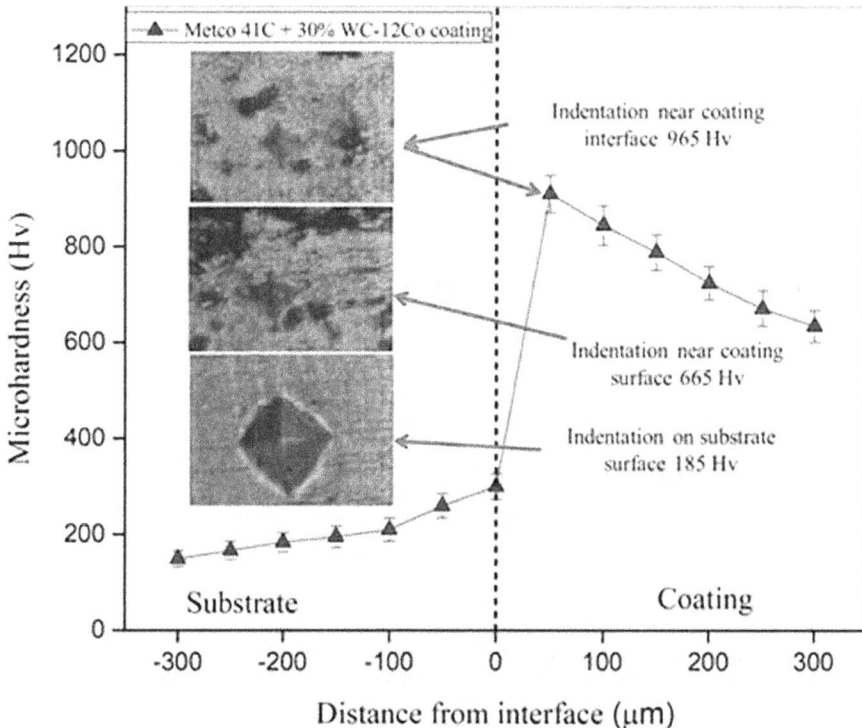

FIGURE 6.4   Microhardness plot of Metco 41C+WC-12Co coating across the interface.

observed to be too small due to the stress built resulting in a higher hardness of 965 Hv. Similarly, indentation near the coating surface region exhibits little more enlargement in size due to the presence of unmelted/partial-melted particles and the formation of oxide stringers. This results in decrease in microhardness of coating value to 665 Hv. As the load was applied on the substrate surface, the impact of the indentation was very high due to a lower hardness of 185 Hv.

6.2.4 EXAMINATION OF WEAR MORPHOLOGY

The worn surface morphologies of Metco 41C+WC-12Co coating are shown in Figure 2.5a and b. The wear tracks and crack were noticed in Figure 6.5a. Debris formed from the worn-out surface was detected as being loaded and adhered to the surface. This pit-shaped valley becomes the accumulation point to some of the debris produced in the path of the wear causing the loading of the coating surface. The wear surface has scratches caused by exfoliated carbide particles, and the wear mechanism was adhesive wear [10, 12, 18, 22]. The surface of the coating was smooth and flat, with a small amount of oxidative wear. Some region of the coating were peeled off, mainly due to adhesive wear, accompanied by a little oxidative wear, and also splats detachment under 20 N load was identified, which was the reason for the slight increase in material removal, as shown in Figure 6.5b. The coating was peeled off mainly around the tiny holes, because the crack around the tiny holes of the coating continued to expand until it broke. This phenomenon was caused by repeated rubbing of the steel disc against the coating, and the residual stress was released [10–14, 28–31]. This thus results in higher wear rate compared to 10 N normal load.

6.2.5 FRICTION AND WEAR RATE

Figure 6.6a and b depicts the plots of friction coefficient (μ) of coating and substrate with respect to testing temperature and loads. The substrate possesses low strength and hardness at all test conditions, resulting in high coefficient of friction (COF) [12, 24–27]. Even for an increase in the load from 10 to 20 N, it was noticed that the COF of the substrate tends to increase drastically.

FIGURE 6.5 SEM wear scar micrographs of Metco 41C+WC-12Co coating.

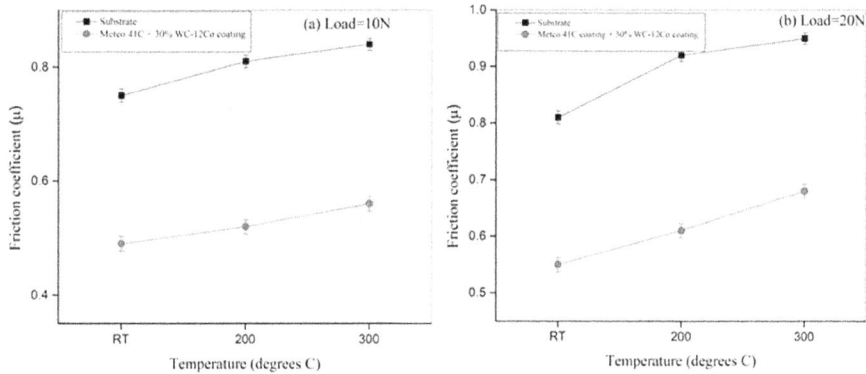

FIGURE 6.6   Frictional profiles of coating and substrate: (a) 10 N, (b) 20 N.

The coating reveals linear increasing trend in COF with a rise in temperature, but stated lower values when compared with the substrate. As compared to the substrate, the coating exhibits 3 times lower COF under all test conditions. With the rise in temperature and normal load, coating surface layer subjected to oxidative and debris were formed due to the breaking of splats, resulting in slight increase in COF. Wear debris was continuously in contact with the surface layer of the coating and the counter disc, which led to an indication of rise in COF at 300°C under 20 N, as presented in Figure 6.6b. The existence of WC in Fe-based matrix coating gives a cushion effect against hard sliding disc, resulting in decrease in COF [1–3, 10, 12, 18]. With continued friction, wear debris particles acts as scratching pits on the surface. The frictional heat generated during abrasion leads to localized increase in the temperature, resulting in the formation of oxides of Fe, Si, Cr, and Ni. This tribo-oxide film with low shear strength was responsible for the reduction in COF. Evaluated friction values of coating and substrate are given in Table 6.2.

Figure 6.7a and b illustrates the coating and substrate volume loss graphs. The substrate registered higher wear volume loss and was observed to increase with rise in normal load due to poor surface properties against the sliding action. The surface layer of the substrate was deformed completely under 10 and 20 N at 300°C temperature; it was observed that the substrate experienced severe oxidation, as shown in

TABLE 6.2
Values of COF with Respect to Normal Load and Temperatures

Specimens	Room Temperature		200°C		300°C	
	10N	20N	10N	20N	10N	20N
Substrate	0.71	0.82	0.86	0.93	0.90	0.96
Metco 41C+WC-12Co coating	0.49	0.55	0.52	0.61	0.56	0.68

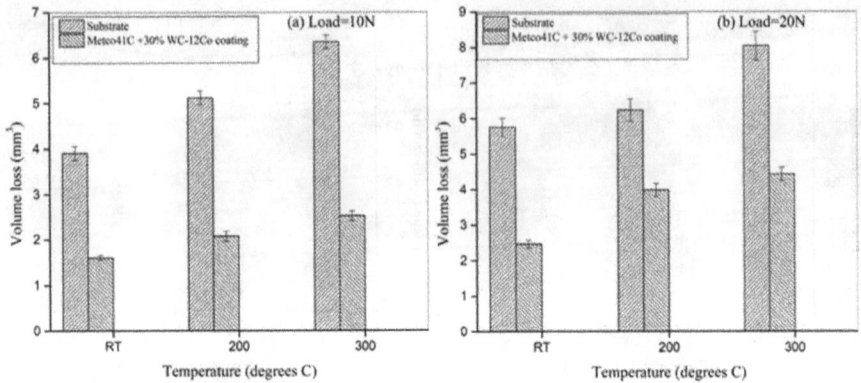

FIGURE 6.7 Volume loss profiles of coating and substrate: (a) 10 N, (b) 20 N.

Figure 6.7b. The Metco 41C+WC-12Co coating exhibited lowest volume loss at all test temperatures and loads compared to the substrate. It was noticed that nearly 70% of material loss is reduced by the coating than the substrate. This was since hard carbides phases such as Cr_3C_2, Ni_3C, W_2C, and Mo_2C were suspended in rich FeNiCr matrix region, which strengthens the layers and avoids loss of material [22, 31–34]. The values of volume loss of the substrate and the coating are mentioned in Table 6.3.

Figure 6.8 presents the plots of the wear rates of the coating and the substrate with respect to temperature and normal load. It was observed that the boiler steel substrate presents higher wear rate at all tested parameter conditions due to lower strength and hardness. During sliding test, substrate produced high vibrations and noise due to oxidation resulting in more friction [35–37]. The top layer of substrate was deformed severely due to continuous contact with the hard sliding disc. As noticed in Figure 6.8b, the wear rate of the substrate reached 0.0411 under 20 N normal load at 300°C test conditions which was a drastic increment. However, the coating exhibits 2.5–3 times lower rate at test conditions. There was little rise in coating wear rate when the test conditions were changed from room temperature to 200°C, as observed in Figure 6.8a. The formation of hard carbides and intermetallic phase like Co_3W_3C (see Figure 2.3) imparts the high-temperature strength and

TABLE 6.3

Values of Volume Loss with Respect to Normal Load and Temperature

Specimens	Room Temperature		200°C		300°C	
Volume Loss (mm³)	10N	20N	10N	20N	10N	20N
Substrate	3.9073	5.7565	5.1205	6.2301	6.3509	8.0399
Metco 41C +WC-12Co coating	1.6011	2.4641	2.0754	3.9815	2.5204	4.4361

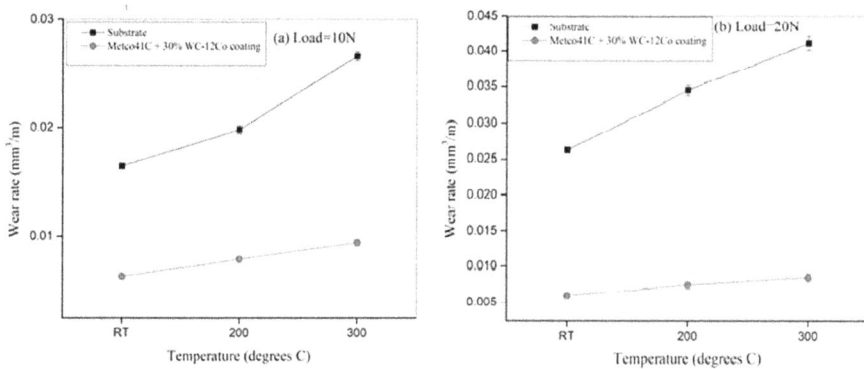

FIGURE 6.8 Wear rate profiles: (a) 10 N, (b) 20 N.

TABLE 6.4

Values of Wear Rate with Respect to Normal Load and Temperature

Specimen Wear Rate (mm³/m)	Room Temperature		200°C		300°C	
	10N	20N	10N	20N	10N	20N
Substrate	0.0165	0.0262	0.0198	0.0345	0.0266	0.0411
Metco 41C +WC-12Co coating	0.0063	0.0060	0.0079	0.0075	0.0094	0.0085

---

microhardness of the coating to withstand the deformation of coating splats [38, 39]. Further, a rise in parameters was traced to a slight rise in coating wear rate and this was due to the removal of unmelted/partial-melted particles by localized deformation [4, 10, 26–30]. Later, the wear rate reached steady state as the oxidative wear restricts further loss of the material. The values of wear rate of coating and substrate are listed in Table 6.4.

6.3 CONCLUSION

- The Metco 41C +WC-12Co coating was successfully deposited on the boiler steel substrate by HVOF spray method. The coating exhibited compact structure and uniform thickness in the range 330–360 μm. The coating also produced a surface roughness of Ra 7.274 μm due to the existence of unmelted particles.
- Carbide phases such as Cr_3C_2, Ni_3C, W_2C, Fe_2C, and SiC, along with intermetallic phases like Mo_2C and Co_3W_3C were formed, resulting in higher microhardness of coating 965 Hv than substrate.

- Sliding wear of the substrate reveals thermal softening and breaking of pits under all test conditions. At 300°C, coatings tend to show deformation of hard layers due to abrasive wear mechanism.
- The coating exhibits less wear rate due to a harder layer and also at 300°C, oxide films were formed, which cover the underlying surface and restrict the damage.
- The appearance of a stable and continuous hard layer is responsible for the lower friction coefficient measured for the coating. When it was not stable, slight rise in friction coefficients were noticed.
- The coating can improve the product life by 2.66 times of the existing value. This was evaluated in terms of loss of volume of the coating during the wear test.

ACKNOWLEDGMENTS

The authors wish to thank Aum Techno Spray, Bangalore, India for providing high-velocity oxy fuel coating facility.

REFERENCES

1. Bolelli, G., Cannillo, V., Lusvarghi, L., and Ricco, S. Mechanical and tribological properties of electrolytic hard chrome and HVOF-sprayed coatings. *Surface and Coatings Technology*, 200, (2009), 2995–3009.
2. Zhang, H., Hu, Y., Hou, G., An, Y., and Liu, G. The effect of high-velocity oxy-fuel spraying parameters on microstructure, corrosion and wear resistance of Fe-based metallic glass coatings. *Journal of Non-Crystalline Solids*, 406(2014), 37–44.
3. Wielage, B., Wank, A., Pokhmurska, H., Grund, T., Rupprecht, C., Reisel, G., and Friesen, E. Development and trends in HVOF spraying technology. *Surface and Coatings Technology*, 201(5), (2006), 2032–2037. doi:10.1016/j.surfcoat.2006.04.049.
4. Murugan, K., Ragupathy, A., Balasubramanian, V., and Sridhar, K. Optimizing HVOF spray process parameters to attain minimum porosity and maximum hardness in WC-10Co-4Cr coatings. *Surface and Coatings Technology*, 247, (2014), 90–102.
5. Pukasiewicz, A.G.M., De Boer, H.E., Sucharski, G.B., Vaz, R.F., and Procopiak, L.A.J.The influence of HVOF spraying parameters on the microstructure, residual stress and cavitation resistance of FeMnCrSi coatings. *Surface and Coatings Technology*, 327, (2017), 158–166.
6. Dejuna, K. and Tianyuan, S. Wear behaviors of HVOF sprayed WC-12Co coatings by laser remelting under the lubricated condition. *Optics & Laser Technology*, 89, (2017), 86–91.
7. Navas, C., Colaço, R., de Damborenea, J., and Vilar, R. Abrasive wear behaviour of laser clad and flame sprayed-melted NiCrBSi coatings. *Surface andCoatings Technology*, 200, (2006), 6854–6862.
8. Li, G.J., Li, J., and Luo, X. Materials characterization effects of high temperature treatment on microstructure and mechanical properties of laser-clad NiCrBSi /WC coatings on titanium alloy substrate. *Materials Characterization*, 98, (2014)., 83–92.
9. Ramesh, M.R., Prakash, S., Nath, S.K., Sapra, P.K., and Venkataraman, B. Solid particle erosion of HVOF sprayed WC-Co/NiCrFeSiB coatings. *Wear*, 269(3–4), (2010), 197–205. doi:10.1016/j.wear.2010.03.019.

10. Hong, S., Wu, Y., Wang, B., Zhang, J., Zheng, Y., and Qiao, L. The effect of temperature on the dry sliding wear behavior of HVOF sprayed nanostructured WC-CoCr coatings. *Ceramics International*, 43, (2017), 458–462.

11. Prasad, C.D., Joladarashi, S., Ramesh, M.R., Srinath, M.S., and Channabasappa, B.H.Effect of microwave heating on microstructure and elevated temperature adhesive wear behavior of HVOF deposited CoMoCrSi-Cr$_3$C$_2$composite coating. *Surface and Coatings Technology*, 374, (2019) 291–304.

12. Karaoglanli, A.C., Oge, M., Doleker, K.M., and Hotamis, M. Comparison of tribological properties of HVOF sprayed coatings with different composition. *Surface and Coatings Technology*, 318, (2017), 299–308.

13. Kim, H.J., Hwang, S.Y., Lee, C.H., and Juvanon, P. Assessment of wear performance of flame sprayed and fused Ni-based coatings. *Surface and Coatings Technology*, 172, 2–3, (2003), 262–269.

14. Chaliampalias, D., Andronis, S., Pliatsikas, N., Pavlidou, E., Tsipas, D., Skolianos, S., Chrissafis, K., Stergioudis, G., Patsalas, P., and Vourlias, G. Formation and oxidation resistance of Al/Ni coatings on low carbon steel by flame spray. *Surface and Coatings Technology*, 255, (2014), 62–68.

15. Sidhu, H.S., Sidhu, B.S., and Prakash, S. Mechanical and microstructural properties of HVOF sprayed WC-Co and Cr$_3$C$_2$-NiCr coatings on the boiler tube steels using LPG as the fuel gas. *Journal of Materials Processing Technology*, 171(1), (2006), 77–82. doi:10.1016/j.jmatprotec.2005.06.058.

16. Kaur, M., Singh, H., and Prakash, S. High-temperature corrosion studies of HVOF-sprayed Cr$_3$C$_2$-NiCr coating on SAE-347h boiler steel. *Journal of Thermal Spray Technology*, (2009), 18 (4), 619–632. doi:10.1007/s11666-009-9371-9.

17. Fernández, E., Cadenas, M., González, R., Navas, C., Fernández, R., and de Damborenea, J. Wear behaviour of laser clad NiCrBSi coating. *Wear*, 259(7–12), (2005), 870–875. doi:10.1016/j.wear.2005.02.063.

18. Chen. J, An, Y., Yang, J., Zhao, X., Yan, F., Zhou, H., and Chen, J. Tribological properties of adaptive NiCrAlY-Ag-Mo coatings prepared by atmospheric plasma spraying. *Surface and Coatings Technology*, 235, (2013), 521–528.

19. Prasad, C.D., Joladarashi, S., Ramesh, M.R., Srinath, M.S., and Channabasappa, B.H., Microstructure and tribological behavior of flame sprayed and microwave fused CoMoCrSi/CoMoCrSi-Cr$_3$C$_2$ coatings. *Materials Research Express*, 5, (2018), 086519.

20. Buytoz, S., Ulutan, M., Islak, S., Kurt, B., and NuriÇelik, O. Microstructural and wear characteristics of high velocity oxygen fuel (HVOF) sprayed NiCrBSi-SiC composite coating on SAE 1030 steel. *Arabian Journal for Science and Engineering*, 38(6), (2013), 1481–1491. doi:10.1007/s13369-013-0536-y.

21. Jalali Azizpour, M. and Tolouei-Rad, M. The effect of spraying temperature on the corrosion and wear behavior of HVOF thermal sprayed WC-Co coatings. *Ceramics International*, 45(11), (2019)., 13934–13941. doi:10.1016/j.ceramint.2019.04.091

22. Fang, W., Cho, T.Y., Yoon, J.H., Song, K.O., Hur, S.K., Youn, S.J., and Chun, H.G. Processing optimization, surface properties and wear behavior of HVOF spraying WC–CrC–Ni coating. *Journal of Materials Processing Technology*, 209, (2009), 3561–3567.

23. Prasad, C.D., Joladarashi, S., Ramesh, M.R., Srinath, M.S., and Channabasappa, B.H.Development and sliding wear behavior of Co-Mo-Cr-Si cladding through microwave heating. *Silicon*, (2019). doi:10.1007/s12633-019-0084-5.

24. Prasad, C.D., Joladarashi, S., Ramesh, M.R., Srinath, M.S., and Channabasappa, B.H. Influence of microwave hybrid heating on the sliding wear behaviour of HVOF sprayed CoMoCrSi coating. *Materials Research Express*, 5, (2018), 086519.

25. Yang, Q., Senda, T., and Ohmori, A. Effect of carbide grain size on microstructure and sliding wear behavior of HVOF-sprayed WC-12% Co coatings. *Wear*, 254(1–2), (2003), 23–34. doi:10.1016/S0043-1648(02)00294-6.

26. Asl, S.K., Sohi, M.H., Hokamoto, K., and Uemura, M. Effect of heat treatment on wear behavior of HVOF thermally sprayed WC-Co coatings. *Wear*, 260, (2006), 1203–1208.

27. Prasad, C.D., Joladarashi, S., Ramesh, M.R., and Sarkar, A. High temperature gradient cobalt based clad developed using microwave hybrid heating. *American Institute of Physics*, 1943, (2018), 020111. doi:10.1063/1.5029687.

28. Uyulgan, B., Dokumaci, E., Celik, E., Kayatekin, I., Azem, N.A., Ozdemir, I., and Toparli, M. Wear behaviour of thermal flame sprayed FeCr coatings on plain carbon steel substrate. *Journal of Materials Processing Technology*, 190, (2007), 204–210.

29. Prasad, C.D., Joladarashi, S., Ramesh, M.R., Srinath, M.S., and Channabasappa, B.H. Comparison of high temperature wear behavior of microwave assisted HVOF sprayed CoMoCrSi-WC-CrC-Ni/WC-12Co composite coatings. *Silicon*, (2020), 1–19. doi:10.1007/s12633-020-00398-1.

30. Gisario, A., Puopolo, M., Venettacci, S., and Veniali, F. Improvement of thermally sprayed WC-Co/NiCr coatings by surface laser processing. *International Journal of Refractory Metals and Hard Materials*, 52, (2015), 123–130.

31. Ma, N., Guo, L., Cheng, Z., Wu, H., Ye, F., and Zhang, K. Improvement on mechanical properties and wear resistance of HVOF sprayed WC-12Co coatings by optimizing feedstock structure. *Applied Surface Science*, 320, (2014), 364–371.

32. Jafari, M., Enayati, M. H., Salehi, M., Nahvi, S. M., and Park, C. G. Microstructural and mechanical characterizations of a novel HVOF-sprayed WC-Co coating deposited from electroless Ni-P coated WC-12Co powders. *Materials Science and Engineering A*, 578, (2013), 46–53. doi:10.1016/j.msea.2013.04.064.

33. Bolelli, G., Berger, L.M., Bonetti, M., and Lusvarghi, L. Comparative study of the dry sliding wear behavior of HVOF-sprayed WC–(W, Cr) 2C–Ni and WC–CoCr hard metal coatings. *Wear*, (2014), 309, 96–111.

34. Hou, G.L., Zhou, H.D., An, Y.L., Liu, G., Chen, J.M., and Chen, J. Microstructure and high-temperature friction and wear behavior of WC-(W, Cr)2C-Ni coating prepared by high velocity oxy-fuel spraying. *Surface and Coatings Technology*, (2011), 206, 82–94.

35. Kandeva, M., Grozdanova, T., Karastoyanov, D., and Assenova, E. Wear resistance of WC/Co HVOF-coatings and galvanic Cr coatings modified by diamond nanoparticles. *Materials Science and Engineering*, 174, 2017, 012060.

36. Sahariah, B.J., Vashishtha, N., and Sapate, S.G. Effect of abrasive particle size on friction and wear behaviour of HVOF sprayed WC-10Co-4Cr coating. *Materials Research Express*, 5, (2018), 066424.

37. Prasad, C.D., Jerri, A., and Ramesh, M.R. Characterization and sliding wear behavior of iron based metallic coating deposited by HVOF process on low carbon steel substrate. *Journal of Bio and Tribo-Corrosion*, 6, (2020), 69. doi:10.1007/s40735-020-00366-7

38. Kong, D. and Zhao, B. Effects of loads on friction wear properties of HVOF sprayed NiCrBSi alloy coatings by laser remelting. *Journal of Alloys and Compounds*, 705, (2017), 700–707.

39. Fervel, V., Normand, B., Liao, H., Coddet, C., Beche, E., and Berjoan, R. Friction and wear mechanisms of thermally sprayed ceramic and cermet coatings. *Surface and Coatings Technology*, 111, (1999), 255–262.

7 Performance Status of Different Advanced Thermal Barrier Coatings against Failure Mechanisms in Gas Turbines

Amrinder Mehta, Hitesh Vasudev, and Sharanjith Singh

CONTENTS

7.1   INTRODUCTION

The multi-layered coated structure has been utilized to provide thermal insulation for the gas turbine components against the high-temperature working environments in the thermal barrier coating (TBCs) processes [1]. In thermal barrier coatings, the NiCoCrAlY is used as a bond-coat (BC) and ceramic-based materials as a top-coat. The thermally grown oxide (TGO) was produced between the top-coat and bond-coat microstructure during a high-temperature oxidation test performed on the coated structure. The coating microstructure mainly depended on the quality of the material and the process used for depositing the coating material on the base metal to help protect them from failure at high-temperature and corrosive environmental conditions [2, 3]. The coating material provided thermal resistance to the base metal by applying different TBC methods, since the steam turbine worked at high temperatures. The concurrent air cooling systems in the turbine components were utilized since this technique is beneficial for reducing the surface temperature to approximately 300°C and can also protect the surface of the material and improve the service life [4]. On employing the TBCs, the thermal insulation advantages outweigh all different material-science accomplishments. The single-crystal NiCoCrAlY base alloys have been used for a long time in surface-engineering applications. Continuous improvement has occurred to overcome the demand for heat and corrosion-resistant materials. Modern TBC techniques make it possible to coat the intricate design component [5]. The aim behind the development of TBCs is to decrease the base metal heating temperature, heat transfer coefficient, cooling effects from the lower side, the geometry of intricate parts, the thickness of the coated structure and thermal conductivity. Continuous efforts have been made in advanced coating techniques to increase the thermal and hot-corrosion efficiency of diesel engines and aerospace engine components, working at high temperatures and corrosive environmental conditions. An innovative cooling mechanism has been developed that significantly supported the thicker-coated structure by modifying the ceramic powder materials' chemistry before deposition on the substrate materials [6]. The coated components' temperature decreased from 4°C to 9°C per 25 μm in this surface coating process that results in the need to increase the thickness of the coating from 600 to 2000 μm for improving the thermal resistance performance [7]. TBCs are used to achieve higher working temperatures [8].

Figure 7.1 shows that the TBCs have played a significant function in regulating the thermal efficiency and improving the higher working temperature of gas turbine blades deteriorating toward some outside boundary, such as turbine blades' sharp edges [9]. TBCs are usually deposited on the piston's top surface near the aperture in SI engines to reduce knocking. The coating is also often employed in automotive braking systems to protect hydraulic parts from high-temperature thermal oxidation.

7.2   SINTERING IMPACTS AND RESIDUAL STRESSES

7.2.1   SINTERING IMPACTS ON THE COATING MICROSTRUCTURE

Sintering effect principal causes failure of TBC's mechanisms at high working temperatures. There has been a detailed investigation into the sintering activity of several material deposition methods such as Air Plasma Spray (APS) and Suspension Plasma Spray (SPS) in TBCs [10]. The different microstructure properties affects the coating because of the sintering effect produced inside the structure due to ceramic splats and bond-coat interfacial connection during working at high temperatures [11]. Throughout sintering, hole reduction and tiny-crack recovery can improve stiffness and reduce strain resistance, which is unacceptable for the surface coating process.

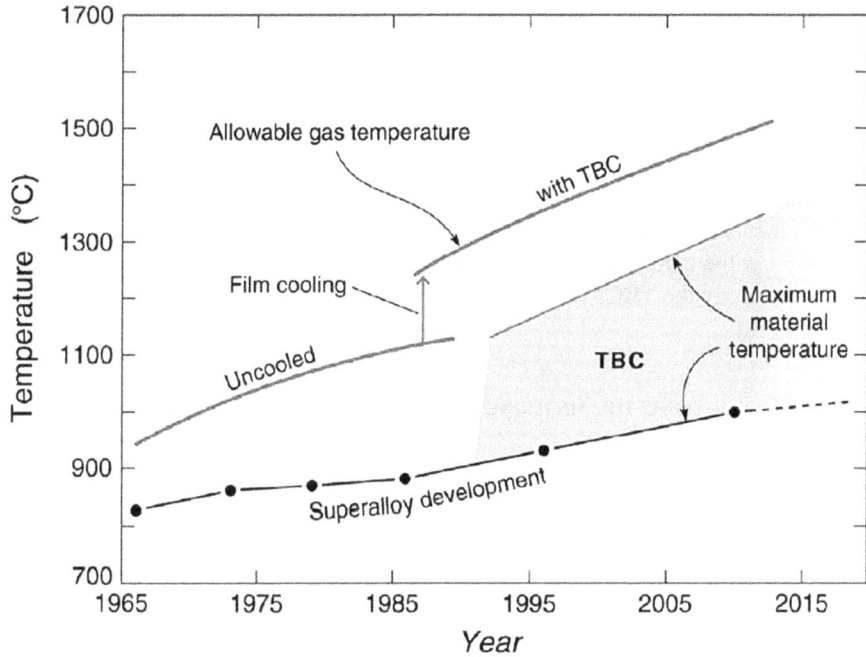

FIGURE 7.1   Historical development of a gas turbine's high pressure and temperature shows the role of super-alloy metal growth, air cooling and thermal barrier (TBC) coatings [9].

The sintering effect can change the coated structure's mechanical and microstructural properties that also help decide the working component's strength and thermal resistance [12]. Unlike the assessment of the entire coated surface's mechanical and thermal properties tested through oxidation and bending test, nano-indentation experiments can provide the hardness value of the coated structure. The complex $La(Al_{1/4}Mg_{1/2}Ta_{1/4})O_3$ (LAMT) rare-earth powders were deposited on the substrate metal by SPS method to investigate the service life of the deposited material. The LAMT coating structure provides lower thermal conductivity as compared to the 3YSZ at 1000°C [13]. The SPS coating method is used to improve some important properties of the coating, such as porosity level and microhardness fracture durability with crystalline/amorphous conditions at 1000°C, 1300°C and 1500°C before and after sintering. This coating process includes higher amorphous states with lesser hardness and fracture toughness, discovered to be inadequate sintering stability [14]. In SPS coating, the thermal aging action occurs in the SPS system and the temperature of the supply air is varied from 1200°C to 1600°C for 24 hrs, and kept at 1550°C for 20 to 100 hrs. Nano-indentation experimental technique examines the hardness and toughness value that has increased due to improving the sintering effect [15]. The SPS martial deposition process gives precise-quality coating structure for complex parts that improve the work performance of the coating at high temperatures and in corrosive environments. The results also show that the Young's modulus and hardness value of the coating structure have increased on applying the isothermal heat

treatment process at 1150°C for 100hrs and this heating technique can play a significant role in improving the surface texture [16]. An examination of the vertical cracks
formed and the mechanical strength inside the coating microstructure using nano-
indentation experiment and annealing process under heat treatments at1000°C,
1200°C and 1400°C showed a reduction of hardness value [17]. The typical APS
YSZ layer suffers from a quick formation sintering effect while working at high
temperatures. Over the past three decades, researcher has continuously worked for
improvements in coating strength and durability. The obtained novel coating microstructure, on the other hand, exhibited a sintering effect improvement for protection
design, whereby the TBCs were able to maintain superior strain toleration for extra
duration [18].

7.2.2 COUNTERING THE SINTERING EFFECT USING NANO-SIZE COATING STRUCTURE

The nano-level coating structure existed sufficiently, to work in high-temperature
and erosive environments. On the other hand, YSZ nano-size powder also provided
better hardness, toughness and strength compared to conventional ones. In these
types, the coating structure was obtained from nano-size powder and the measured
diameters of the powder particles were less than 100 nm before deposition of these
particles onto substrate metal [19]. The air plasma spray (APS) coating process is
most generally used to deposit YSZ powder particles on the substrate metal. This has
provided low operating cost and excellent thermal resistance for the coated structure
compared to the HVOF method [20]. The key problem of thermal spray technique is
to deposit YSZ nano-size grains without altering the coating structure's thermal insulation and mechanical properties. During the thermal spray process, the powder particles are continuously supplied by a feedstock mechanism that can work at high
temperatures. The SPS is more influential than the traditional thermal spay method
because it protects the YSZ ceramic molten droplets from overheating before depositing on the base material. The partially melted molten droplet required plays a significant role in retaining the powder's existing property [21]. APS is carried out by
the successively depositing the ceramic powder molten droplet on a base metal of
fully or semi-melted to grains, accompanied by similar flattening, fast crystallization
formation and temperature reduction [22]. The partially melted powder grains help
form the nanostructure on the base metal and a feedstock mechanism is used to supply nano-size powder grains into the plasma jet regularly. This coating technique also
exhibits the solidification, nucleation and growth behaviour, which results in the formation of the typical lamellar region inside the coating [23]. The sintering effect on
the nano-size coating structure rapidly affected the fully molten and semi-molten
droplet density that also provided sufficient strength and fatigue against the failure of
the coating nanostructure [24]. The researchers worked to combat sintering impres-
sions by engineering APS coating method and producing the nanostructure on base
metal from nano-size ceramic powder grains. In this study, the outcomes show how
the coarse porosity of the nano-size coating structure can be increased by applying
the heat-treatment process. Subsequently, the heat-treatment process was done on the
coated structure at 1400°C and time taken was 20 hrs. The outcomes showed that 3.5

times higher porosity was obtained in coated nanostructure than the traditionally coated microstructure. The essential activity happens inside the coating structure because of the various matrix and nano-zone sintering concentrations that improve the improved nanostructured coating. The higher-magnification level on SEM shows the sintering effect inside the coating surface texture, and this effect has been reduced by improving the coating quality. The nano-powders play a significant role in achieving the required property of the coating structure during the use of thermal spray depositing techniques, such as a faster shrinkage rate, improved thermal stability and hardness of material after the solidification. The coated components make enables it to work in high-temperature and erosive environments.

Figure 7.2 shows the sintering consequence: areas in the as-spray layer defined as the nanostructure were examined after 20 hrs of heat treatment at a more powerful magnification. After the heat-treatment process was completed, the porosity zones were still present inside the nanostructured coating. It is important to remember that during the densification process (among other things), grain size and velocity of the molten particles before deposition on the substrate (i.e., the nanostructure) play a significant role.

Figure 7.3 displays thermal diffusivity values for this nanostructured and traditional YSZ coatings process as-sprayed, and the effect of heat treatment. These thermal diffusivity rates from both sides of the coating have increased continuously, and then after heating of the coating structure for 1 hr at 1400°C, a 45% decrease in the gap was observed [24]. The porosity plays a significant role in the nanostructured coating and it is found that as the porosity level rises, thermal conductivity gradually decreases, hence improving thermal stability. Sintering for this coating can also significantly enhance the microstructure after the heat-treatment process has been performed. The particle diameter for this YSZ-based nanostructure coating was continually increased by extending the annealing time [25].

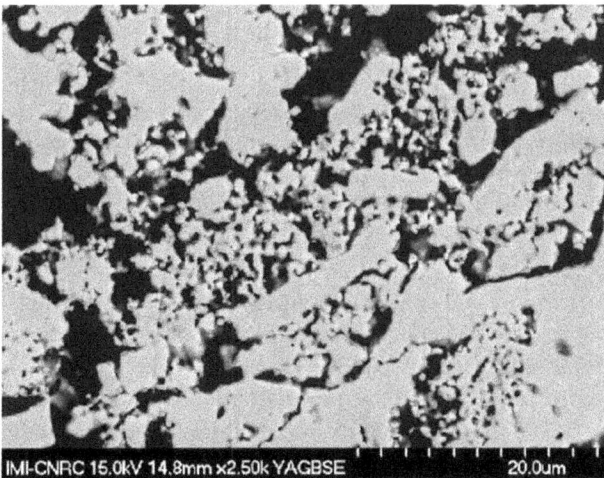

FIGURE 7.2 At high magnification, an SEM image of the nanostructured YSZ coating after heat-treatment at 1400°C is shown, and the time taken for completion process is 20 hrs [24].

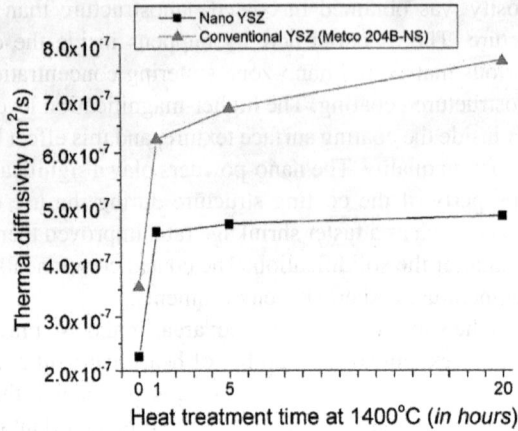

FIGURE 7.3 The difference in thermal diffusivity of the as-sprayed coating structure and the heat-treatment process for the nano- and traditional YSZ coatings for 1, 5 and 20 hrs at 1400°C [24].

The sintering effect inside the coating structure helps provide more thermal insulation and durability of the nanostructure than traditional ones. The nanostructure coating has to provide higher thermal insulation and lower stress generation inside the coating than the traditional coating.

7.3   INFLUENCE OF THE RESIDUAL STRESSES ON COATING FEATURES

The residual stresses inside the coating structure depend upon the thickness of the coating and selection parameters of the thermal spray. The coating modulus may be distinguished through some axial forces inside the layered structure. Axial strength has been used to analyse the quality of the coating structure [26]. The quenching process can produce better strength throughout the ceramic material deposition process due to continuous shrinkage and reduced molten splats. The grit blast operation was performed on the base metal before deposition of the bond layer to provide sufficient surface roughness to improve grip strength. In the APS coating process, stresses generated inside the microstructure during the work can be measured and regulated using the in situ curvature method for stress evaluation [27]. The ceramic nanostructure's adhesive and cohesive strengths were measured using a fundamental material resistance method, and material deposition was performed using laser technique. For measuring the crucial load corresponding to the crack initiated, load-displaced areas were taken throughout this four-point bending experiment. Various factors influence the surface coating method, and parameter variances such as laser scan velocity for the coating's cohesive intensity have been studied [28]. The preheating temperature can influence the distribution of the residual stresses for the coating and is therefore an effective thermal spray method. Before the coating starts, preheating to the base metal at 250°C is optimized; it can continuously improve and distribute heating temperature to reduce mechanical and thermal stresses. The coated microstructure were analysed for adhesive and cohesive strengths by preheating the substrate metal at 200°C–1000°C before deposition of the ceramic material. The temperature range for the ceramic powder material spraying was selected as 400°C–600°C to improve the coating material's overall mechanical properties [29]. In the thermal spray process, conventional YSZ ceramic powers have been used as a topcoat, MCrAlY as a bond-coat and SS-304 as a base metal. To analyse the adhesive, cohesive and other mechanical properties of the material, the base metal was preheated at three different temperatures and it was found that 100°C gave the best results [30]. The improved coating microstructure was developed by deposition of the YSZ powder grain in full- and semi-molten droplet phases before solidifying on the surface of super-alloy-based bond-coat at high temperatures by the APS coating method. This research discovered that the coating's stiffness, durability, interface bond strength and protection from the cavitation effect are highest when the substrate's preheat temperature was 800°C [31].

Figure 7.4 shows that due to the excess amount of residual stresses generated inside the coating structure, the coating material starts peeling off slowly from the substrate metal. In the TBC system, the coating operating under high temperature

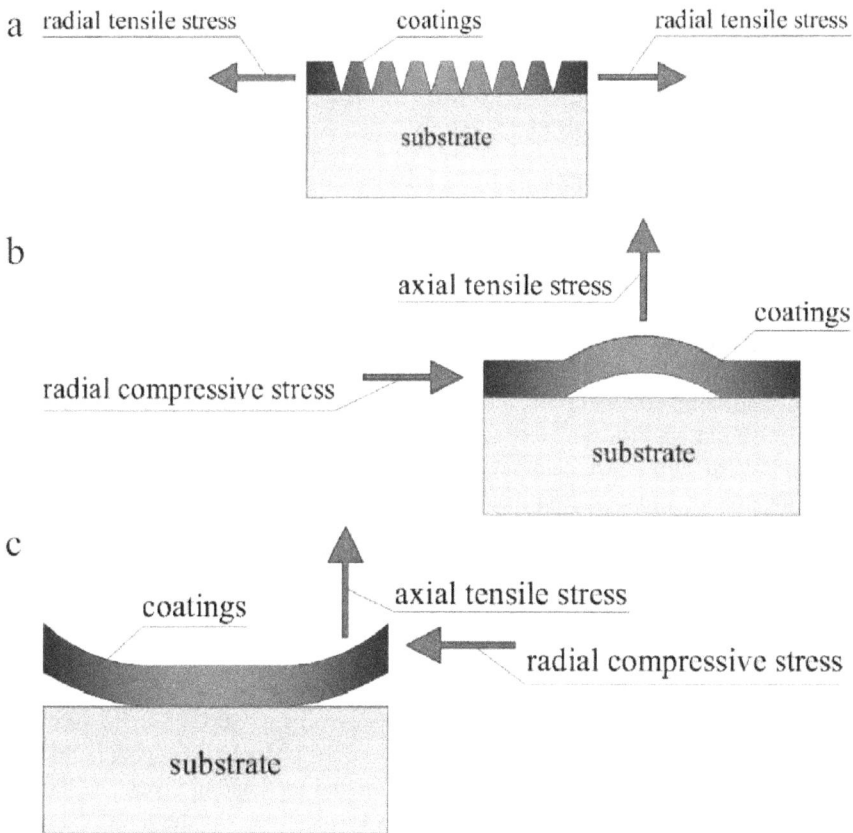

FIGURE 7.4   Due to the formation of residual stresses inside the coating structure, the coating starts to deteriorate: (a) groove fracturing, (b) internal arch and (c) edge warping [32].

and corrosive environment experiences a residual stress. The layer is likely to crack, but improving the porosity level at the same time can overcome this challenge. The main reasons for the formation of groove crackers inside the coating structure are the excess amount of radial tensile stresses generated on the coating, as shown in Figure 7.4a of the crack propagation. The coating's thermal shock resistance due to the induced vertical-directional cracks in the coating microstructure repairs the thermal expansion during working at high temperatures. The coated components' service life get reduced if the generated vertical cracks are deeper than half of the top-coat thickness inside the coating microstructure with the possibility of increasing the oxidation and corrosion rate for the coating surface. Figure 7.4b shows the internal arch grated inside the coating surface due to the combined impact of the axial tensile stress and radial residual compressive stress. As shown in Figure 7.4c, the edge warping defects occurred in the coating microstructure due to the radial compressive and axial tensile stress's combined effect [32]. In this research, the YSZ-based ceramic material was deposited on the SS-316-based substrate metal at a different heating temperature by using the thermal spray method. For this coating method, the heating temperature used for material deposition plays a small role in the phase composition because it can be less than t-ZrO_2 to m-ZrO_2 transformation temperatures [33].

7.4   OXIDATION PERFORMANCE

7.4.1   AT HIGH-TEMPERATURE OXIDATION

A bilayer TBC system comprises two layers: one is a metallic bond-coat as an oxidation-resistant layer with the general composition of MCrAlY (M = Ni and Co), a ceramic top-coat acting as a thermal insulation layer. Higher thermal stability and oxidation resistance can be obtained from 8YSZ ceramic powder and it is widely used as a top-coat for high-temperature applications [34].

Figure 7.5a considers the interface of the CoNiCrAlY bond-coat and YSZ top-coat, at which the TGO layer formed. In the figure, the oxidized layer's formation at the top- and bond-coat's interface can help protect the base metal by serving as a diffusion shield against oxidation. Figure 7.5b shows that the oxidized composite TBC microstructure, in which oxide scales may be observed intermittently scattered along the coating cross section instead of a uniform oxide layer, is primarily composed of Al_2O_3 at the top of the bond-coat. The development of TGO in Figure 7.5c presents the effect of the particles existing in the CoNiCrAlY bond-coat. Figure 7.5d presents the persistent TGO layer growing at the interface within the CoNiCrAlY-based bond-coat and gradient top-coat and the unusual development of the oxide layer [34].

The formation of TGO inside the coating structure in between the top-coat and bond-coat starts after the thermal oxidation test perform at 950°C. Two major factors play a significant role in developing TGO between the top- and bond-coat in the TBC method, like the coating material's composition and the coating layer's depth [35]. For working at higher-temperature-engineering applications, the thermally expanded interface mismatch of the splats in between the ceramic top-coat and the metallic bond-coat of the traditional bi-layer coating process mainly creates more

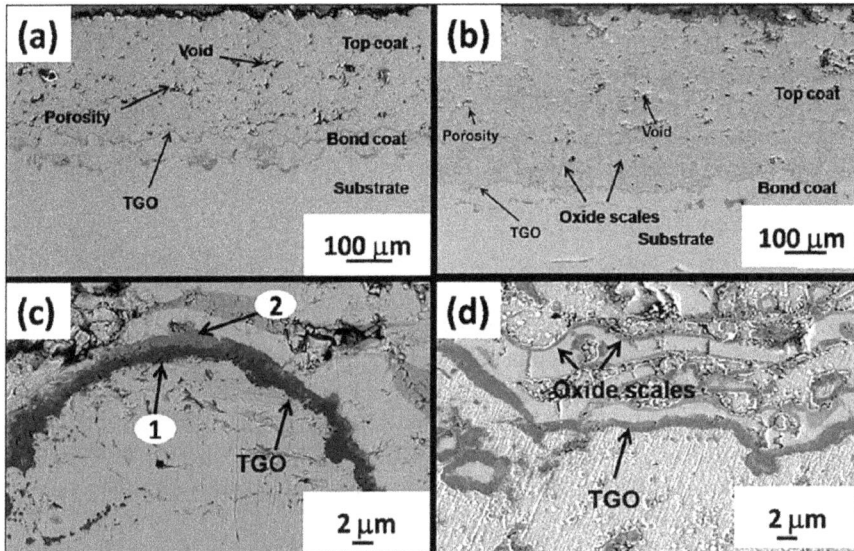

FIGURE 7.5 Cross-section SEM of (a, c) duplex and (b, d) composite TBC oxidized at 1000°C for 96 hrs at (a, b) low magnification and (c, d) high magnification, showing the TGO rate between the bond- and top-coat layer interface [34].

comprehensive heating stress at the coating interface. The spallation of coating microstructure is due to residual stresses generated since the coated metal primarily works at higher temperatures [36]. Numerous studies have been carried out to reduce the higher temperature oxidation outcomes, such as thermal shocks producing residual stresses inside the bond-coat oxidation and sintering effect [37]. In recent decades, the composite coatings have been widely used in the thermal spray process due to its superior quality microstructure, improved bond strength and better corrosion resistance performance [38]. On the other hand, TGO motion was lowered throughout the thermal oxidation by employing compositionally graded coating compared to the traditional bi-layer coating [39]. In recent decades, the researcher developed a compositionally graded layer (CGL) material deposition technique that helps to reduce the residual stress and improves the coating structure's thermal stability. The nanopowder-based coating structure has given better interactive mechanical and thermal resistance properties than the conventional ones [40]. To prepare the nanosize, YSZ powder-based coating under the TBCs process composed of bimodal nanostructure that consists of densifying YSZ nano-zones such as a matrix. The nanopowder-based coating structure has provided better thermal insulation and reduced the sintering effect that can help improve the coating's surface texture. The nano-zones start to heal uncovered gaps due to a mismatch of splats between the bond- and the top-coated nanostructures during working at a higher temperatures. This phenomenon is used to reduce the oxygen infiltration within the bond-coat during high-temperature work. The corrosive environment plays a significant role in removing the thermal growth rate [41]. Some researchers work to analyse the performance of thermal oxidation behaviour of the TBC by applying a nanostructured CGL-based coating. The outcomes of the research confirmed that the CGL deposition method under nanostructured base coating was effectively performed and reduced the TGO rate during working at high temperatures and corrosive environmental conditions. The nano-zones' appearance inside the layer has benefited the decrease of oxygen movement at the nanostructured base top-coat and bond-coat interface [42].

7.4.2 IMPORTANCE OF WATER VAPOUR IN THE OXIDATION PERFORMANCE

The bond-coat's purpose is to prevent base metal from oxidation, apart from implementing an equal thermal development between the top-coat and the base metal. At a heating temperature of 1050°C, the isothermal oxidation test was conducted for investigating the effect of water vapour on the TBC's oxidation behaviour. Figure 7.6 displays that isothermal oxidation of TBC with O_2 + 5% H_2O. In the TBC method, oxidation in pure O_2 is relatively slow at 1050°C and it was shown that the oxidation flow was parabolic in nature.

As shown in Figure 7.6, the exposure to O_2 + 5% H2O during the testing of specimens, weight gain during the oxidation test has reduced and improved the oxidation resistance. The effect of water vapour on the oxidation of the NiCrAlY coating can be due to the increased movement for the Ni and Cr cation [43].

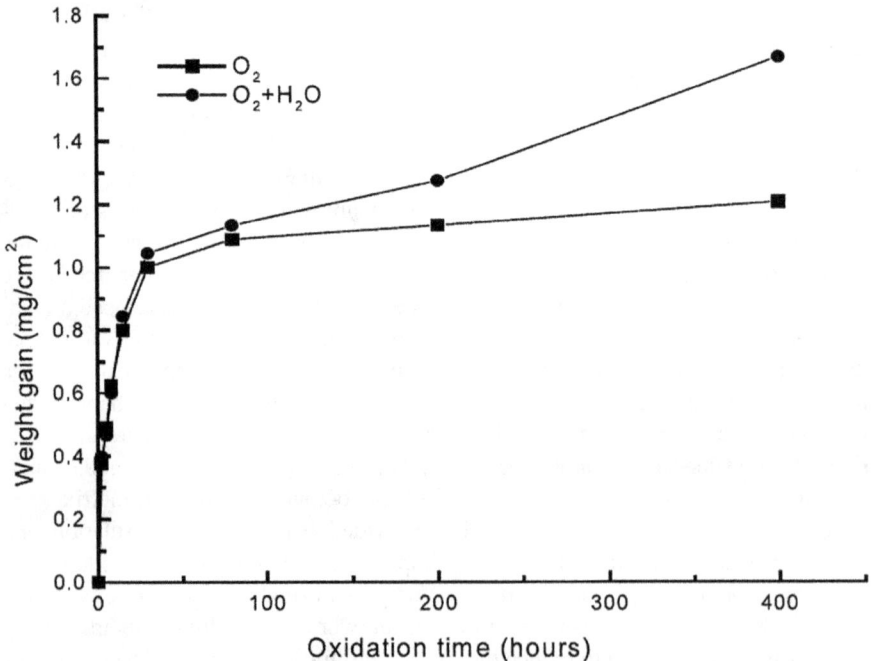

FIGURE 7.6 Isothermal oxidation kinetics of TBCs at 1050°C in O_2 and mixture of O_2+5%H_2O [43].

In marine engineering applications, the turbine blades always work under a high temperature and hot corrosion environment condition inside the combustion chamber of the marine engine. The NiCrAlY/YSZ-based coating performed under a hot corrosion test, and the temperature range was set around at 1050°C for 1 hr and air cooling was for 10 minutes [44]. The results show the relationship between water impression and top-coat spall to check the coating's postponed deterioration caused by moisture [45]. The acoustic emission was not present in the as-sprayed sample, which was in agreement with earlier reports on hydrogen generation by the reaction of δ-Al_2O_3-modified Al powder grains [46].

7.4.3 OXIDATION PERFORMANCE IMPROVEMENT BY EMPLOYING GRAPHENE PROTECTIVE COATING MATERIAL

The refined metal alloys are extensively used in various modern manufacturing and industrial applications such as marine, pipelines and automotive industries. The surface coating technique is used to improve alloy metals' service life because it can protect the surface of the metal from failure by using a ceramic layer on the upper body of the substrate metal that works at high-temperature, corrosive environmental conditions [47]. To date, several different techniques have been commonly used in shielding carbon steel from harsh corrosive conditions, such as cathodic safety [48]. A passive corrosion protection coating (PCP coating) is a protective barrier that keeps corrosive media away from the coating's surface [49]. Compared to other coating methods, APS coatings are generally used to effectively protect the ferrous portion from corrosion in numerous applications [50]. Both one-dimensional (1D) and two-dimensional (2D) nano-size powder grains are used to get better-quality water-based coatings for adequate corrosion safety [51]. Particles of SiO_2 [52], Al_2O_3 [53], ZnO [54], TiO_2 [55], talc [56], CeO_2 [57], graphene [58], etc., have been included in water-based surface coating arrangements to efficiently improve their hardness, strength, high-temperature inertness and corrosion and wear-resistance performances. As a 2D material, graphene has played a significant role due to its incomparable mechanical, chemical and thermal properties [59]. The researchers have discovered that graphene material protects oxidation and corrosion of Cu and Si base metal for a short time duration [60]. Many researchers have enhanced the dispersion properties by depositing the graphene layer at the material surfaces' outer parts and improving hardness and corrosion resistance [61]. It can be utilized to improve the interactions between GO sheets and epoxy resin (EP) as modifiers for diamine and p-phenylenediamine (PPDA) to make functionalized graphene oxide (GO) [62].

Polyisocyanate resin functionally GO can also be utilized as fillers for protection from corrosion and improving the working efficiency of polyurethane (PU) coatings observed in this research. Researchers discovered that encapsulating graphene sheets in protected material is an effective way of enhancing barrier and corrosion resistance properties, in contrast to the techniques explained earlier [63]. It employed (3-aminopropyl)-triethoxysilane (APTES) as a protecting substance to encapsulate GO sheets to examine the corrosion resistance efficiency. It has been observed that APTES can work as a spacer that impedes graphene–metal/graphene connections to disrupt the electron transference pathway and efficiently inhibit corrosion activity [64]. Some researchers have found that polyaniline–graphene composites (PAGCs) have superior corrosion-protective properties to polyaniline (PANI) which is one of a class of organic polymers and polyaniline-clay composites (PACCs) with the same filling. To enhance the corrosion resistance and thermal resistance properties of the coated material, the addition of functionalized graphene is much preferable, based on earlier investigations [64, 65]. The hydrophilic accumulations on the graphene exterior section (resistance film) preferentially transport corrosive electrolytes through the nano-layers and thereby protect the bottleneck from long-term corrosion [66]. Since defects are the crucial explanation for the degradation of the coatings' physical barrier properties, it is desirable to solve this issue from a molecular-level design perspective [67]. Graphene oxide (GO) C_2OH is used for extraordinary thermal resistance and hot corrosion efficiency, and therefore prevents substrate material from degradation.

Figure 7.7 presents and elaborates the systematic procedure applied for the preparation of the SPANI-rGO-based coating structure.

As shown in Figure 7.8, the schematic procedure applies for the corrosion protection device of EP/SPANI-rGO-4-based layer on the outside of Q235 substrate metal. The traditional epoxy coating (EP) material from corrosion is usually called an insulating layer that can help to prevent the components against corrosion. Prolonged time immersion in eroding mechanisms appears to happen at the coated-and base-material interface during the thermal oxidation and contraction effects. As compared to the pure form of EP, the corrosion-protective performance of epoxy resin (EP) coatings with 0.5% mechanical properties of rGO improved from 44.3% to 98.0%. Besides, the expansion of the immersion span, epoxy resin and phase position replaced significantly, can provide an opportunity for high-performance corrosion-safety applications [68].

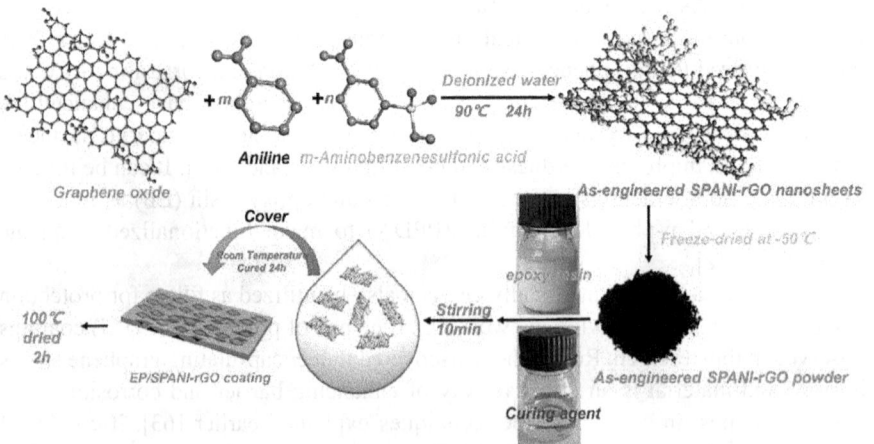

FIGURE 7.7 Schematic diagram of the development process of SPANI-rGO that holds on to the epoxy resin composite layers.

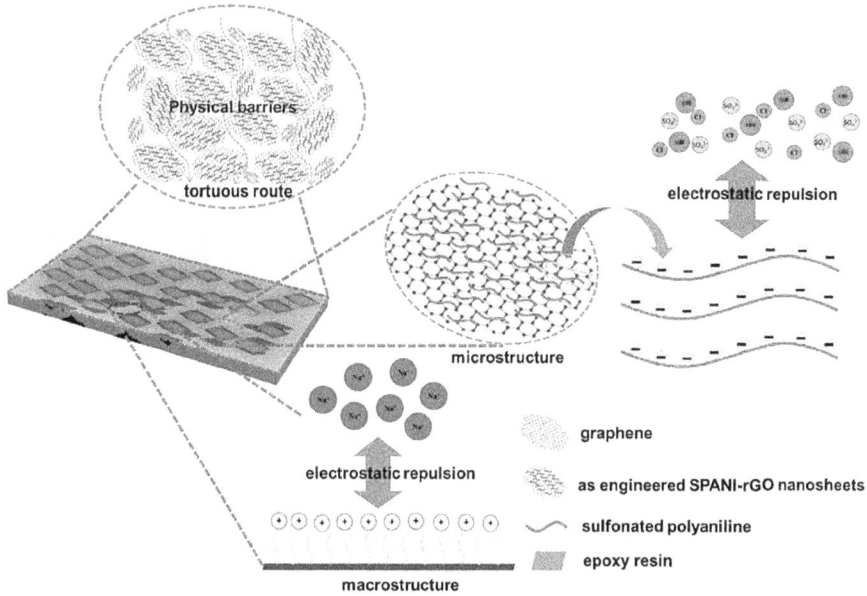

FIGURE 7.8   Schematic diagram based on the Q235 steel surface of rGO-SPANI at nano-level sheets, schematic design of a non-corrosion device for coatings.

7.5   CMAS PROTECTIVE SURFACE COATINGS

In combination with air-filled external elements like airborne/volcanic/concrete dust, sand fly ash and fuel residue, TBCs are used in aerospace turbine blades that are exposed to hostile oxidization atmosphere [69]. These powder grains are deposited in molten calcium – magnesium – alumina – silicate (CMAS) glass on the heated substrate surface, which starts penetration into the porous layer [70]. In the deposition, these crystallize the material and cause thermal increment difference between the CMAS and the coating structure's degradation. In the thermal spray process, lots of research have been done for the previous two decades to reduce CMAS attacks on the coating surface, that improved the working component's service life [71].

Some micro-level fractures and tiny pores were generated inside the coating microstructure after deposition of the ceramic material on to the substrate metal. Throughout the deposition method, the latest splats covered the previous splats, and crystal structures were formed between various splats because the neighbouring splats do not entirely overlap. Each depositing splat has an existing different melting point temperature; this causes the difference of the shrinkage amount obtained after impinging at previous splats. Figure 7.9a presents the differences in shapes, size and arrangements of the splats formed inside the top-coat surface microstructures. During an aircraft engine's operation at elevated temperature in a hostile combusted environment, the CMAS is deposited steadily on the ceramic top-coat surface, as shown in Figure 7.9b. In Figure 7.9c shows that when the temperature of the heated gas of this oxidization rises to the CMAS melting point usually beyond 1200°C, the solid phase

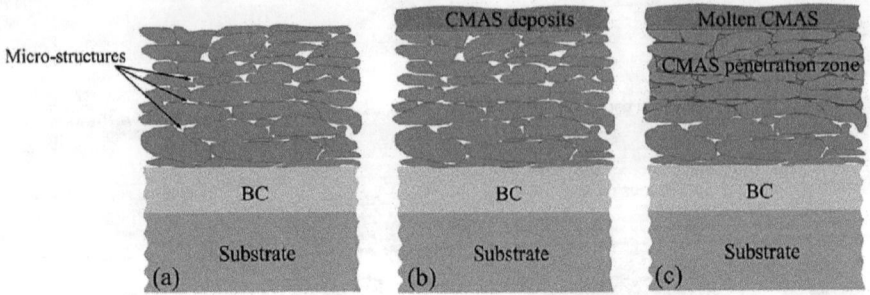

FIGURE 7.9   The penetration process of CMAS into APS TBC [72].

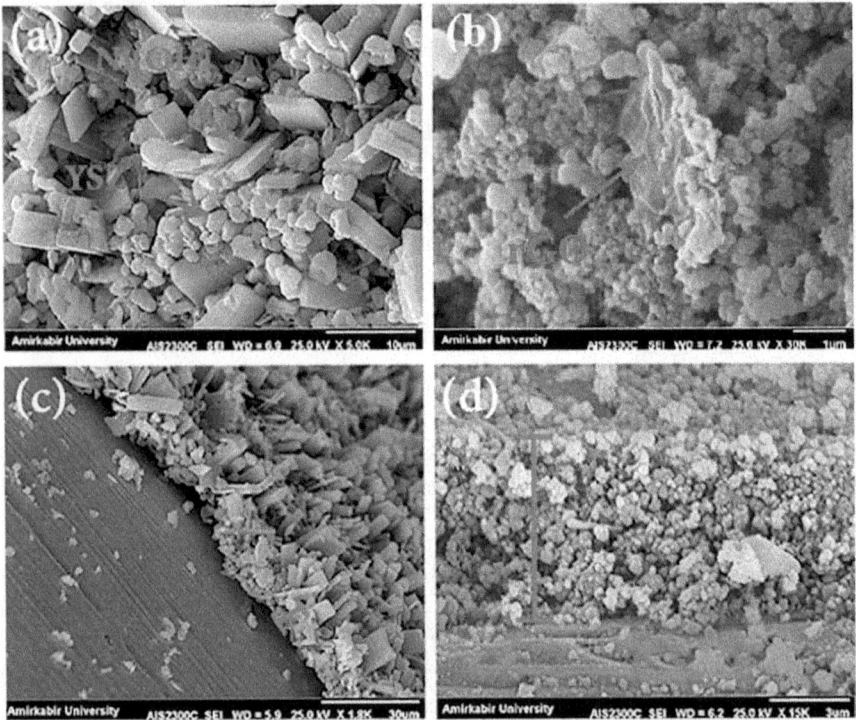

FIGURE 7.10   FE-SEM images of the outside and cross-sectional layers of the Ni(OH)$_2$-YSZ-based composite coating at a higher-magnification level [73].

CMAS during a deposition gets converted to a liquid phase, and then gradually starts penetrating into the coated surface [72].

During the Electrophoretic Deposition (EPD) process, Figure 7.10a presents Ni(OH)$_2$ films with a depth of 60 nm in Ni(OH)$_2$-YSZ powder will be interconnecting with all different in the electric field. This cross-sectional FE-SEM image of the composite layer Ni(OH)$_2$-YSZ have been observed in Figure 7.10c with a coarse surface and a 15 μm depth coated on 316L. This is also one of the principal understandings for decreasing corrosion protection of the $Ni(OH)_2$-YSZ layer consisting of proprietary square $Ni(OH)_2$ sheet and micron-level round-shaped YSZ agglomerated powder. To decrease the depth of the coated structure seen in the cross-section FE-SEM image of the uniform $Ni(OH)_2$-YSZ/rGO and in Figure 7.10 (b, d), the coating results in compressed composition and the lower level of porosity for the coating structure [73]. The atmospheric CMAS is melted at an elevated temperature and starts deteriorating from the upper side of the coating microstructure in the thermal spray [74]. Each ceramic coating layer's properties can gradually change because of the penetration of the melted CMAS into the developed coating structure [75]. The melted form of CMAS glass within the upper coating surface texture may repeatedly generate a considerably harder varnished solidification to the CMAS [76]. In the thermal spray coating process, thermal expansion coefficient plays a significant role in starting the mismatch between the upper coating microstructure and CMAS deposited layer. This situation can also play a leading role in initiating the crack propagation inside the coating microstructure [77]. During repeated thermal oxidation and cooling cycles, the strain toleration in this coating structure might degrade from within to the turbine blades, causing cracks inside the coating microstructure [78].

7.6   OUTLOOK AND CONCLUSIONS

The TBC methods have mainly been used for providing thermal and corrosive insulation for the diesel, marine and steam turbine engines at high-temperature environmental condition. The lamellar microstructure created by the quick solidification of impinging liquid droplets and cohesiveness among splats distinguishes thermal spray coatings. Some principal causes in the TBCs, such as the sintering effect, TGO and CMAS inside the coating microstructure, play a significant role in losing the coating material. All these causes generation of harmful effects inside the coating structure, such as increasing the stress level and rapid phase transformation of the coating material. In the TBC method, the thermal conductivity, modulus of elasticity and hardness can increase because of the shrinkage and the sintering effect acting on the surface layer. Due to the thermal expansion coefficient, the mismatch beginning between top-coat/CMAS and top-coat/TGO leads to the start of crack propagation inside the coating structure. In this coating process, the crack propagation is created inside the coating because of the reduction of strain tolerance and the rise of splash generation. The sintering effect in the coating can be reduced by replacing the conventional YSZ ceramic powder to be deposited on the substrate metal with the nano-size powder grains. The feedstock mechanism used to control the supply of powder grains is at an optimum level. The nanostructure-based TBC has improved and better work performance compared to conventional ones, for example, improved corrosion protection, reduced thermal conductivity and decreased elastic modulus. Extraordinary-quality oxidation protection has been displayed for the Pt aluminide coating, reducing the stress produced due to the increase of the TGO rate. The TBC is mainly related to the high working temperature composition obtained at a reduced price, and CMAS work also plays a critical role in advanced coating developments.

REFERENCES

1. Feuerstein A, Knapp J, Taylor T, Ashary A, Bolcavage A, Hitchman N. Technical and economical aspects of current thermal barrier coating systems for gas turbine engines by thermal spray and EBPVD: a review. *Journal of Thermal Spray Technology*, 2008; 17: 199–213.
2. Karaoglanli AC, Ogawa K, Turk A, Ozdemir I. Thermal shock and cycling behavior of thermal barrier coatings (TBCs) used in gas turbines. *Progress in Gas Turbine Performance*, 2013; 2013: 237–260.
3. Mehta A, Vasudev H, Singh S. Recent developments in the designing of deposition of thermal barrier coatings–A review. *Materials Today: Proceedings*, 2020; 26: 1336–1342.
4. Myoung S-W, Lee S-S, Kim H-S, Kim M-S, Jung Y-G, Jung S-I, et al. Effect of post heat treatment on thermal durability of thermal barrier coatings in thermal fatigue tests. *Surface and Coatings Technology*, 2013; 215: 46–51.
5. Amagasa S, Shimomura K, Kadowaki M, Takeishi K, Kawai H, Aoki S, et al. Study on the turbine vane and blade for a 1500 C class industrial gas turbine. 1994: 597–604. 1994.
6. Mogro-Campero A, Johnson C, Bednarczyk P, Dinwiddie R, Wang H. Effect of gas pressure on thermal conductivity of zirconia thermal barrier coatings. *Surface and Coatings Technology*, 1997; 94: 102–105.
7. Bhatia T, Ozturk A, Xie L, Jordan EH, Cetegen BM, Gell M, et al. Mechanisms of ceramic coating deposition in solution-precursor plasma spray. *Journal of Materials Research*, 2002; 17: 2363–2372.
8. Liu Q, Huang S, He A. Composite ceramics thermal barrier coatings of yttria stabilized zirconia for aero-engines. *Journal of Materials Science & Technology*, 2019; 35: 2814–2823.
9. Zhao H, Levi CG, Wadley HN. Molten silicate interactions with thermal barrier coatings. *Surface and Coatings Technology*, 2014; 251: 74–86.
10. Lv B, Mücke R, Fan X, Wang T, Guillon O, Vaßen R. Sintering resistance of advanced plasma-sprayed thermal barrier coatings with strain-tolerant microstructures. *Journal of the European Ceramic Society*, 2018; 38: 5092–5100.
11. Cocks A, Fleck N, Lampenscherf S. A brick model for asperity sintering and creep of APS TBCs. *Journal of the Mechanics and Physics of Solids*, 2014; 63: 412–431.
12. Lima RS, Marple BR. Toward highly sintering-resistant nanostructured ZrO 2–7wt.% Y 2 O 3 coatings for TBC applications by employing differential sintering. *Journal of Thermal Spray Technology*, 2008; 17: 846–852.
13. Guignard A, Mauer G, Vaßen R, Stöver D. Deposition and characteristics of submicrometer-structured thermal barrier coatings by suspension plasma spraying. *Journal of Thermal Spray Technology*, 2012; 21: 416–424.
14. Tarasi F *Suspension Plasma Sprayed Alumina-Yttria Stabilized Zirconia Nano-Composite Thermal Barrier Coatings: Formation And Roles Of The Amorphous Phase*. Montreal, QC: Concordia University; 2010.
15. Zhao Y, Wang L, Yang J, Li D, Zhong X, Zhao H, et al. Thermal aging behavior of axial suspension plasma-sprayed yttria-stabilized zirconia (YSZ) thermal barrier coatings. *Journal of Thermal Spray Technology*, 2015; 24: 338–347.
16. Zou Z, Donoghue J, Curry N, Yang L, Guo F, Nylén P, et al. A comparative study on the performance of suspension plasma sprayed thermal barrier coatings with different bond coat systems. *Surface and Coatings Technology*, 2015; 275: 276–282.
17. Mahade S, Li R, Curry N, Björklund S, Markocsan N, Nylén P. Isothermal oxidation behavior of Gd2Zr2O7/YSZ multi-layered thermal barrier coatings. *International Journal of Applied Ceramic Technology*, 2016; 13: 443–450.

18. Huang J, Wang W, Li Y, Fang H, Ye D, Zhang X, et al. Improve durability of plasma-splayed thermal barrier coatings by decreasing sintering-induced stiffening in ceramic coatings. *Journal of the European Ceramic Society*, 2020; 40: 1433–1442.

19. Lu Y, Liaw PK. The mechanical properties of nanostructured materials. *JOM*, 2001; 53: 31–35.

20. Zeng Y, Lee S, Gao L, Ding CJ. Atmospheric plasma sprayed coatings of nanostructured zirconia. *Journal of the European Ceramic Society*, 2002; 22: 347–351.

21. Lima R, Kucuk A, Berndt CJ. Integrity of nanostructured partially stabilized zirconia after plasma spray processing. *Materials Science and Engineering: A*, 2001; 313: 75–82.

22. Yang G-J, Li C-X, Hao S, Xing Y-Z, Yang E-J, Li CJ, et al. Critical bonding temperature for the splat bonding formation during plasma spraying of ceramic materials. *Surface and Coatings Technology*, 2013; 235: 841–847.

23. Cipitria A, Golosnoy I, Clyne TJ. A sintering model for plasma-sprayed zirconia TBCs: Part I: Free-standing coatings. *Acta Materialia*, 2009; 57: 980–992.

24. Lima R, Marple BJ Nanostructured YSZ thermal barrier coatings engineered to counter-act sintering effects. *Materials Science and Engineering: A*, 2008; 485: 182–193.

25. Kai W, Hui P, Hongbo G, Shengkai GJ. Effect of sintering on thermal conductivity and thermal barrier effects of thermal barrier coatings. *Chinese Journal of Aeronautics*, 2012; 25: 811–816.

26. Zhang X, Watanabe M, Kuroda SJ. Effects of processing conditions on the mechanical properties and deformation behaviors of plasma-sprayed thermal barrier coatings: Evaluation of residual stresses and mechanical properties of thermal barrier coatings on the basis of in situ curvature measurement under a wide range of spray parameters. *Acta Materialia*, 2013; 61: 1037–1047.

27. Zhang X, Xu B, Wang H, Wu YJ. Error analyses on some typically approximate solutions of residual stress within a thin film on a substrate. *Journal of Aapplied Physics*, 2005; 98: 053516.

28. Kadolkar P, Dahotre NB. Effect of processing parameters on the cohesive strength of laser surface engineered ceramic coatings on aluminum alloys. *Materials Science and Engineering: A*, 2003; 342: 183–191.

29. Fomin A, Fomina M, Koshuro V, Rodionov I, Zakharevich A, Skaptsov A. Structure and mechanical properties of hydroxyapatite coatings produced on titanium using plasma spraying with induction preheating. *Ceramics International*, 2017; 43: 11189–11196.

30. Hui L, Benpeng W, Qi W, Xiaojian J, Zhuoxin L. Influence of Plasma Spray Parameters on the Microstructure and Property of ZrO_2–7 wt% Y_2O_3 Coating. *Rare Metal Materials And Engineering*, 2012; 41: 291–295.

31. Deng W, An Y, Hou G, Li S, Zhou H, Chen J. Effect of substrate preheating treatment on the microstructure and ultrasonic cavitation erosion behavior of plasma-sprayed YSZ coatings. *Ultrasonics Sonochemistry*, 2018; 46: 1–9.

32. Pang M, Zhang X-H, Liu Q-X, Fu Y-X, Liu G, Tan W-D, et al. Effect of preheating temperature of the substrate on residual stress of Mo/8YSZ functionally gradient thermal barrier coatings prepared by plasma spraying. *Surface and Coatings Technology.*, 2020; 385: 125377.

33. Li S, An Y, Zhou H, Chen JJ. Plasma sprayed YSZ coatings deposited at different deposition temperatures, part 1: Splats, microstructures, mechanical properties and residual stress. *Surface and Coatings Technology.*, 2018; 350: 712–721.

34. Nath S, Manna I, Majumdar JDJCs. Kinetics and mechanism of isothermal oxidation of compositionally graded yttria stabilized zirconia (YSZ) based thermal barrier coating. *Corrosion Science*, 2014; 88: 10–22.

35. Thibblin A, Jonsson S, Olofsson U. Influence of microstructure on thermal cycling life-time and thermal insulation properties of yttria-stabilized zirconia thermal barrier coatings for diesel engine applications. *Surface and Coatings Technology*, 2018; 350: 1–11.

36. Rico A, Gómez-García J, Múnez C, Poza P, Utrilla V. Mechanical properties of thermal barrier coatings after isothermal oxidation: Depth sensing indentation analysis. *Surface and Coatings Technology*, 2009; 203: 2307–2314.

37. Khor K, Gu YJ, A E. Effects of residual stress on the performance of plasma sprayed functionally graded ZrO2/NiCoCrAlY coatings. *Materials Science and Engineering*, 2000; 277: 64–76.

38. Kawasaki A, Watanabe R. Thermal fracture behavior of metal/ceramic functionally graded materials. *Engineering Fracture Mechanics*, 2002; 69: 1713–1728.

39. Ghasemi R, Vakilifard H. Plasma-sprayed nanostructured YSZ thermal barrier coatings: thermal insulation capability and adhesion strength. *Ceramics International*, 2017; 43: 8556–8563.

40. Jiang K, Liu S, Wang X. Phase stability and thermal conductivity of nanostructured tetragonal yttria–stabilized zirconia thermal barrier coatings deposited by air–plasma spraying. *Ceramics International*, 2017; 43: 12633–12640.

41. Wang J, Sun J, Zhang H, Dong S, Jiang J, Deng L, et al. Effect of spraying power on microstructure and property of nanostructured YSZ thermal barrier coatings. *Journal of Alloys and Compounds*, 2018; 730: 471–482.

42. Sezavar A, Sajjadi SA, Babakhani A, Peng RL, Yuan K. Oxidation behavior of a nano-structured compositionally graded layer (CGL) thermal barrier coating (TBC) deposited on IN-738LC. *Surface and Coatings Technology*, 2019; 374: 374–382.

43. Zhou C, Yu J, Gong S, Xu H. Influence of water vapor on the high temperature oxidation behavior of thermal barrier coatings. *Materials Science and Engineering: A*, 2003; 348: 327–332.

44. Zhou C, Song Y, Wang C, Xu H. Cyclic-oxidation behavior of thermal-barrier coatings exposed to NaCl vapor. *Oxidation of Metals*, 2008; 69: 119–130.

45. Rudolphi M, Renusch D, Schütze M. Verification of moisture-induced delayed failure of thermal barrier coatings. *Scripta Materialia*, 2008; 59: 255–257.

46. Deng ZY, Liu YF, Tanaka Y, Zhang HW, Ye J, Kagawa Y. Temperature effect on hydrogen generation by the reaction of γ-Al_2O_3-modified Al powder with distilled water. *Journal of the American Ceramic Societ*, 2005; 88: 2975–2977.

47. Rao BA, Iqbal MY, Sreedhar BJ. Self-assembled monolayer of 2-(octadecylthio) benzothiazole for corrosion protection of copper. *Corrosion Science*, 2009; 51: 1441–1452.

48. Ye Y, Liu Z, Liu W, Zhang D, Zhao H, Wang L, et al. Superhydrophobic oligoaniline-containing electroactive silica coating as pre-process coating for corrosion protection of carbon steel. *Chemical Engineering Journal*, 2018; 348: 940–951.

49. Liu X, Gu C, Wen Z, Hou B. Improvement of active corrosion protection of carbon steel by water-based epoxy coating with smart CeO_2 nanocontainers. *Progress in Organic Coatings*, 2018; 115: 195–204.

50. Leal DA, Riegel-Vidotti IC, Ferreira MGS, Marino CEB. Smart coating based on double stimuli-responsive microcapsules containing linseed oil and benzotriazole for active corrosion protection. *Corrosion Science*, 2018; 130: 56–63.

51. Li W, Tian H, Hou BJ, Corrosion. Corrosion performance of epoxy coatings modified by nanoparticulate SiO_2. *Materials and Corrosion*2012; 63: 44–53.

52. Yu S, Liu Y, Li W, Liu J, Yuan DJ. The running-in tribological behavior of nano-SiO_2/Ni composite coatings. *Composites Part B: Engineering*, 2012; 43: 1070–1076.

53. Saha R, Khan T. Effect of applied current on the electrodeposited Ni–Al$_2$O$_3$ composite coatings. *Surface and Coatings Technology* 2010; 205: 890–895.
54. Sun M, Chen Z, Bu Y, Yu J, Hou BJ. Effect of ZnO on the corrosion of zinc, Q235 carbon steel and 304 stainless steel under white light illumination. *Corrosion Science*, 2014; 82: 77–84.
55. Al-Maadeed MAS. TiO 2 nanotubes and mesoporous silica as containers in self-healing epoxy coatings. *Scientific Reports*, 2016; 6: 38812.
56. Sathiyanarayanan S, Syed Azim S, Venkatachari GJ. Performance studies of phosphate-doped polyaniline containing paint coating for corrosion protection of aluminium alloy. *Journal of Applied Plymer Science*2008; 107: 2224–2230.
57. Ramezanzadeh B, Bahlakeh G, Ramezanzadeh M. Polyaniline-cerium oxide (PAni-CeO2) coated graphene oxide for enhancement of epoxy coating corrosion protection performance on mild steel. *Corrosion Science*, 2018; 137: 111–126.
58. Chen C, Qiu S, Cui M, Qin S, Yan G, Zhao H, et al. Achieving high performance corrosion and wear resistant epoxy coatings via incorporation of noncovalent functionalized graphene. *Carbon*, 2017; 114: 356–366.
59. Li Z, Zhang W, Wang H, Qin ZJ. Activated pyrene decorated graphene with enhanced performance for electrochemical energy storage. *Chemical Engineering Journal*, 2018; 334: 845–854.
60. Schriver M, Regan W, Gannett WJ, Zaniewski AM, Crommie MF, Zettl AJ. Graphene as a long-term metal oxidation barrier: Worse than nothing. *ACS Nano*, 2013; 7: 5763–5768.
61. Gu L, Liu S, Zhao H, Yu HJ, Facile preparation of water-dispersible graphene sheets stabilized by carboxylated oligoanilines and their anticorrosion coatings. *ACS Applied Materilas & Interfaces*. 2015; 7: 17641–17648.
62. Ramezanzadeh B, Niroumandrad S, Ahmadi A, Mahdavian M, Moghadam MM. Enhancement of barrier and corrosion protection performance of an epoxy coating through wet transfer of amino functionalized graphene oxide. *Corrosion Science*, 2016; 103: 283–304.
63. Ramezanzadeh B, Ghasemi E, Mahdavian M, Changizi E, Moghadam MM. Covalently-grafted graphene oxide nanosheets to improve barrier and corrosion protection properties of polyurethane coatings. *Carbon*, 2015; 93: 555–573.
64. Sun W, Wang L, Wu T, Wang M, Yang Z, Pan Y, et al. Inhibiting the corrosion-promotion activity of graphene. *Chemistry of Materials*, 2015; 27: 2367–2373.
65. Chang C-H, Huang T-C, Peng C-W, Yeh T-C, Lu H-I, Hung W-I, et al. Novel anticorrosion coatings prepared from polyaniline/graphene composites. *Carbon*, 2012; 50: 5044–5051.
66. Tang J, Yao W, Li W, Xu J, Jin L, Zhang J, et al. Study on a novel composite coating based on PDMS doped with modified graphene oxide. *Journal of Coatings Technology and Research*, 2018; 15: 375–383.
67. Zhang M, Ma L, Wang L, Sun Y, Liu Y, Insights into the use of metal–organic framework as high-performance anticorrosion coatings. *ACS Applied Maerials & Interfaces*, 2018; 10: 2259–2263.
68. Zhou C, Hong M, Yang Y, Hu N, Zhou Z, Zhang L, et al. Engineering sulfonated polyaniline molecules on reduced graphene oxide nanosheets for high-performance corrosion protective coatings. *Applied Surface Science*, 2019; 484: 663–675.
69. Borom MP, Johnson CA, Peluso LA, Role of environment deposits and operating surface temperature in spallation of air plasma sprayed thermal barrier coatings. *Surface and Coatings Technology*, 1996; 86: 116–126.

70. Krämer S, Yang J, Levi CG, Johnson CAJJotACS. Thermochemical interaction of ther-
    mal barrier coatings with molten CaO–MgO–Al₂O₃–SiO₂ (CMAS) deposits. *Journal of
    the American Ceramic Society*, 2006; 89: 3167–3175.

71. Mack DE, Wobst T, Jarligo MOD, Sebold D, Vaßen R. Lifetime and failure modes of
    plasma sprayed thermal barrier coatings in thermal gradient rig tests with simultaneous
    CMAS injection. *Surface and Coatings Technology*, 2017; 324: 36–47.

72. Cai Z, Jiang J, Wang W, Liu Y, Cao ZJ. CMAS penetration-induced cracking behavior in
    the ceramic top coat of APS TBCs. *Ceramics International*, 2019; 45: 14366–14375.

73. Salehzadeh D, Sadeghian Z, Marashi PJD, Materials R. Enhanced protective properties
    of hydrothermally synthesized Ni (OH) 2-YSZ/reduced graphene oxide (rGO) nano-
    composite coating. *Diamond and Related Materials*, 2020; 101: 107655.

74. Krämer S, Faulhaber S, Chambers M, Clarke D, Levi C, Hutchinson J, et al. Mechanisms
    of cracking and delamination within thick thermal barrier systems in aero-engines sub-
    ject to calcium-magnesium-alumino-silicate (CMAS) penetration. *Materials Science
    and Engineering: A*, 2008; 490: 26–35.

75. Chen W, Zhao L. Review–volcanic ash and its influence on aircraft engine components.
    Procedia Engineering, 2015; 99: 795–803.

76. Krause AR, Garces HF, Dwivedi G, Ortiz AL, Sampath S, Padture NP. Calcia-magnesia-
    alumino-silicate (CMAS)-induced degradation and failure of air plasma sprayed yttria-
    stabilized zirconia thermal barrier coatings. *Acta Materialia*, 2016; 105: 355–366.

77. Zhang G, Fan X, Xu R, Su L, Wang T. Transient thermal stress due to the penetration of
    calcium-magnesium-alumino-silicate in EB-PVD thermal barrier coating system.
    Ceramics International, 2018; 44: 12655–12663.

78. Yanar N, Pettit F, Meier G. Failure characteristics during cyclic oxidation of yttria stabi-
    lized zirconia thermal barrier coatings deposited via electron beam physical vapor depo-
    sition on platinum aluminide and on NiCoCrAlY bond coats with processing
    modifications for improved performances. *Metallurgical and Materials Transactions A*,
    2006; 37: 1563–1580.

8 Hot Corrosion Characteristics of HVOF-Sprayed Cr_3C_2-25NiCr Protective Coating on Ni-Based Superalloys

M. Sathishkumar, M. Vignesh, V. Sreenivasulu,
M. Nageswara Rao, N. Arivazhagan, and
M. Manikandan

CONTENTS

DOI: 10.1201/9781003213185-8

8.1   INTRODUCTION

8.1.1   Hot Corrosion

In various power generation and high-temperature applications, impurities present in the high-temperature environments such as sulphur, sodium, vanadium, chlorine get converted to compounds like vanadium pentoxide (V_2O_5), sodium chloride (NaCl), sodium sulphate (Na_2SO_4), etc. These salts are often in molten condition, reacting with the underlying metal producing unsafe oxide layers on the components used in power plant, thus causing severe deterioration of the component [1]. This phenomenon of corrosion in the presence of salts at high temperature is called hot corrosion. The salt deposit used for hot corrosion studies are ionic conductors like Na_2SO_4, K_2SO_4, V_2O_5, NaCl and KCl. The acid/base nature of products of their reaction with substrate leads to fluxing of the protective oxide layer on the material surface [2]. The hot corrosion reaction often produces voids and pits and also initiates cracks in the protective oxide layer. This deteriorates the protective nature of oxide layer, facilitating the base metal getting attacked [3]. The areas where the degradation of metals and alloys takes place are boilers, gas turbines, fluidized bed combustion chambers, internal combustion engines, etc.

This chapter addresses the critical role played by hot corrosion in influencing the life of components used for aggressive power plant applications [4]. The commonly used salts for hot corrosion evaluation are $NaVO_3$, NaCl, Na_2SO_4, V_2O_5, etc. The duration and number of cycles for the hot corrosion studies are selected based on the application of the coated material at the place of use.

8.1.2   Procedure for Hot Corrosion Study

The operating conditions of the sample at the work place are selected for the hot corrosion studies. A piece of sample is sliced and cleaned to remove any impurities or scales present on the sample surface. Post cleaning, the sample is preheated to remove any moisture present on it. The sample is coated with the desired molten salt (simulating environment at the place of use) on the specimen surface and the weight of it is measured before the corrosion study. The coated specimen is subjected to hot corrosion, exposure time being based on the duration of the work material at the place of application. The weight gain/loss of the specimen is measured after every cycle and it is utilized for calculating corrosion kinetics and corrosion rate. The schematic procedure of the hot corrosion study is given in Figure 8.1.

The hot corrosion study is performed on the material used at the application site, by maintaining the same environment during testing. The material used could be a raw material or the coated material. The coating material and the process of coating play a major role in deciding the corrosion resistance of the component used. In this study, the hot corrosion testing was performed by keeping the coated sample inside a tubular furnace with conditions such as specimen size of 20 × 15×7 mm, Na_2SO_4 +

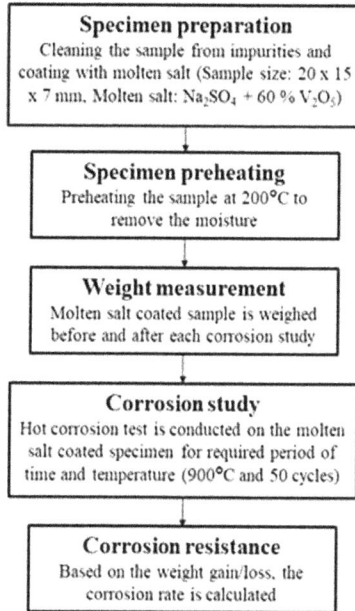

FIGURE 8.1   Schematic of the hot corrosion procedure.

60% V_2O_5 molten salt environment, preheating temperature of 200°C, hot corrosion test temperature of 900°C and 50 cycles (each cycle of 1 hr heating and 30 mins cooling period). Various details about the coating process and coating powders available are discussed in the next section.

8.2   THERMAL SPRAY COATINGS

In the present day of advanced power plant construction, materials should withstand very high operating temperatures, complex mechanical and tribological loading. A single bulk material can't possess these many properties to cater the today's industrial need. Hence, the use of protective coating on the work material is needed to enhance the mechanical, chemical and material properties [5]. Physical vapour deposition and chemical vapour deposition have been important coating processes in the industrial world. Thermal spray coating and laser cladding, with their high efficiency and productivity have occupied an important place in the industry [6]. Out of these two, thermal spraying is found to be more convenient and promising method to coat the components, effectively and economically without affecting properties of the material. The schematic of thermal spray coating is shown in Figure 8.2 [7].

Thermal spray coatings are produced by heating the material in a hot gaseous medium and projecting it at a high velocity on the work material surface where the desired coating is to be done. It comes under the hardfacing techniques, with a flexibility in choosing coating material, low thermal input and no substrate dissolution.

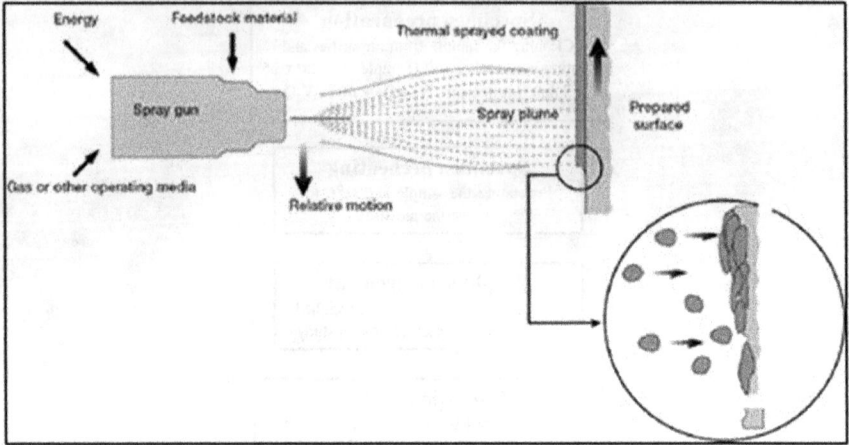

FIGURE 8.2 Schematic of thermal spray coating [7].

These thermal spraying techniques are extensively used on boiler tubes in steam-generating plants to prevent erosion and corrosion problems [8]. The available thermal spraying techniques are detonation-gun, plasma coating and high-velocity oxy-fuel (HVOF) coating. These coatings are applied on the work material substrate to impart resistance against corrosion, erosion and wear at elevated temperatures up to 900°C. Out of the said thermal spraying techniques, HVOF is found to be the most promising and reliable spraying process, because of high gas jet velocity and lower flame temperature [9]. A detailed discussion about the HVOF coating is done in the following section.

8.2.1   HVOF COATING

Out of various thermal spraying techniques, HVOF coating received special attention among the users because it yields sophisticated, wear-resistant cermet coatings with a high-quality microstructure and surface finish. Additionally, this coating features a very low oxidation, low carbide decomposition and low porosity or carbide–matrix dissolution. These attributes result in higher abrasion resistance and higher hardness on the material surface [10].

The process of coating on the work material surface through HVOF process includes the introduction of coating powder in the gas stream path and spraying it on the material surface at very high velocity and at higher temperature. The obtained coating is characterized by low porosity and closely controlled chemistry. The HVOF coatings are continuous in nature and found to be effective at the place of use. The schematic of the HVOF coating is shown in Figure 8.3 [11].

The corrosion resistance of the HVOF process is found higher when compared to other thermal spray techniques like flame or arc spraying, atmospheric plasma spraying (APS), etc. [12]. Also, the properties of the HVOF-coated layer is significantly influenced by the spray process and its parameters, in addition to the powder properties. As far as adhesive property is concerned, the bond between the coated layer and

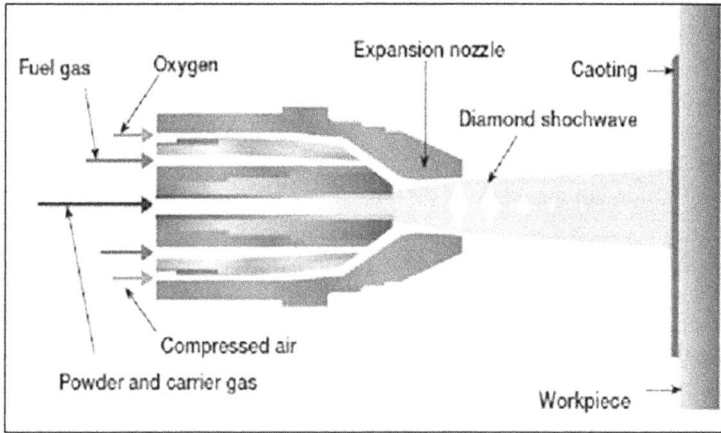

FIGURE 8.3 Schematic of the HVOF coating process [11].

the substrate should be strong enough for the coating to be effective. The HVOF cermet coating conditions are controlled by factors like morphology of the powder, size and distribution of the carbide particles, hardness of the carbide particles relative to the coating process and its characteristics like hardness, bonding strength, density, residual stress, etc. [13].

The microstructural features that improve the corrosion resistance of the HVOF-coated layer are high density, low porosity, fine grain size, freedom from cracks, etc. Also, the surface roughness plays a major role in deciding the corrosion resistance of the coated layer. Higher the surface roughness, more would be the corrosion attack on the coated surface. The various types of powders that could be coated using HVOF process are discussed in the upcoming section.

8.3 CARBIDE COATING

The primary goal of the coating is to effectively protect the surface from oxidation, corrosion, erosion and to have an excellent service behaviour. The thermal conductivity of the coating layer used in heat exchangers, gas turbines should be good enough for its effective utilization. Hence, the coating or the surface treatment should be in such a way, that it should provide an excellent barrier between the coated alloy and the environment to prevent oxidation [14]. The carbide coating on the work material surface provides efficient protection from the said environmental issues. The coating of carbide particles embedded in the metallic matrix, called Cermet coating, is applied using thermal spraying techniques. Cermets combine the advantages of the matrix and the reinforcement – high hardness of carbide and excellent ductility of metallic matrix. Also, the other enhanced properties of cermet are moderate toughness, good chemical stability, excellent adhesion and protection against wear [15]. Some of the commonly used cermet coatings are WC-Co, Fe-CrAlY-Cr_3C_2 and Cr_3C_2-NiCr. Since the coatings are applied on the surface in a powder form, properties of coating depend on the powder characteristics and its manufacturing methods. Some of the powder characteristics are:

- Powder particle shape and size
- Powder particle density
- Powder particle distribution
- Distribution homogeneity of the carbide particle in powder medium, etc.

The size of the carbide powders used in the cermet mixture usually is in the range 10–50 μm and their weight fraction is in the range 10–40%. The size of the cermet powder usually ranges below 5μm. The production of cermet powder with such small particle size is achieved either through "agglomeration and sintering" process or "sinter and crushing" process.

8.4 CERMET PRODUCTION PROCESS

8.4.1 AGGLOMERATION AND SINTERING PROCESS

In this process, the blended raw materials are dispersed in the water along with a polymer binder. A spray drying process is used, in which the dispersion is introduced into the heating chamber in the form of droplets produced through nozzle or the centrifugal atomizer. Since the dispersion passes through the heating chamber, the water present in the mixture gets eliminated, resulting in solid particles or pellets, which are then collected at the bottom.

Followed by agglomeration, sintering is performed on the collected solid particles. Initially, lower temperature is adopted, to remove the polymer binder particles from the collected pellets. Then, the sintering is conducted at higher temperatures of 1100°C–1400°C. Normally, the sintered powder particles are porous in nature and are loosely held together, thus eliminating the need for heavy crushing equipment. This prevents the iron contamination in the sintered powder mixture and the obtained powder is spherical in shape.

8.4.2 SINTER AND CRUSHING PROCESS

In this process, the blended ceramic and metal powders are pressed to form bricks or cylinders, which are then subjected to sintering process at 1100°C–1400°C. Post sintering, the bricks are crushed into smaller particles using hammer jaw or steel crushers, which would result in iron dust contamination. Because of the contamination in the powders obtained, the coating would result in lesser corrosion resistance than the powder obtained from agglomerated and sintered process.

Out of various cermet (WC-Co, Fe-CrAlY-Cr_3C_2 and Cr_3C_2-NiCr) powder particles, Cr_3C_2-NiCr cermet particles are widely used as a coating powder because of its corrosion-, erosion- and abrasion-wear resistance at elevated temperatures up to 850°C. The coating characteristics of the Cr_3C_2-NiCr are discussed in detail in the next section.

8.4.3 Cr₃C₂-NiCr Coating Characteristics

Cr_3C_2-NiCr powder coatings find extensive usage for corrosion resistant applications involving high temperatures. The NiCr matrix shows excellent corrosion resistance, while the carbide ceramic phase shows excellent wear resistance. The carbide powder particles have very high melting point and possess attractive properties like hardness, strength and excellent wear resistance at elevated temperatures. The commonly used materials for high-temperature applications prominently include iron- and nickel-based alloys, whose co-efficient of thermal expansion is $11.4 \times 10^{-6} °C^{-1}$ and $12.8 \times 10^{-6} °C^{-1}$, respectively. The co-efficient of thermal expansion for the selected coating layer is $10.3 \times 10^{-6} °C^{-1}$[5], which is almost equal to that of the iron- and nickel-based alloys. Hence, the chosen coating layer is suitable for its utilization at the high-temperature areas and reduces the stress generation between the coating and the substrate material during thermal cycles. The SEM image of the Cr_3C_2-NiCr powder particle is shown in Figure 8.4.

Based on the above-mentioned properties of the selected cermet particle powder (Cr_3C_2-NiCr), it is evident that this could be a suitable coating layer for the high-temperature materials. Researchers have found that the use of carbide-based cermet coatings provide excellent erosion resistance on the material surface [17]. But, the researchers are not sure, what percentage of the carbide addition would result in maximum erosion resistance. This creates confusion in selection of the carbide-based additions for optimal erosion resistance. To study this, two types of carbide-based cermet with different compositions are produced, Cr_3C_2-20NiCr and Cr_3C_2-25NiCr, respectively. Among them, Cr_3C_2-20NiCr cermet, exhibits excellent wear resistance both at elevated temperature and room temperature and Cr_3C_2-25NiCr coatings show excellent wear- and corrosion-resistance property [16, 18].

With respect to the particle size of the cermet is concerned, smaller the particle size of the coating material, better are the characteristics of the coated layer. If the

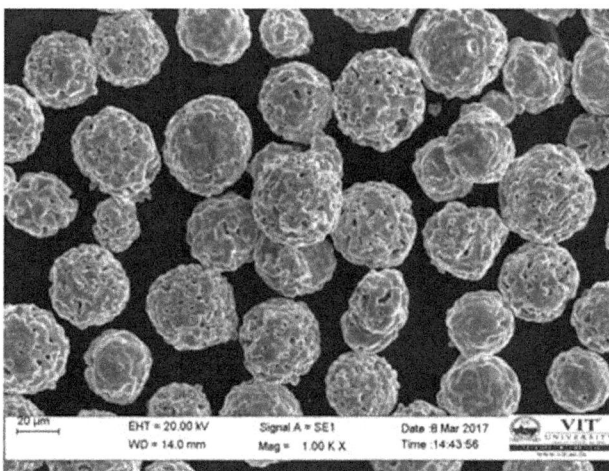

FIGURE 8.4 SEM image of the Cr_3C_2-NiCr powder particle [16].

size of the selected coating powders is reduced to a scale of micro level or to a nano level, the coating technology in the thermal spray processes would produce a novel nanostructured or fine-structured coating layer. These layers show enhanced microstructural, mechanical and functional properties. Also, these layers show very low porosity when compared to the layer coated with coarse particles. But the production of very small particles for coating would be a challenging one.

Hitherto, details of the coating materials and coating processes have been discussed. Details of the substrate onto which the coatings are deposited will be covered in the following section. The commonly used materials for high-temperature applications are iron-, nickel- and cobalt-based alloys. Coating processes are carried out on the said materials to enhance the service life. Nickel-based alloys have received higher attention and attraction in the industrial and aerospace applications.

8.5   NICKEL-BASED SUPERALLOYS

Superalloys are generally known as heat-resistant or high-temperature alloys, which are categorized as nickel-based, iron-based, cobalt–based superalloys, etc. Out of the said superalloys, Nickel-based superalloys are most predominantly used in hazardous and extreme applications like aircraft engines, turbochargers, power plants, gas turbines, etc., because of their superior properties like high-temperature corrosion resistance, creep resistance, oxidation resistance, fatigue resistance, etc. There are many grades of nickel-based superalloy used at different places based on application requirements. Some of the commonly used nickel-based superalloys are, Inconel 718, Inconel 625, Inconel 825, Inconel 100, Nimonic C-263, Nimonic – 80A, Nimonic – 105, Hastelloy X, Hastelloy C, etc. Components/systems, where nickel-based superalloys come into picture are illustrated in Figure 8.5 [19].

Though the materials exhibit high performance levels, the low quality fuel usage at the site of application results in the release of impurities like sodium, vanadium,

FIGURE 8.5   Applications of nickel-based superalloy [19].

sulphur, etc. These impurities gets settled down in the form of salts like Na_2SO_4, V2O5 and NaCl on the surface of the material, resulting in deterioration of the material life at the site of interest. In order to overcome the said shortcoming and to prevent the effect of salt layer on the material surface, coating is done over the material substrate to enhance its service life. Many researchers have used various coating techniques like cold spraying technique, physical vapour deposition, HVOF coating, etc. [20–22], to coat the nickel-based superalloy for its increased service efficiency. The HVOF coating is the most widely used coating technique for superalloys and a detailed discussion of it is given in the next section.

8.5.1 HVOF COATING OVER NICKEL-BASED SUPERALLOY

HVOF is one of the thermal spraying techniques used for coating materials with less porosity, lesser decarburization, more hardness and more bonding strength than any other thermal spraying technique. With these advantages, people started coating material substrate with various powder particles on Ni-based superalloy (Superni 75). The objective of coating powders on the surface is to protect the material from oxidation caused by various salt compounds like NaCl, Na_2SO_4, etc. NiCrBSi powders are found very effective in reducing the oxidation and increases the corrosion resistance on the selected nickel-based superalloy at 900°C molten salt environment [23].

Instead of studying the coating effect on a single material, it is effective to study the single coating performance on multiple materials. IN 100, M247LC SX, CMSX-4 nickel-based superalloy was coated with NiCoCrAlYTa powder through HVOF process at 1100°C and studies were carried out on microstructural and chemical evolution [24]. Another Ni-based superalloy named Inconel 792 is coated with MCrAlY powders through HVOF process. On performing the oxidation study, the diffusion of Al particles into the material changes the microstructure, which in turn enhances the performance of the work material at extreme conditions [25].

Study on the performance of different coatings with various powder particles on the 19.5Cr-3Fe-0.3Ti-0.1C-Ni material substrate was also conducted. The different powder particles used were Cr_3C_2-NiCr, Ni-20Cr, NiCrBSi and Stellite-6 and were coated by the HVOF process. On the coated material surface, hot corrosion studies are conducted in the molten salt environment at 900°C. Out of the selected coating material, Ni-20Cr exhibits maximum corrosion resistance than other coatings. The maximum hot corrosion resistance is attributed to the oxide and spinel formation of nickel and chromium [26, 27].

Comparison of cermet coating (Cr_3C_2-25NiCr) with metallic coating (NiCrMoNb) is conducted by coating it on the surface of alloy 80A at 900°C and studied in molten salt environment. The results proved that the presence of chromium and nickel reduced the corrosion capability of the material in the molten salt environment [16]. The corrosion kinetics of the Superni 600 material is given in Figure 8.6 [28].The plot inferred that the weight gain values of Superni 600 material is found to be higher when compared to the coated material. Out of the two coating layers, NiCrbSi coating shows higher protection with lesser weight gain/area, than the Stellite 6 coating.

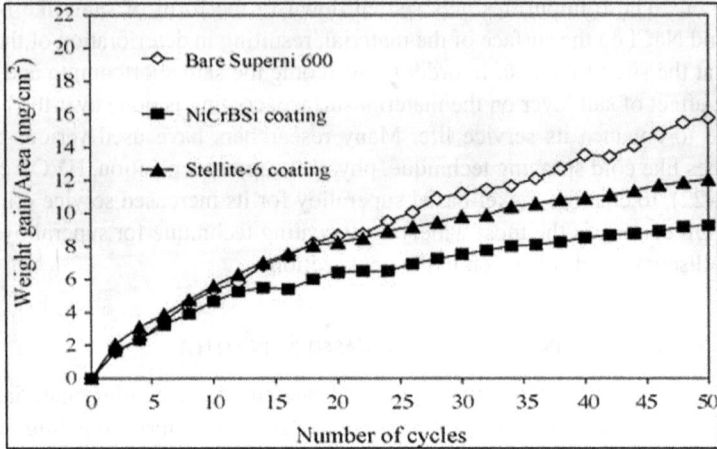

FIGURE 8.6 Plot showing weight gain/area vs number of cycles [28].

Hence, the coating of nickel-based superalloy through the HVOF process would result in excellent resistance in hot corrosion and oxidation of the material on its surface. The various types of characterization (microstructure, porosity, etc.) available for the HVOF-coated material is discussed in detail in the next section. And also, the effect of hot corrosion behaviour of the coated material on the nickel-based superalloy is discussed in the next section.

8.6   HOT CORROSION BEHAVIOUR OF NI-BASED SUPERALLOYS

8.6.1   COATING THICKNESS

The coating thickness has a strong influence on preventing the corrosion and erosion rate of the substrates. The coating thickness was measured using the inverted optical microscope embedded with image analysis software (Serial No: 1034509 and Make: Metascope). Too thick a coating allows the electrolyte to easily penetrate into the substrate due to the generation of stresses and crack development during the coating deposition. Too thin a coating also allows the electrolyte to go through easily due to its low thickness. Hence, the coating thickness is optimized, with the optimum value lying in the range 300–500 μm [29]. The Ni-based substrates are often coated with widely available Cr_3C_2-25NiCr powder. The coating powder has a particle size in the range 15–45 μm with a spheroidal shape which is manufactured through agglomeration and sintering process. The coating is performed all over the surfaces with the uniform thickness. Sidhu et al. [27] studied the hot corrosion behaviour of HVOF-coated Superni 600 in Na_2SO_4-60%V_2O_5 environment at 900°C. The thickness of Cr_3C_2-25NiCr coating on Superni 600 is 290 μm. Sreenivasulu and Manikandan [6] also reported studies on air-plasma-sprayed Cr_3C_2-25NiCr coating over alloy 80A. The average thickness of the coating is 160 μm.

8.6.2 Coating Porosity

The open type of porosity makes the sample highly susceptible to penetration in molten salt condition. The porosity of the coating was determined as per ASTM B276 using material plus image analysis software 4.2. The thermal conductivity of the coating material has significant effects on porosity size, and it is inversely proportional to the porosity. Further, the denser coating reduces the corrosion loss [30]. The porosity values should be in the range of 1–3.5% for Cr_3C_2-25NiCr coating, and this value needs to be maintained during the coating of materials used for power plant application. The high kinetic energy of Cr_3C_2-25NiCr powder develops low-porosity coating (dense coating) which are supposed to give better corrosion resistance than porous coating. The corrosion species easily penetrate into the coating through pores and cause severe damage in case of porous coating. Hence, the rate of corrosion is higher for low porosity-coating than high-porosity coating [31, 32].

8.6.3 Coating Bond Strength

The bond strength of the coating was measured using universal testing machine controlled by advanced microprocessor (Model No: 5969, Make: Instron) as per ASTM C633. The testing was performed by mating the uncoated and coated substrate with HTK ultra-bond epoxy glue with the cylindrical sample size of 25.4×38.1 mm [16].

8.6.4 Coating Microhardness

The performance of Cr_3C_2-25NiCr coating highly depends on the microhardness properties. The coating microhardness was measured on Vickers scale as per ASTM E384 [33, 34]. It is observed that Cr_3C_2-25NiCr coating gives high microhardness due to existence of hard Cr_3C phase in NiCr matrix [6]. Besides, the high cohesive strength, density of individual splats and high kinetic energy obtained by Cr_3C_2-25NiCr powder significantly increase the microhardness [32, 35].Also, the non-uniform variation in the microhardness over the cross section of the coating is primarily attributed to the changes in the microstructure of the coatings along the cross-sectional area. The existences of the oxidized, un-melted, semi-melted and melted particles in the coated substrates mainly cause these changes in the microstructure of the coating [36].

8.6.5 Visual Observation

The coating's appearance is determined through visual observation, which broadly reveals corrosion behaviour of the substrate, after each cycle of hot corrosion analysis. During this analysis, the changes in the colour of the substrates, existence of spallation, irregular scale formation, porosities, white spots and peeling off in the coated substrate are noted. Sreenivasulu and Manikandan [6] reported the changes in the colour of the substrates from dark grey to brown for air-plasma-sprayed Cr_3C_2-25NiCr coating over alloy 80A in the $Na_2SO_4 + 60\%$ V_2O_5 environment. The authors also pointed out the minimum spallation and porosities in Cr_3C_2-25NiCr coating compared to other powder coatings.

8.6.6 CORROSION KINETICS

The rate of corrosion is determined through corrosion kinetics analysis. The change in the mass of the sample is evaluated after every cycle using a Metler Toledo electronic balance with ± 0.01 mg of accuracy. The difference in the weight between cycles reveals weight gain or weight loss of hot corroded substrates. Further, these results are plotted between a number of cycles and weight gain (mg/cm^2) or weight gain square. From this plot, the parabolic (k_p) rate is determined through linear least square methods as given in Equation 8.1. A high K_p value indicates severe hot corrosion, whereas a low K_p value is indicative of mild corrosion. Guangyan et al. [37] reported the hot corrosion demeanour of Ni and Ni-10Cr-Al alloy coated with a molten salt film of 75% Na$_2$SO$_4$ and 25% NaCl condition at 900°C for 24 hrs. The alloy Ni-10Cr-Al shows less weight gain compared to Ni.

8.6.7 METALLURGICAL STUDIES

The microstructural control of the coating (of the Ni-based substrate) is as important as the elemental composition of the coating. In HVOF coating techniques, substantial changes in microstructure and compositions are possible when exposing the substrate materials to high-temperature conditions. Besides, HVOF coating causes voids and oxides, primarily originating from the spraying process at the splat boundaries, where corrosive environment easily attacks the substrate material. Hence, the coating has to be sprayed carefully over the substrate material, so that corrosion caused by molten salts does not penetrate into the coated layer. The HVOF spray coating develops relatively less porosity and dense microstructure due to the high impact, low temperature and high velocity of the powder particles over the substrates during coating. The HVOF coating's microstructure shows a typical splat morphology with different boundaries because of the addition and re-solidification of molten droplets. The splat is an individual impacted particle/droplet, which is developed when accelerated, molten droplets impact the surface of the substrate or already created coating material.

The microstructure is majorly lamellae with splat boundaries, and it is aligned through the substrate surface. It also shows small amounts of un-melted particles, oxide inclusions and voids which are the distinct characteristics of the HVOF process. The oxides are developed because of in-flight oxidation in the existing powder materials [32]. Further, the high density Cr$_3$C$_2$-25NiCr powder develops some pores (black spot) in the coating.

8.7 CASE STUDIES ON HOT CORROSION CHARACTERISTICS OF CR$_3$C$_2$-25NICR COATING OVER ALLOY 80A

Alloy 80A is a Ni-based superalloy mainly produced for high-temperature environment. This alloy primarily consists of Ni and Cr elements, and it is strengthened by precipitation mechanism with the addition of Ti, Al and C. It is widely used in steam turbine and gas turbine components like a disc, rings and power plants' blades. The main issues associated with alloy 80A are hot corrosion, oxidation and erosion due to the development of low melting constituents from the mixture of Na, S and V available in the fuels used in gas turbine applications. Therefore, the developments of these low melting constituents accelerate the corrosion rate when exposed to high temperatures.

In the present study, the effects of Cr_3C_2-25NiCr powder coating on alloy 80A through the HVOF spray technique are studied in aggressive molten salt conditions $(Na_2SO_4 + 60V_2O_5)$ at 900°C. The results are characterized through measurement of thickness, porosity, microhardness, corrosion kinetics and metallurgical properties using SEM/EDS observation.

The coating thickness plays a significant role in protecting the substrate material from erosion and corrosion and is measured through microstructure. The HVOF-sprayed Cr_3C_2-25NiCr coating shows a thickness of 200 μm. The porosity percentage of the HVOF coating should be kept lesser than 1% for getting desired outcomes. The optical microstructure of alloy 80A coated with HVOF-sprayed Cr_3C_2-25NiCr coating as displayed in Figure 8.7. It is observed from this study that HVOF-coated substrate shows the porosity of 0.6%, which is well within 1%. Besides, it also revealed that denser HVOF-sprayed Cr_3C_2-25NiCr coating.

The surface morphology of uncoated and Cr_3C_2-25NiCr-coated alloy 80A exposed to molten salt condition is shown in Figure 8.8. The colour of uncoated alloy 80 turned into blackish-grey from metallic colour after a few cycles; then, there is no such changes in the colour till the end of the 50th cycle. Besides, the development of white colour spots is observed after the 2nd cycle. Whereas, the Cr_3C_2-25NiCr-coated substrate is turned into grey after the 10th cycle and maintains the same colour until the end of the 50th cycle. The surface morphology ofCr_3C_2-25NiCr-coated substrate is revealed to be crack-free.

Further, no peeling-off and significantly less spallation are observed on both substrates [6, 16]. The Cr_3C_2-25NiCr coating reveals the microhardness of 853±30 $HV_{0.3}$. Therefore, the microhardness value of coated (Cr_3C_2-25NiCr) substrate confirmed 189% higher hardness than alloy 80A. The high kinetic energy of

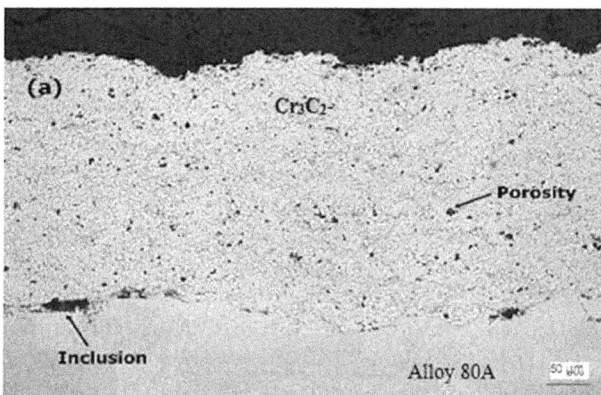

FIGURE 8.7 Optical microstructure of alloy 80A-coated HVOF-sprayed Cr_3C_2-25NiCr coating [16].

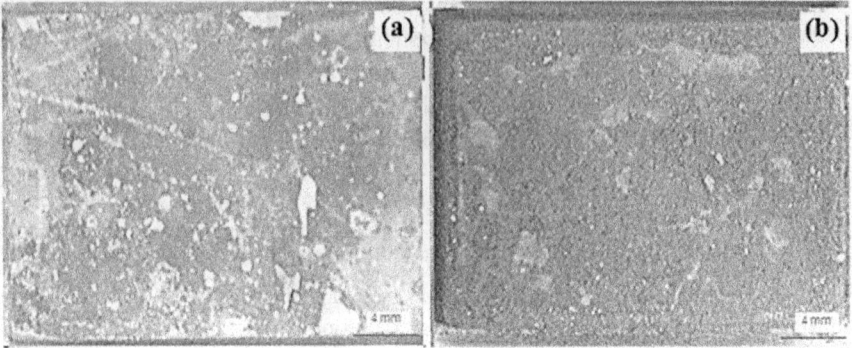

FIGURE 8.8 Surface morphology of alloy 80A exposed to molten salt: (a) uncoated alloy 80A, (b) Cr_3C_2-25NiCr-coated [16].

Cr_3C_2-25NiCr powder and the occurrence of Cr_3C_2inNiCr matrix are the main reasons for this enhanced microhardness. The results of Cr_3C_2-25NiCr coating follow the same trend with other studies reported by Żórawski and Kozerski [38] and Murthy et al. [39]. The alloy 80A substrate coated with HVOF spray technique shows the bond strength of 71 MPa and the observed strength is in agreement with work reported by Somasundaram et al. [40] and Sreenivasulu and Manikandan [16].

The corrosion kinetic plot is drawn between a number of cycles and weight gain square for the coated substrate and base substrate, as displayed in Figure 8.9. The graphs follow the parabolic rate law for both coated and uncoated substrates. During

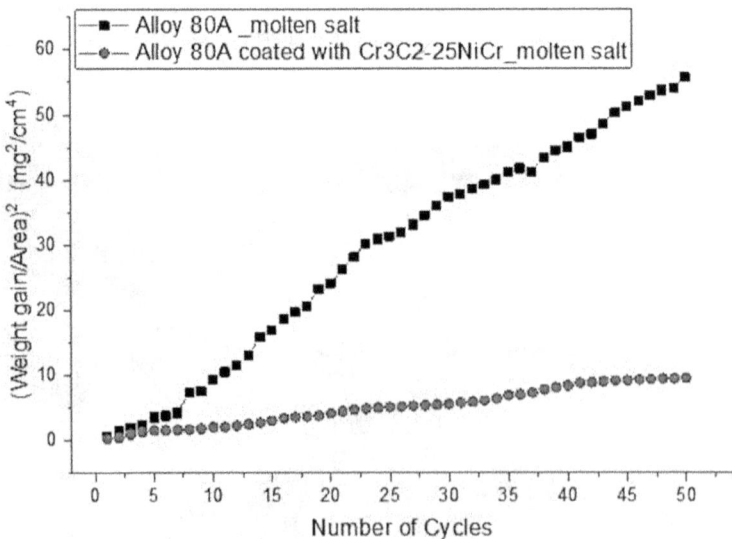

FIGURE 8.9 Plot of number of cycles vs weight gain square for uncoated alloy 80A and HVOF-coated Cr_3C_2-25NiCr exposed to molten salt.

the first 20–25 cycles, weight gain increases drastically due to the rapid development of oxides. After that, the weight gain becomes gradual till 50th cycle. The Kp value was calculated from the parabolic rate law (Equation 8.1).

$$\left(\frac{\Delta w}{S_A}\right)^2 = K_p \times t \qquad (8.1)[41, 42]$$

where S_A is the surface area (cm^2), K_p the parabolic constant (mg^2/(cm^4.s)), t the time (s), Δw the difference of final and initial weight of the substrate (mg).

This study observed that the cumulative weight gain of the Cr$_3$C$_2$-25NiCr-coated sample is 3.06 mg/cm^2, whereas the uncoated sample shows 7.51 mg/cm^2. Further, the K_p value of Cr$_3$C$_2$-25NiCr-coated substrate is 0.00000436 mg^2/(cm^4.s), whereas uncoated substrate shows 0.0000278 mg^2/(cm^4.s). Therefore, the HVOF-sprayed Cr$_3$C$_2$-25NiCr-coated alloy 80A has comparatively low K_p and weight gain than bare alloy 80A, when exposed to molten salt condition at 900°C. The low weight gain in HVOF carbide coating is mainly attributed to the high level of Cr in the composition of Cr$_3$C$_2$-25NiCr powder. The high level of Cr$_2$O$_3$ oxides prevents the sulphidation process by reducing the ion level of Na$_2$SO$_4$. Figure 8.10 shows the SEM/EDS analysis of uncoated and Cr$_3$C$_2$-25NiCr-coated alloy 80A exposed to the molten salt [16]. The sponge nodule shape is observed for uncoated substrate (Figure 8.10a), whereas

FIGURE 8.10 SEM/EDS analysis of alloy 80A exposed to molten salt: (a) uncoated alloy 80A, (b) Cr$_3$C$_2$-25NiCr-coated [16].

the Cr_3C_2-25NiCr-coated alloy 80A revealed a combination of sponge nodules and flake scales (Figure 8.10b).

Around 400°C, the development of NiO phase (Equation 8.2) starts on the substrate and Cr_2O_3 (Equation 8.3) forms between 500°C and 600°C range under molten salt deposition (Na_2SO_4 + 60 V_2O_5). The above oxides are effectively retained in HVOF-coated alloy 80A due to the high level of Cr and Ni formed on the coated alloy 80A than the uncoated substrate. Therefore, below 900°C, Na_2SO_4 dissolves as NiO oxide (Equations 8.4 and 8.5) and also forms $NaNiO_2$ (Equation 8.6) and $NiSO_4$ (Equation 8.7), which has a melting point of 840°C. The XRD plot of coated and uncoated samples of alloy 80A are illustrated in Figure 8.11. The thin layer's development over the alloy 80A through HVOF coating act as a hindrance to the penetration of corrosion products such as $NaNiO_2$ which improves the hot corrosion resistance of alloy 80A.

Below 900°C,

$$Ni + O = NiO \tag{8.2} [43]$$

$$4Cr + 3O_2 = 2Cr_2O_3 \tag{8.3} [43]$$

$$Na_2SO_4 = Na_2O + SO_3 \tag{8.4} [6]$$

$$4Ni + Na_2SO_4 = 3NiO + NiS + Na_2O \tag{8.5} [44, 45]$$

$$2NiO + Na_2O + \frac{1}{2}O_2 = 2NaNiO_2 \tag{8.6} [6]$$

$$NiO + SO_3 = NiSO_4 \tag{8.7} [6]$$

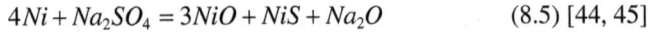

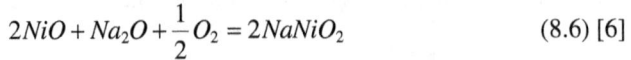

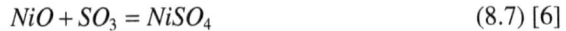

The rate of corrosion of Ni-Cr alloy is better than Cr-free Ni in the fused salt condition. The reaction of V_2O_5 and Na_2SO_4 at 900°C, form $NaVO_3$ (Equation 8.8), and

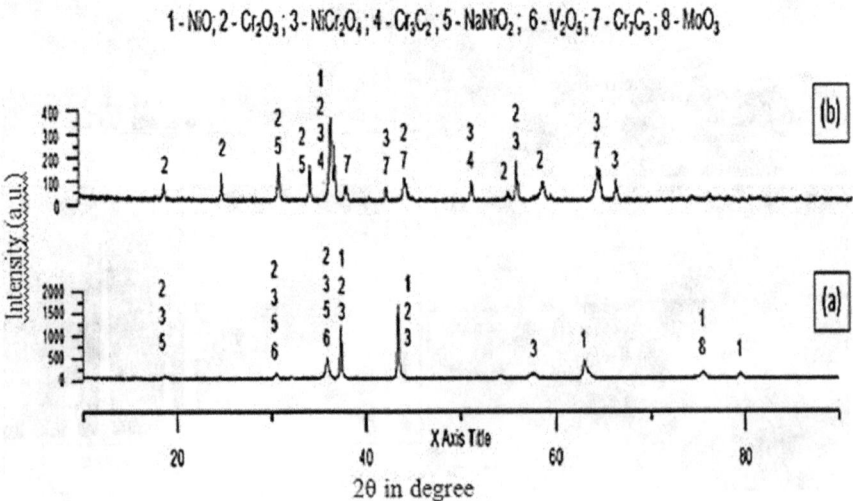

FIGURE 8.11 XRD of alloy 80A exposed to molten salt: (a) uncoated alloy 80A, (b) Cr_3C_2-25NiCr-coated.

melts at 610°C, as pointed out by Kolta et al. [46]. At 900°C, the formed $NaVO_3$ is entirely in a molten liquid state that attacks the existing protective oxides layer like NiO and Cr_2O_3, severely damaging the substrate. The developed $NaVO_3$ reacts with Cr_2O_3 protective oxides and forms Na_2CrO_4 (Equation 8.9) and $CrVO_4$ (Equation 8.10), which evaporate at 900°C, as pointed by Fryburg [47].

At 900°C,

$$Na_2SO_4 + V_2O_5 = 2NaVO_3 + SO_2 + \frac{1}{2}O_2 \qquad (8.8)\ [45]$$

$$Cr_2O_3 + 4NaVO_3 + \frac{3}{2}O_2 = 2Na_2CrO_4 + 2V_2O_5 \qquad (8.9)\ [6]$$

$$Cr_2O_3 + 2NaVO_3 = 2CrVO_4 + Na_2O \qquad (8.10)\ [45]$$

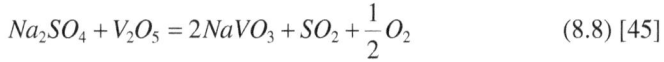

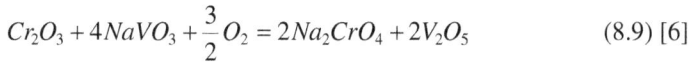

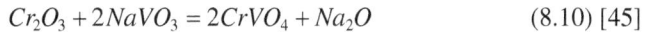

The development of non-protective Na_2CrO_4 salt evaporation causes Cr depletion in uncoated substrates under molten salt condition. This results in decreased hot corrosion resistance of alloy 80A compared to HVOF-coated alloy 80A. The presence of high Cr in HVOF coating compensates these Cr depletions. Goebel and Pettit [48] discussed the salt of fluxing mechanism. Based on this, protective oxides scale on the substrate is dissolved due to the formation of alloying elements from $NiSO_4$, V_2O_5 and MoO_3. This kind of loss in the protective scale of the hot corroded substrate is called acidic fluxing. The occurrence of these acidic fluxing over the substrate alloy 80A in a molten salt condition significantly improves the weight gain and K_p value. Whereas, HVOF-sprayed Cr_3C_2-25NiCr-coated alloy 80A controls these salt fluxing mechanisms by retaining the protective oxides throughout the hot corrosion processes.

8.8   SUMMARY

The chapter discusses the effect of Cr_3C_2-25NiCr coating on the hot corrosion behaviour of nickel-based superalloys. In addition, the chapter also reviews the different types of coatings tried out on superalloys for high-temperature applications and their effect on hot corrosion. The techniques for production of coating materials in the form of powders have been described. The different mixed salt environments that are encountered during service have been covered. Important attributes relating to coatings such as porosity, thickness, microstructure and mechanical properties have been covered in some detail. A specific study on alloy 80A alloy with Cr_3C_2-25NiCr coating, coated through the HVOF process, is discussed in the latter part of the chapter. Effect of the coating on hot corrosion behaviour of alloy 80A in an aggressive $Na_2SO_4 + 60V_2O_5$ salt environment at 900°C was studied and the findings are reported in the form of a case study. The HVOF spray Cr_3C_2-25NiCr-coated alloy shows lower K_p value (0.00000436 mg^2/(cm^4.s)), compared to that for uncoated alloy (0.0000278 mg^2/(cm^4.s)). The high presence of protective oxides such as Cr_2O_3, $NiCr_2O_4$ (spinel) and NiO at the surface in case of the coated alloy 80A improves its hot corrosion resistance compared to that of uncoated material.

REFERENCES

1. Jegadeeswaran N, Udaya Bhat K, Ramesh MR. Improving hot corrosion resistance of cobalt based superalloy (Superco-605) using HVOF sprayed oxide alloy powder coating. *Trans Indian Inst Met* 2015; 68: 309–316. doi:10.1007/s12666-015-0605-x.
2. Rapp RA. Hot corrosion of materials: A fluxing mechanism? *Corros Sci* 2002; 44: 209–221.
3. Muthu SM, Arivarasu M. Oxidation and hot corrosion studies on Fe-based superalloy A-286 pulsed current GTA weldments in gas turbine environment. *Mater Res Express* 2019; 6. doi:10.1088/2053-1591/ab49cb.
4. Sathishkumar M, Subramani P, Natesh M, Venkateshkannan M, Arivazhagan N, Manikandan M. Effect of hot corrosion demeanour on aerospace-grade Hastelloy X made by pulsed and constant current arc welding in molten salts at 820°C. *IOP Conf Ser Mater Sci Eng* 2020; 912: 1–14. doi:10.1088/1757-899X/912/3/032060.
5. Chatha SS, Sidhu HS, Sidhu BS. Characterisation and corrosion-erosion behaviour of carbide based thermal spray coatings. *J Miner Mater Charact Eng* 2012; 11: 569–586. doi:10.4236/jmmce.2012.116041.
6. Sreenivasulu V, Manikandan M. High-temperature corrosion behaviour of air plasma sprayed Cr3C2-25NiCr and NiCrMoNb powder coating on alloy 80A at 900°C. *Surf Coatings Technol* 2018; 337: 250–259. doi:10.1016/j.surfcoat.2018.01.011.
7. Ahmad Z, Khan AU, Farooq R, Saif T, Mastoi NR. Mechanism of corrosion and erosion resistance of plasma-sprayed nanostructured coatings. *High Temp Corros* 2016: 123–146. doi:10.5772/64316.
8. Lashmi PG, Ananthapadmanabhan PV, Chakravarthy Y, Unnikrishnan G, Aruna ST, Ananthapadmanabhan PV, et al. Hot corrosion studies on plasma sprayed bi-layered YSZ/La2Ce2O7 thermal barrier coating fabricated from synthesized powders. *J Alloys Compd* 2017; 711: 355–364. doi:10.1016/j.jallcom.2017.04.022.This.
9. Jithesh K, Arivarasu M. Comparative studies on the hot corrosion behavior of air plasma spray and high velocity oxygen fuel coated Co-based L605 superalloys in a gas turbine environment. *Int J Miner Metall Mater* 2020; 27: 649–659. doi:10.1007/s12613-019-1943-1.
10. Tillmann W, Vogli E, Baumann I, Kopp G, Weihs C. Desirability-based multi-criteria optimization of HVOF spray experiments to manufacture fine structured wear-resistant 75Cr3C2-25 (NiCr20) coatings. *J Therm Spray Technol* 2010; 19: 392–408. doi:10.1007/s11666-009-9383-5.
11. Chandra Yadaw R, Kumar Singh S, Chattopadhyaya S, Kumar S, Singh RC Tribological behavior of thin film coating: A review. *Int J Eng Technol* 2018; 7: 1656–1663. doi:10.14419/ijet.v7i3.11788.
12. Wuxi Z, Kesong Z, Chunming D, Keli Z, Yuxi L. Hot corrosion behavior of HVOF-sprayed Cr3C2- WC-NiCoCrMo coating. *Ceram Int* 2017; 43: 9390–9400.
13. Zhang SH, Cho TY, Yoon JH, Fang W, Song KO, Li MX, et al. Characterization of microstructure and surface properties of hybrid coatings of WC-CoCr prepared by laser heat treatment and high velocity oxygen fuel spraying. *Mater Charact* 2008; 59: 1412–1418. doi:10.1016/j.matchar.2008.01.003.
14. Sivakumar G, Banerjee S, Raja VS, Joshi SV. Hot corrosion behavior of plasma sprayed powder-solution precursor hybrid thermal barrier coatings. *Surf Coatings Technol* 2018; 349: 452–461. doi:10.1016/j.surfcoat.2018.06.021.
15. Mannava V, Rao AS, Paulose N, Kamaraj M, Kottada RS. Hot corrosion studies on Ni-base superalloy at 650°C under marine-like environment conditions using three salt mixture (Na_2SO_4 + NaCl + $NaVO_3$). *Corros Sci* 2016; 105: 109–119. doi:10.1016/j.corsci.2016.01.008.

16. Sreenivasulu V, Manikandan M. Hot corrosion studies of HVOF sprayed carbide and metallic powder coatings on alloy 80A at 900°C. *Mater Res Express* 2018; 6: 36519. doi:10.1088/2053-1591/aaf65d.

17. Subramani P, Padgelwar N, Shetty S, Pandit A, Sreenivasulu V, Arivazhagan N, et al. Hot corrosion studies on detonation-gun-sprayed NiCrAlY and 80Ni–20Cr coatings on alloy X22CrMoV12-1 at 600°C. *Trans Indian Inst Met* 2019; 72: 1639–1642. doi:10.1007/s12666-019-01567-6.

18. Goyal K, Goyal R. Improving hot corrosion resistance of Cr3C2–20NiCr coatings with CNT reinforcements. *Surf Eng* 2020; 36: 1200–1209. doi:10.1080/02670844.2019.166 2645.

19. INCONEL Alloy 718: Properties and applications. 2018. Corrotherm International

20. Silvello A, Cavaliere P, Rizzo A, Valerini D, Dosta Parras S, Garcia Cano I. Fatigue bending behavior of cold-sprayed nickel-based superalloy coatings. *J Therm Spray Technol* 2019; 28: 930–938. doi:10.1007/s11666-019-00865-1.

21. Shaolin L, Xiaoguang Y, Hongyu Q, Guoqiang X, Duoqi S. Influence of MCrAlY coating on low-cycle fatigue behavior of a directionally solidified nickel-based superalloy in hot corrosive environment. *Mater Sci Eng A* 2016; 678: 57–64.

22. Zhang H, Dong X, Chen S. Solid particle erosion-wear behaviour of Cr3C2-NiCr coating on Ni-based superalloy. *Adv Mech Eng* 2017; 9: 1–9. doi:10.1177/1687814017694580.

23. Sidhu TS, Prakash S, Agrawal RD. Hot corrosion behaviour of HVOF-sprayed NiCrBSi coatings on Ni- and Fe-based superalloys in Na_2SO_4-60% V_2O_5 environment at 900°C. *Acta Mater* 2006; 54: 773–784. doi:10.1016/j.actamat.2005.10.009.

24. Mora-García AG, Ruiz-Luna H, Mosbacher M, Popp R, Schulz U, Glatzel U, et al. Microstructural analysis of Ta-containing NiCoCrAlY bond coats deposited by HVOF on different Ni-based superalloys. *Surf Coatings Technol* 2018; 354: 214–225. doi:10.1016/j.surfcoat.2018.09.025.

25. Yuan K, Lin Peng R, Li XH, Johansson S, Wang YD. Some aspects of elemental behaviour in HVOF MCrAlY coatings in high-temperature oxidation. *Surf Coatings Technol* 2015; 261: 86–101. doi:10.1016/j.surfcoat.2014.11.053.

26. Sidhu TS, Prakash S, Agrawal RD. Evaluation of hot corrosion resistance of HVOF coatings on a Ni-based superalloy in molten salt environment. *Mater Sci Eng A* 2006; 430: 64–78. doi:10.1016/j.msea.2006.05.099.

27. Sidhu TS, Prakash S, Agrawal RD. Hot corrosion studies of HVOF sprayed Cr3C2–NiCr and Ni–20Cr coatings on nickel-based superalloy at 900°C. *Surf Coatings Technol* 2006; 201: 792–800. doi:10.1016/j.surfcoat.2005.12.030.

28. Sidhu TS, Prakash S, Agrawal RD. A comparative study of hot corrosion resistance of HVOF sprayed NiCrBSi and Stellite-6 coated Ni-based superalloy at 900°C. *Mater Sci Eng A* 2007; 445–446: 210–218. doi:10.1016/j.msea.2006.09.015.

29. Guilemany JM, Fernandez J, Delgado J, Benedetti A V, Climent F. Effects of thickness coating on the electrochemical behaviour of thermal spray Cr3C2 – NiCr coatings. *Surf Coatings Technol* 2002; 153: 107–113. doi:10.1016/S0257-8972(01)01679-6.

30. Lawrence J, Li L. Augmentation of the mechanical and chemical resistance characteristics of an Al_2O_3-based refractory by means of high power diode laser surface treatment. *J Mater Process Technol* 2003; 142: 461–465. doi:10.1016/S0924-0136(03)00626-5.

31. Singh B, Prakash S. Studies on the behaviour of stellite-6 as plasma sprayed and laser remelted coatings in molten salt environment at 900°C under cyclic conditions. *J Mater Process Technol* 2006; 172: 52–63. doi:10.1016/j.jmatprotec.2005.08.018.

32. Sidhu TS, Prakash S, Agrawal RD. Characterisations of HVOF sprayed NiCrBSi coatings on Ni- and Fe-based superalloys and evaluation of cyclic oxidation behaviour of some Ni-based superalloys in molten salt environment. *Thin Solid Films* 2006; 515: 95–105. doi:10.1016/j.tsf.2005.12.041.

33. Sathishkumar M, Bhakat YJ, Kumar KG, Giribaskar S, Oyyaravelu R, Arivazhagan N, et al. Investigation of double-pulsed gas metal arc welding technique to preclude carbide precipitates in aerospace grade hastelloy X. *J Mater Eng Perform* 2020. doi: 10.1007/s11665-020-05360-1.

34. Vigneshwara Kumaran TA, Nithin Joseph Reddy SA, Jerome S, Anbarasan N, Arivazhagan N, Manikandan M, et al. Development of pulsed cold metal transfer and gas metal arc welding techniques on high-strength aerospace-grade AA7475-T761. *J Mater Eng Perform* 2020; 29: 7270–7290. doi:10.1007/s11665-020-05240-8.

35. Verdon C, Karimi A, Martin J-L. A study of high velocity oxy-fuel thermally sprayed tungsten carbide based coatings. Part 1: Microstructures. *Mater Sci Eng A* 1998; 246: 11–24. doi:10.1016/S0921-5093(97)00759-4.

36. Matthews SJ, James BJ, Hyland MM. Microstructural influence on erosion behaviour of thermal spray coatings. *Mater Charact* 2007; 58: 59–64. doi:10.1016/j.matchar.2006.03.014.

37. Guangyan F, Zeyan Q, Jingyu C, Qun L, Yong S. Hot corrosion behavior of Ni-Base alloys coated with salt film of 75%Na$_2$SO$_4$+25%NaCl at 900°C. *Rare Met Mater Eng* 2015; 44: 1112–1115. doi:10.1016/S1875-5372(15)30077-1.

38. Żórawski W, Kozerski S. Scuffing resistance of plasma and HVOF sprayed WC12Co and Cr3C2-25 (Ni20Cr) coatings. *Surf Coatings Technol* 2008; 202: 4453–4457. doi:10.1016/j.surfcoat.2008.04.045.

39. Murthy JKN, Bysakh S, Gopinath K, Venkataraman B. Microstructure dependent erosion in Cr3C2–20(NiCr) coating deposited by a detonation gun. *Surf Coatings Technol* 2007; 202: 1–12. doi: 10.1016/j.surfcoat.2007.03.017.

40. Somasundaram B, Kadoli R, Ramesh MR. Hot corrosion behaviour of HVOF sprayed (Cr3C2–35 % NiCr) + 5 % Si coatings in the presence of Na$_2$SO$_4$–60% V$_2$O$_5$ at 700°C. *Trans Indian Inst Met* 2015; 68: 257–268. doi:10.1007/s12666-014-0453-0.

41. Sathishkumar M, Subramani P, Arivazhagan N, Gokulkumar K, Naiju C, Jerome S, et al. Hot corrosion demeanour of key-hole plasma arc welded aerospace grade hastelloy X in molten salts environment. *SAE Tech Pap* 2020; 2020-28–04. doi: 10.4271/2020-28-0422.

42. Sathishkumar M, Manikandan M. Hot corrosion behaviour of continuous and pulsed current gas tungsten arc welded Hastelloy X in different molten salts environment. *Mater Res Express* 2019; 6:126553. doi:10.1088/2053-1591/ab562a.

43. Sreenivas KS, Radhakrishnan VM. Oxidation and hot corrosion behaviour of Nimonic-75 superalloy. *Indian J Eng Mater Sci* 1998; 5: 295–301.

44. Li W, Liu Y, Wang Y, Han C, Tang H. Hot corrosion behavior of Ni–16Cr–xAl based alloys in mixture of Na$_2$SO$_4$–NaCl at 600°C. *Trans Nonferrous Met Soc China* 2012; 21: 2617–2625. doi:10.1016/s1003-6326(11)61100-x.

45. Salehi Doolabi M, Ghasemi B, Sadrnezhaad SK, Habibollahzadeh A, Jafarzadeh K. Hot corrosion behavior and near-surface microstructure of a "low-temperature high-activity Cr-aluminide" coating on inconel 738LC exposed to Na$_2$SO$_4$, Na$_2$SO$_4$ + V$_2$O$_5$ and Na$_2$SO$_4$ + V$_2$O$_5$ + NaCl at 900°C. *Corros Sci* 2017; 128: 42–53. doi:10.1016/j.corsci.2017.09.004.

46. Kolta GA, Hewaidy IF, Felix NS. Reactions between sodium sulphate and vanadium pentoxide. *Thermochim Acta* 1972; 4: 151–164. doi:10.1016/S0040-6031(72)80029-7.

47. Fryburg GC, Kohl FJ, Stearns CA. Chemical reactions involved in the initiation of hot corrosion of IN-738. *J Electrochem Soc* 1984; 131: 2985–2997. doi:10.1149/1.2115455.

48. Goebel JA, Pettit FS. Na$_2$SO$_4$-induced accelerated oxidation (hot corrosion) of nickel. *Metall Trans* 1970; 1: 1943–1954. doi:10.1007/BF02642794.

9 Evaluation of Microstructural and Dry Sliding Wear Resistance of Iron-Based SiC-Reinforced Composite Coating by HVOF Process

Wear Resistance of Fe-Based Coatings

C. Durga Prasad, Akhil Jerri, and M. R. Ramesh

CONTENTS

DOI: 10.1201/9781003213185-9

9.1 INTRODUCTION

At first glance, it might seem that the boiler and petrochemical applications have little in common. The components under these sectors are critical, among dozens of other industrial sectors, and face constant and challenging exposure to extreme conditions. These conditions include high temperatures and extremely abrasive and corrosive environments [1–3]. Humid environments, with the presence of suspended particles and chemical atmospheres require components with specific characteristics. Besides that, extreme environmental conditions can trigger the combined action of degassing mechanisms. The environment conditions where the component operates must also be observed [4–7]. A high wear resistance coating in ambient temperature conditions can perform poorly at high working temperatures [8]. To safeguard boiler tubes and petrochemical valves from such conditions, manufacturers use what is known as thermal spray coatings [9–11].

Thermal coatings utilize advanced material systems that can be sprayed on to metallic or composite surfaces [12]. Once in place, the coatings form an integral part of a critical component's exhaust heat management system and environmental protection shield. Critical for design engineers, coatings allow equipment to operate at higher operating temperatures [13–15]. For example, advanced thermal coatings can allow temperatures higher than the melting point of metal tubes in boiler applications. This is an important consideration as demand increases for efficient supply of steam/water through tubes that can operate at high temperatures while still delivering high durability and long lifetimes [16–20]. To be most effective, a thermal spray coating offers the following characteristics: higher deposition rate, thick layers, good mechanical bonding, no thermal distortion and no effects on substrate bulk properties [21–24].

Thermal sprayed Cr_3C_2-NiCr coatings to protect components under elevated temperature wear due to the high wear resistance imparted by the hard carbide particles and the high-temperature oxidation resistant nature of the Cr_2O_3 oxide were formed over both phases [9, 25–27]. These combinations of coatings were well established and accepted in boiler applications. But these coatings exhibit slight lower hardness and strength at high-temperature conditions. Also, nickel and chromium are little expensive to develop coatings for components having large area [16, 28–31].

In order to reduce the cost of secondary process, iron (Fe)-based powders can be sprayed on the surface of boiler tubes, petrochemical valves and heat exchangers by HVOF method [1, 4]. Fe-based powders are commercially available which is very economical and has similar advantages compared to other metallic and carbide-based powders [24, 25]. The iron-based metallic powder exhibits excellent physical, chemical and mechanical properties, such as high strength, high hardness, excellent wear resistance and corrosion resistance. Fe-based exhibits various feedstock's depending on the applications suitable powders need to select.

In this paper, 70% wt. Metco 41C and 30% wt SiC feedstock were mixed mechanically using ball milling and further deposited on ASTM-SA213-T11 boiler tube base metal through HVOF process. The coating was subjected to metallurgical and mechanical characterizations and then later high temperature as well as room condition sliding wear behavior was investigated using pin on disc apparatus.

9.2 EXPERIMENTAL PROCEDURE

ASTM-SA213-T11 grade was used as a substrate having $30 \times 30 \times 5$ mm size for the coatings. To ensure suitable adhesion of the coating, the substrates were grit-blasted and cleaned with ethanol prior to the coating application. Metco 41C iron-based and SiC powders were selected as feedstock. The chemical composition of Metco 41C is Fe 12Ni17Cr2.5Mo2.3Si0.03C. Metco 41C powder was produced by water-atomized technique with particle sizes from -106 to $+45\mu m$, whereas SiC exhibits -45 to $+15\mu m$ size particles. The powders in a ratio of (70%) Metco 41C and (30%) SiC were mixed mechanically using ball milling method. Further, the prepared powders were sprayed on the substrate. For the coating application, the HVOF diamond jet 2100 gun was used. The coating was deposited with the parameters given in Table 9.1.

During spraying, high gas fluxes and gas velocities were achieved. After exiting the nozzle, the gas flow was mainly directed toward the substrate. Near the substrate, this gas flow was strongly deflected. The strength of this impingement flow increases with higher gas velocities. However, the high gas velocity is advantageous for the application of dense coatings.

9.2.1 CHARACTERIZATION OF COATING

Prior to development of coating, powder morphology was examined through scanning electron microscope (SEM). Further, the developed coating samples were trimmed into size of $12 \times 12 \times 5$ mm using wire electro-discharge machining to accommodate samples for various characterization techniques. The coated samples were prepared by performing polishing on both surface and cross-section side of coating using emery papers. Then, the coating sample was taken for surface and cross-sectional microstructural analysis using SEM and its chemical composition was evaluated using energy dispersive spectroscopy (EDS). However, the identification of phases in coating was performed using X-ray diffraction (XRD). The surface roughness (Ra) of the coating was evaluated using surface roughness tester (SJ-201P, Mitutoyo, Japan). At 5 different locations on the coating, the test was performed and the average value was reported. Vickers microhardness of coating was estimated using OMNITECH microhardness tester. On the direction of cross section of coating and substrate, the microhardness was measured under a load of 300 g with duration

TABLE 9.1
HVOF Spray Parameters

Parameter	Value
Flow rate N_2	50 lpm
Flow rate O_2	250 lpm
Flow rate H_2	700 lpm
Powder feed rate	20 g/min
Spray distance	300 mm
Fuel pressure	6 KPa
Spray pass/thickness	13–15 microns

of 10 sec. Selected 10 various indentation spots were employed and the mean value was stated. According to ASTM B276-05 standard, porosity of the coating was investigated. To test the porosity of the coating, Biovis Materials Plus software, along with image analyzer facility was utilized. By considering 20 various spots on the coating region, the average porosity value was stated.

According to ASTM C135-96 standard, the density of coating was measured with the help of water immersion using pycnometer test. To determine the density, a small layer of coating was mechanically pealed from the sample. The pealed samples are in the range of 1–2 g, having minimum weight.

9.2.2 DRY SLIDING WEAR

According to ASTM G99-05, the wear test of coating and substrate were performed without supply of lubrication using pin on disc apparatus (TR-20LE-PHM 400-CHM 600, Ducom Instruments Pvt. Ltd., Bangalore, India). Samples were tested at ambient and at 200°C, 300°C temperature under normal loads of 10 and 20 N. The sliding velocity and sliding distance were kept constant of 1.3 m/sec and 3000 m, respectively. The coated sample steel disc was used as the counter body for wear tests. The system having inbuilt heating option which transfers heat to the sample attached in the holder and thermocouple was inserted to measure the temperature.

With the help of computer specimen, height loss was estimated. Further, the system generates data to calculate friction, wear rate and volume loss. Characterization of tested specimens like microstructural analysis was done through SEM to understand the wear mechanism. Similar steps were adopted in wear studies and reported [6, 14, 18].

9.3   RESULTS AND DISCUSSION

9.3.1 MICROSTRUCTURAL ANALYSIS

The morphologies of Metco 41 C and SiC were observed to be uniform from the SEM images as shown in Figure 9.1a and b. As noticed in (Figure 9.1a), Metco 41C feedstock exhibits dendritic structure whereas the particles of SiC was mostly spherical and partially distorted from the spheroidal and elliptical shape depicted in Figure 9.1b.

The cross-section SEM image of the 30% SiC-reinforced coating is depicted in Figure 9.2a. The mean thickness of 30%-reinforced SiC coating was 275–300 μm. Nevertheless of the evident distinction between the substrate and the coating, bonding was good, as seen from the micrograph as there were no crack observed along the interface, indicating strong bonding. Dense-packed structure was apparent from the micrograph as it can be detected that the splat units forming the layers are well mixed up, explaining the increase in the coating hardness.

Coating constituents were distributed uniformly with fully melted splats and there were minor signs of semi-melted particles trapped in the pores and voids in the cross section. SEM micrograph of the as-sprayed coating along the cross section shows the lamellar and uniform structure, bonded well with the substrate, indicating layer-by-layer coating with every pass of HVOF spray process.

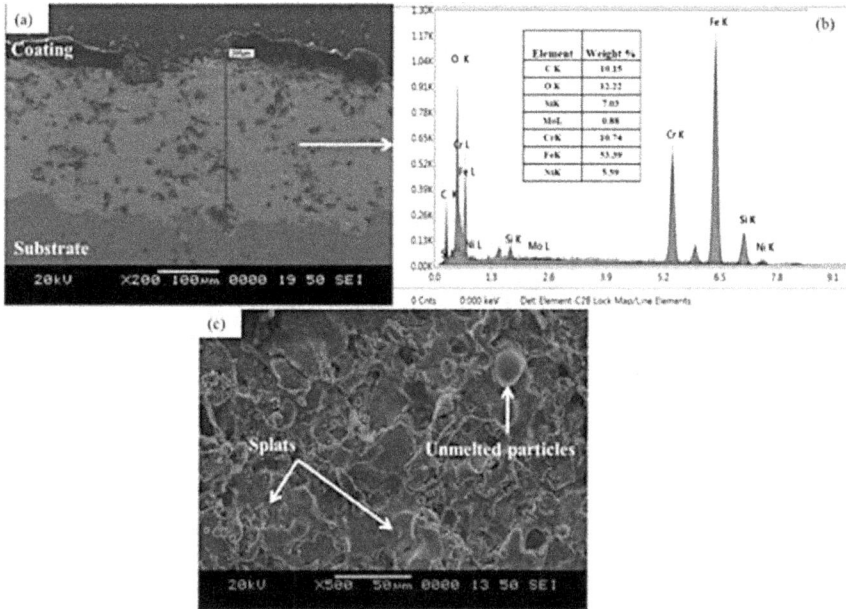

FIGURE 9.1 Morphology of powders: (a) Metco 41C, (b) SiC.

EDS analysis was performed on the coating surface of the selected region shown in Figure 9.2b. As observed from EDS pattern, strong peaks of Fe, Cr and O were generated. Oxide stringers were detected in coating area due to oxidation of inflight particles during deposition of coating as shown in Figure 9.1a. Coating elements were uniformly composed with C, Ni, Si and Mo. This results in the formation of carbides and oxide phases which imparts high-temperature strength.

Coating surface microstructure exhibited lamellar structure with presence of partial-melted particles shown in Figure 9.2c. Also, splats were formed with unusual size, marked on microstructure in Figure 9.2c. A few superficial voids were evident in the coating microstructure. The surface roughness value was indicated to be R_a = 8.232 µm. The density of the coatings was 5.910 g/cm³, as measured using pycnometer as per water immersion technique.

9.3.2 XRD ANALYSIS

Figure 9.3 presents the XRD plot of Metco 41C+SiC coating. The as-sprayed coating spectrum has a sharp crystalline structure with lower intensity. The pattern exhibits new carbide phases such as FeC and Cr_7C_3. At 42°, 50° and 75° showed FeC peak, whereas Cr_7C_3 peak was observed at 58°, 62° and 72°. In addition to these phases, SiC and $MoSi_2$ peaks were noticed in the coating. These phases strongly influence the coating properties such as hardness, high-temperature stability and wear resistance reported by several researchers [24, 25]. However, the presence of Cr_7C_3 has been confirmed by this analysis, which ensures that the decarburization of Cr_3C_2 has taken place.

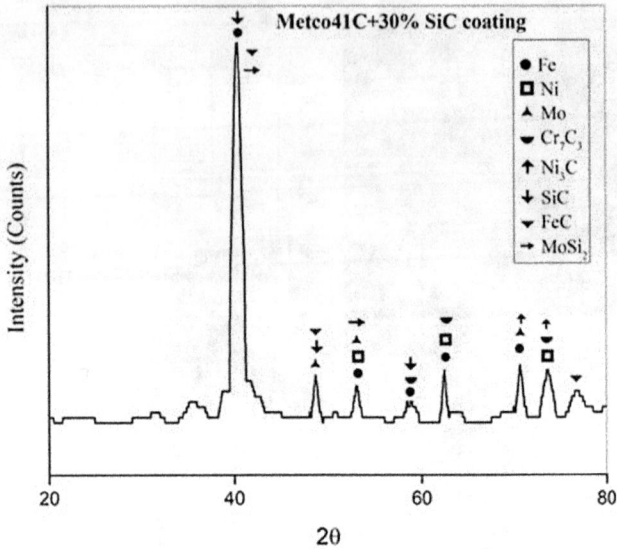

FIGURE 9.2 (a) Metco 41 C+SiC coating cross section, (b) compositional EDS of the coating, and (c) Metco 41 C+SiC coating surface.

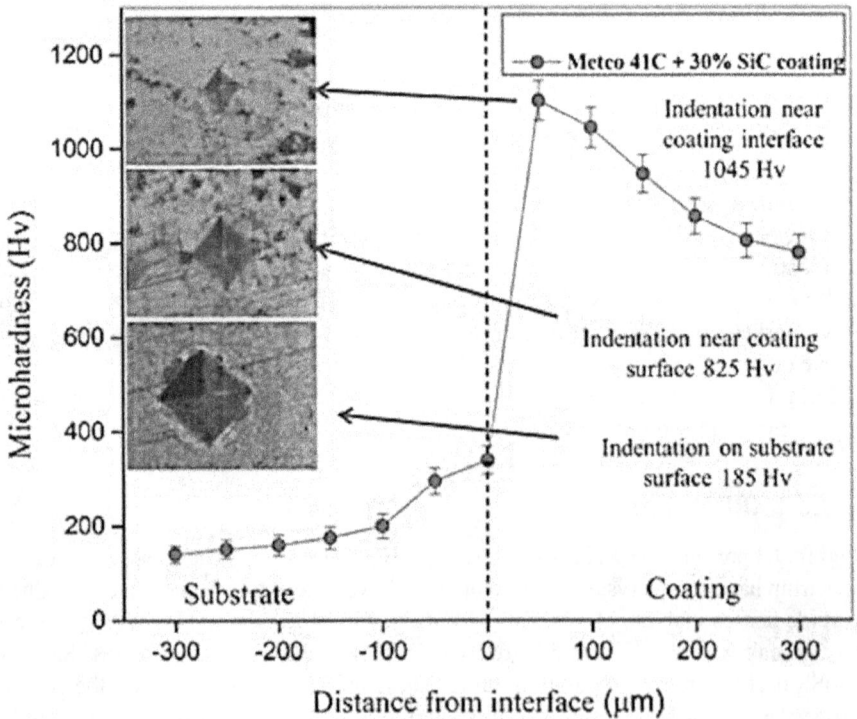

FIGURE 9.3 XRD diffractogram of HVOF coating.

9.3.3 MICROHARDNESS ANALYSIS

The microhardness plot of the coating is depicted in Figure 9.4. Coating hardness test was performed in transverse direction from substrate end to coating end. The SiC-reinforced coatings have exhibited the highest average microhardness of 1045 ± 30 Hv. Substrate hardness was in the range of 180–194 Hv. The average microhardness of the substrate was 189 Hv. The coating possesses heterogeneous structure results in variation of hardness across the thickness. Coating revealed higher hardness at near the interface location. The splats formed at this region exhibits greater compressive strength due to deposition of molten particles. This result in higher hardness due to compressive stress has induced into splats. Further, the developing of layers one after the other led to tensile stress produced in the coating, resulting in decrease in hardness. This significant variation in microhardness along the thickness of the SiC-reinforced coatings might be due to the distribution of the SiC hard phase in Fe-based alloy matrix. The Vickers hardness increases with increasing amount of SiC particles in the Fe-based coatings.

9.3.4 EXAMINATION OF WEAR MORPHOLOGY

The post-wear SEM images of Metco 41C+SiC coating are shown in Figure 9.5a–d. The images (Figure 9.5a–d) of sample with the wear conditions of 200°C and 300°C

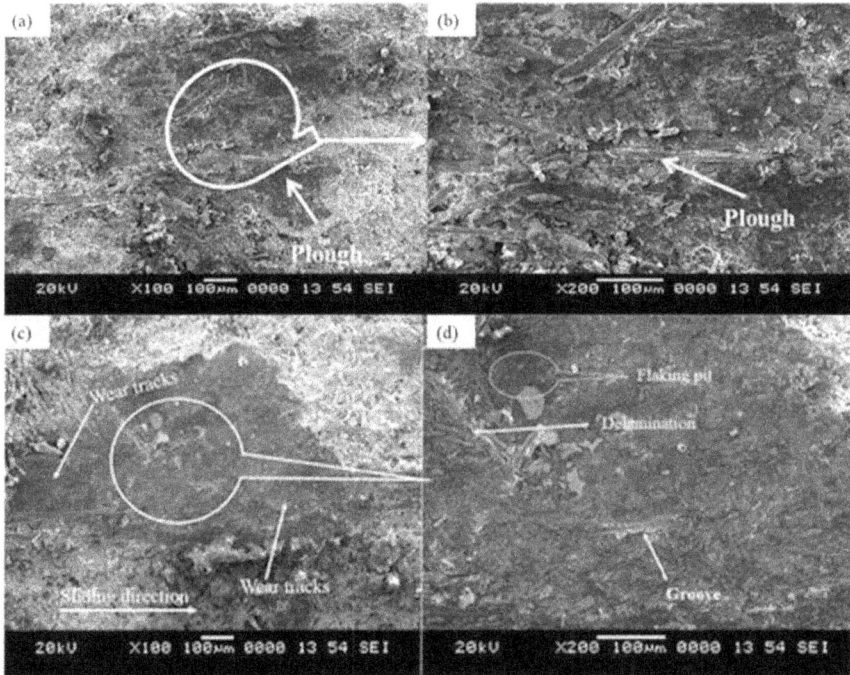

FIGURE 9.4 Microhardness plot of Metco 41C+SiC coating across the interface.

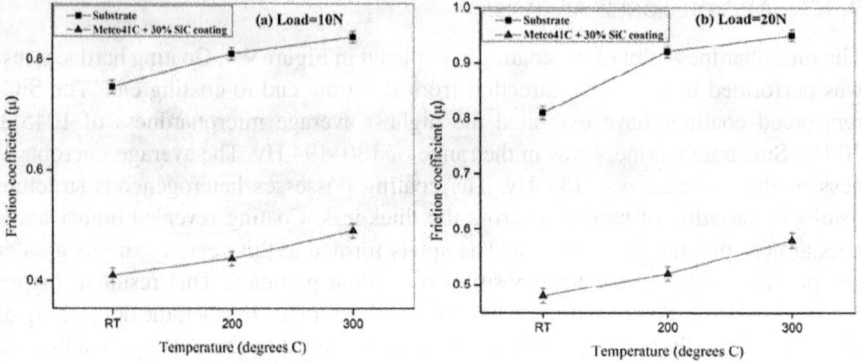

FIGURE 9.5   SEM wear scar micrographs of Metco 41C+SiCcoating under 20 N (a) 200°C (b) 300°C.

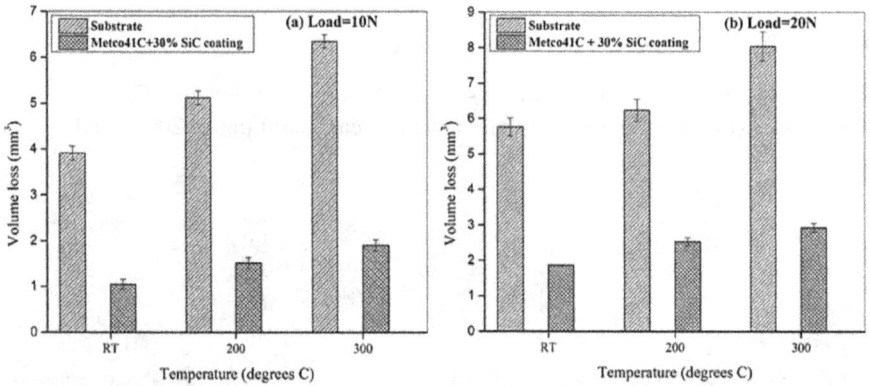

FIGURE 9.6   Frictional profiles of coating and substrate: (a) 10 N, (b) 20 N.

under 20N normal load were being used for analysis. The surface of the coating was smooth and flat, with a small amount of oxidative wear at 200°C under 20 N normal load. Some regions of the coating were peeled off, mainly due to adhesive wear, accompanied by a little oxidative wear as shown in Figure 9.5a–b. The delamination on the worn-out surface can be observed at 300°C under 20N normal load. There was a flaking pit produced besides the delamination on the top of the wear surface. A grooving produced at the center can also be detected as shown in Figure 9.5c–d. But this groove was not extended throughout the length of the surface, making it an intermittent groove.

9.3.5   FRICTION AND WEAR RATE

Figure 9.6 a and b depicts the plots of friction coefficient (μ) of coating and substrate with respect to testing temperature and loads. Substrate possesses low strength and hardness at all test conditions results in high coefficient of friction (COF) [21]. Also, at temperatures 200°C and 300°C, the substrate was exposed to severe oxidation, showing higher COF. It was noticed that even for an increase in normal load from 10 to 20 N, the COF of the substrate tends to increase drastically.

The SiC-reinforced coating revealed lower coefficient of friction at all the test temperatures and loads compared to the substrate. The coating exhibited 4 times lower COF under all performed conditions. As the load increases, the friction-induced temperature at the interface raises, which results in softening, leading to lower friction coefficient as shown in Figure 9.6a and b. The existence of SiC in Fe-based matrix coating acts as cushion effect against hard sliding disc, resulting in decrease in COF. On the surface of the coating, formation of minor scratches was noticed due to wear debris resulting in a slight variation of the friction curve. The frictional heat generated during abrasion leads to localized increase in the temperature, resulting in the formation of oxides of Fe, Si, Cr and Ni. This tribo-oxide film with low shear strength was responsible for the reduction in COF. Evaluated friction values of the substrate and the coating are recorded in Table. 9.2.

Figure 9.7a and b illustrates the coating and substrate volume loss graphs. The substrate registered higher wear volume loss and was observed to increase with rise in normal load due to poor surface properties against to sliding action. The surface layer of the substrate was completely deformed under 10 and 20 N normal loads and at 300°C temperature, the substrate experienced severe oxidation, as shown in Figure 9.7b. The Metco 41C+SiCcoating exhibited lower volume loss at all test

TABLE 9.2

Values of COF with Respect to Normal Load and Temperatures

Specimens	Room Temperature		200°C		300°C	
	10 N	20 N	10 N	20 N	10 N	20 N
Substrate	0.71	0.82	0.86	0.93	0.90	0.96
Coating	0.41	0.48	0.44	0.52	0.49	0.58

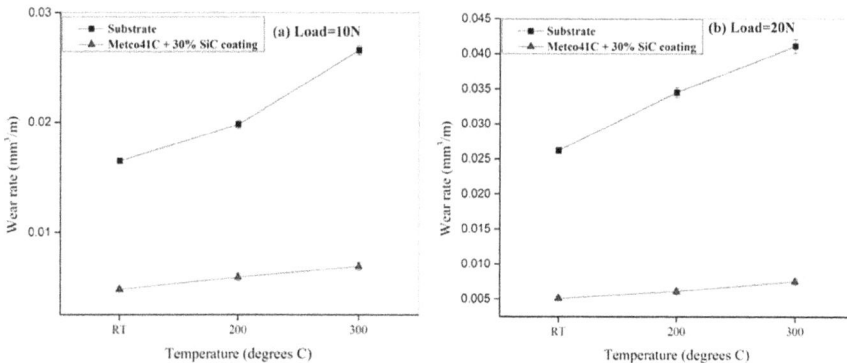

FIGURE 9.7 Volume loss profiles of coating and substrate: (a) 10 N, (b) 20 N.

TABLE 9.3

Values of Volume Loss with Respect to Normal Load and Temperature

Specimens Volume Loss (mm³)	Room Temperature		200°C		300°C	
	10 N	**20 N**	**10 N**	**20 N**	**10 N**	**20 N**
Substrate	3.9073	5.7565	5.1205	6.2301	6.3509	8.0399
Coating	1.045	1.8623	1.509	2.5236	1.898	2.9205

temperatures and loads, compared to the substrate. It was noticed that the nearly 80% of material loss was avoided by coating than substrate. This was due to hard carbide phases such as Cr_7C_3, Ni_3C, FeC, SiC and $MoSi_2$ suspended in rich FeNiCr matrix region which strengthens the layers and avoids loss of material [23–25]. The values of volume loss of coating and substrate are mentioned in Table 9.3.

Figure 9.8a and b illustrates the coating and substrate wear rate plots. It was observed that the boiler steel substrate exhibits higher wear rate at all the temperature conditions due to lower hardness and strength. During the sliding test, the substrate produced high vibrations and noise due to oxidation results in more friction. These parameters were the main cause for increase material loss. The wear rate for substrate was increased as the temperature was raised. Increase in wear rate with increase in normal load from 10 to 20N as noticed during the test.

The SiC-reinforced coating exhibited lowest wear rate at all test parameters compared to the substrate. However, the coating exhibits 3.5–4 times lower wear rate at test conditions. The formation of hard carbide phases such as SiC, FeC and Cr_7C_3 (see Figure 9.3) imparts high-temperature strength and microhardness to the coating, and withstands the deformation of coating splats. Wear rate of the coating was slightly increased under 20 N normal load due to detachment of splats, as observed in the microstructure shown in Figure 9.5b. This was due to the removal of unmelted/ partial-melted particles caused by localized deformation [1, 2, 24–26]. Later, the wear rate reached steady state as oxidative wear restricts further loss of the material. The values of wear rate of coating and substrate are listed in Table 9.4.

FIGURE 9.8 Wear rate profiles: (a) 10 N, (b) 20 N.

TABLE 9.4
Values of Wear Rate with Respect to Normal Load and Temperature

Specimens Wear Rate (mm³/m)	Room Temperature		200°C		300°C	
	10N	20N	10N	20N	10N	20N
Substrate	0.0165	0.0262	0.0198	0.0345	0.0266	0.0411
Coating	0.0048	0.0051	0.0059	0.0061	0.0069	0.0075

9.4 CONCLUSION

- The feedstock Metco 41C and SiC were blended properly by using ball-milling process with a ratio of 70: 30.
- The Metco 41C+SiC coating was successfully deposited on the boiler steel substrate by HVOF spray method. The coating exhibited compact structure and uniform thickness in the range 275–300 μm. The coating also produced a surface roughness R_a of 8.232 μm due to the existence of unmelted particles.
- Carbide phases such as Cr_7C_3, Ni_3C, FeC and SiC along with intermetallic phases like $MoSi_2$ were formed, resulting in higher microhardness of coating, 1045 ± 30 Hv, than the substrate.
- Recorded values of wear rate and friction of coating were lower than that of the substrate due to higher microhardness strengthening the underlying surface at high temperatures.
- The coating can improve the product life by 3.66 times of the existing part. This was evaluated in terms of loss of volume of coating during wear test.

ACKNOWLEDGMENTS

The authors are grateful to Aum Techno Spray, Bengaluru, India, for providing the HVOF coating system.

REFERENCES

1. A.G.M. Pukasiewicz, H.E. de Boer, G.B. Sucharski, R.F. Vaz, & L.A.J. Procopiak, The influence of HVOF spraying parameters on the microstructure, residual stress and cavitation resistance of FeMnCrSi coatings, *Surface and Coatings Technology*, 327, (2017), 158–166.
2. P.D. Miller, J.W. Holladay, Friction and wear properties of titanium, *Wear*, 2 (2), (1958), 133–140.
3. E. Bemporad, M. Sebastiani, F. Casadei, & F. Carassiti, Modelling, production, and characterization of duplex coatings (HVOF and PVD) on Ti–6Al–4V substrate for specific mechanical applications, *Surface and Coatings Technology* 201, (2007), 7652–7662.
4. S. Němeček, L. Fidler, & P. Fišerová, Corrosion resistance of laser clads of Inconel 625 and Metco 41C, *Physics Procedia*, 56 (C), (2014), 294–300. doi:10.1016/j.phpro.2014.08.174.

5. S. Buytoz, M. Ulutan, S. Islak, B. Kurt, & O. Nuri Çelik, Microstructural and wear characteristics of high velocity oxygen fuel (HVOF) sprayed NiCrBSi-SiC composite coating on SAE 1030 steel, *Arabian Journal for Science and Engineering*, 38 (6), (2013), 1481–1491. doi:10.1007/s13369-013-0536-y

6. C. D. Prasad, S. Joladarashi, M. R. Ramesh, & A. Sarkar, High temperature gradient cobalt based clad developed using microwave hybrid heating, *American Institute of Physics*, 1943, (2018), 020111. doi: 10.1063/1.5029687.

7. Z. Bergant, U. Trdan, JanezGrum, Effect of high-temperature furnace treatment on the microstructure and corrosion behavior of NiCrBSi flame-sprayed coatings, *Corrosion Science*, 88, (2014), 372–386.

8. C. D. Prasad, S. Joladarashi, M. R. Ramesh, M. S. Srinath, & B. H. Channabasappa, Effect of microwave heating on microstructure and elevated temperature adhesive wear behavior of HVOF deposited CoMoCrSi-Cr_3C_2 composite coating, *Surface and Coatings Technology*, 374, (2019), 291–304.

9. A. C. Karaoglanli, M. Oge, K. M. Doleker, & M. Hotamis, Comparison of tribological properties of HVOF sprayed coatings with different composition, *Surface and Coatings Technology*, 318, (2017), 299–308. doi:10.1016/j.surfcoat.2017.02.021.

10. Q. Yang, T. Senda, & A. Ohmori, Effect of carbide grain size on microstructure and sliding wear behavior of HVOF-sprayed WC-12% Co coatings, *Wear*, 254 (1–2), (2003), 23–34. doi:10.1016/S0043-1648(02)00294-6.

11. S. Sharma, Parametric study of abrasive wear of Co–CrC based flame sprayed coatings by Response Surface Methodology, *Tribology International*, 75, (2014), 39–50.

12. J. Li, H. Liao, & C. Coddet, Friction and wear behavior of flame-sprayed PEEK coatings, *Wear*, 252, (2002), 824–831.

13. R. Gonzalez, M. Cadenas, R. Fernandez, J. L. Cortizo, & E. Rodrıguez, Wear behaviour of flame sprayed NiCrBSi coating remelted by flame or by laser, *Wear*, 262, (2007), 301–307.

14. C. Navas, R. Colaco, J. de Damborenea, & R. Vilar, Abrasive wear behaviour of laser clad and flame sprayed-melted NiCrBSi coatings, *Surface and Coatings Technology*, 200, (2006), 6854–6862.

15. C. D. Prasad, S. Joladarashi, M. R. Ramesh, M. S. Srinath, & B. H. Channabasappa, Development and sliding wear behavior of Co-Mo-Cr-Si cladding through microwave heating, *Silicon*, (2019). doi:10.1007/s12633-019-0084-5.

16. A. Gisario, M. Puopolo, S. Venettacci, & F. Veniali, Improvement of thermally sprayed WC-Co/NiCr coatings by surface laser processing, *International Journal of Refractory Metals and Hard Materials*, 52, (2015), 123–130.

17. N. Ma, L. Guo, Z. Cheng, H. Wu, F. Ye, & K. Zhang, Improvement on mechanical properties and wear resistance of HVOF sprayed WC-12Co coatings by optimizing feedstock structure, *Applied Surface Science*, 320, (2014), 364–371.

18. G. Bolelli, L.-M. Berger, M. Bonetti, & L. Lusvarghi, Comparative study of the dry sliding wear behavior of HVOF-sprayed WC–(W, Cr)2C–Ni and WC–CoCrhard metal coatings, *Wear*, 309, (2014), 96–111.

19. S. Harsha, D. K. Dwivedi, & A. Agrawal, Influence of WC addition in Co–Cr–W–Ni–C flame sprayed coatings on microstructure, microhardness and wear behavior, *Surface and Coatings Technology*, 201, (2007), 5766–5775.

20. C. D. Prasad, S. Joladarashi, M. R. Ramesh, M. S. Srinath, & B. H. Channabasappa, Influence of microwave hybrid heating on the sliding wear behaviour of HVOF sprayed CoMoCrSi coating, *Materials Research Express*, 5, (2018), 086519.

21. C. D. Prasad, S. Joladarashi, M. R. Ramesh, M. S. Srinath, & B. H. Channabasappa, Microstructure and tribological behavior of flame sprayed and microwave fused CoMoCrSi/CoMoCrSi-Cr_3C_2 coatings, *Materials Research Express*, 6, (2019), 026512.

22. S. Honga, Y. Wu, B. Wang, J. Zhang, Y. Zheng, & L. Qiao, The effect of temperature on the dry sliding wear behavior of HVOF sprayed nanostructured WC-CoCr coatings, *Ceramics International*, 43, (2017), 458–462.

23. S. Houdkova, E. Smazalova, M. Vostrak, & J. Schubert, Properties of NiCrBSi coating, as sprayed and remelted by different technologies, *Surface and Coatings Technology*, 253, (2014), 14–26.

24. B. Uyulgan, E. Dokumaci, E. Celik, I. Kayatekin, N.F. AkAzem, I. Ozdemir, & M. Toparli, Wear behaviour of thermal flame sprayed FeCr coatings on plain carbon steel substrate, *Journal of Materials Processing Technology*, 190, (2007), 204–210.

25. O. Redjdal, B. Zaid, M. S. Tabti, K. Henda, & P. C. Lacaze, Characterization of thermal flame sprayed coatings prepared from FeCr mechanically milled powder, *Journal of Materials Processing Technology*, 213, (2013), 779–790.

26. H. L. Yu, W. Zhang, H. M. Wang, Y. L. Yin, X. C. Ji, K. B. Zhou, Comparison of surface and cross-sectional micro-nano mechanical properties of flame sprayed NiCrBSi coating, *Journal of Alloys and Compounds*, 672, (2016), 137–146.

27. C. D. Prasad, S. Joladarashi, M. R. Ramesh, M. S. Srinath, & B. H. Channabasappa, Comparison of high temperature wear behavior of microwave assisted HVOF Sprayed CoMoCrSi-WC-CrC-Ni/WC-12Co composite coatings, *Silicon*, 6, (2020) 1–19.

28. H. Asgari, G. Saha, & M. Mohammadi, Tribological behavior of nanostructured high velocity oxy-fuel (HVOF) thermal sprayed WC-17NiCr coatings, *Ceramics International*, 43, (2017), 2123–2135.

29. D. Kong, B. Zhao, Effects of loads on friction wear properties of HVOF sprayed NiCrBSi alloy coatings by laser remelting, *Journal of Alloys and Compounds*, 705, (2017), 700–707.

30. C. Durga Prasad, S. Lingappa, S. Joladarashi, M. R. Ramesh, & B. Sachin, "Characterization and sliding wear behavior of CoMoCrSi+Flyash composite cladding processed by microwave irradiation", *Materials Today Proceedings*, 46, (2021), pp. 2387–2391, https://doi.org/10.1016/j.matpr.2021.01.156

31. C. D. Prasad, A. Jerri, & M. R. Ramesh, Characterization and sliding wear behavior of iron based metallic coating deposited by HVOF process on low carbon steel substrate, *Journal of Bio and Tribo-Corrosion*, 6, (2020), 69.

10 Application of Thermal Spray Coatings for Protection against Erosion, Abrasion, and Corrosion in Hydropower Plants and Offshore Industry

Jashanpreet Singh

CONTENTS

DOI: 10.1201/9781003213185-10

10.1 INTRODUCTION

Hydropower is one of the renewable sources of energy which makes its development even more important over fossil fuels and other non-renewable resources. Countries like India, which have a latent potential for hydroelectric power, should invest in this resource to meet their country's enormous energy needs. In India, after thermal energy, hydroelectricity is the second most important source of energy of the total 225,793.10 MW of installed power capacity, hydropower accounts for 17.55% [1]. The net installed capacity of hydroelectricity globally is 1267 GW, with 10 countries accounting for 66% of the capacity. India ranks 7th in the list [2]. In India, the latent potential for hydroelectric power should invest in this resource to meet their country's enormous energy needs. The hydropower potential of India is around 145,000 MW and at a 60% load factor, it can meet the demand of around 88,537 MW [1]. As per the latest data available, India has 39,896.40 MW of installed capacity. An additional 5921.755 MW of hydropower is achieved in 2019, as approved by the Central Electricity Authority of India (CEA). The Northern region boasts of 15,569.75 MW of installed capacity of hydropower utilities in India, making it the region with the maximum installed capacity. Punjab is the state with the maximum installed capacity (constitutes 19.42% of the installed capacity in the Northern region). Among the states, the two Himalayan states of Uttarakhand and Himachal Pradesh (H.P.) have the most hydroelectricity generation capacity as a percentage of total electricity produced. Hydropower still faces many challenges in its way. One of those challenges is silt erosion.

The protection of machinery, equipment, and other systems installed at hydropower plant is recognized by many engineers, researchers, and scientists. This requires a high level of process-based understanding that involves in economic solution of such failure problems. Most common surface failure processes are fatigue failure, scaling, abrasion, adhesion, fatigue, erosion, corrosion, and erosion–corrosion. Sediment erosion in the hydro-turbines and its components is a key challenge in the Northern Indian hydropower plants. The rivers of Himalayas constitute hard particles like quartz, feldspar, etc. During the monsoon season, a large amount of sediment is swept along with the runoff into the hydro–turbines, leading to severe damage to the turbine components. Silt is very hard and large concentrations of it can result in wear of hydro-turbine components like blades, runners, seals, liners, passages, etc. [3]. The silt contains more than 50% of quartz content, which causes several problems in turbines, such as sediment erosion, leakage flow, disturbances in secondary flow, structural failures, deformation of runner and blades, erosion in impeller and casings, reduced power outputs, etc. [4–6]. In hydropower plants, the two different schemes are utilized, namely pumped storage and run-of-the-river (ROR). Sediment erosion due to the sedimentation of particles is observed in both these schemes [7]. Especially, the sediment erosion is pronounced in the ROR type small hydropower stations. Some of the power plants facing the silt erosion problem are Maneri Bhali Stage-I, Maneri Bhali Stage-II, Nathpa Jhakri, Bariasul, Dehar, Salal, Chilla, Khatima, etc. [8]. Hydropower project sites in the Himalayan range in northern and northeastern regions of the country face severe silt erosion problems. All these damages add to the operation and maintenance (O&M) cost and lower the overall efficiency of the hydropower plant. Erosion damages occurring in hydropower plants raises the operational and maintenance cost associated with problems like the damage of guide vanes, runner blades, leakage flow, choking of strainers of the turbines and thus affects the overall efficiency of the hydropower plant. Thus, the sediment erosion rate decreases the service life of the turbine and its components thus, affects the overall efficiency of the hydropower plant.

10.2   CLASSIFICATION OF HYDRO-TURBINES

The amount of power that could be produced from a source of water depends primarily on two key criteria for hydroelectric power stations, i.e., (a) quantity of water and (b) head conditions. The hydro-turbines are built to match the different river conditions that differ according to the region. The selection of hydro-turbines is focused primarily on the concept of conversion of energy, the water head available on the unit, the specific speed of the turbine, and the volume of water usable in continuous electricity generation. Figure 10.1 presents the selection criteria of hydro-turbines

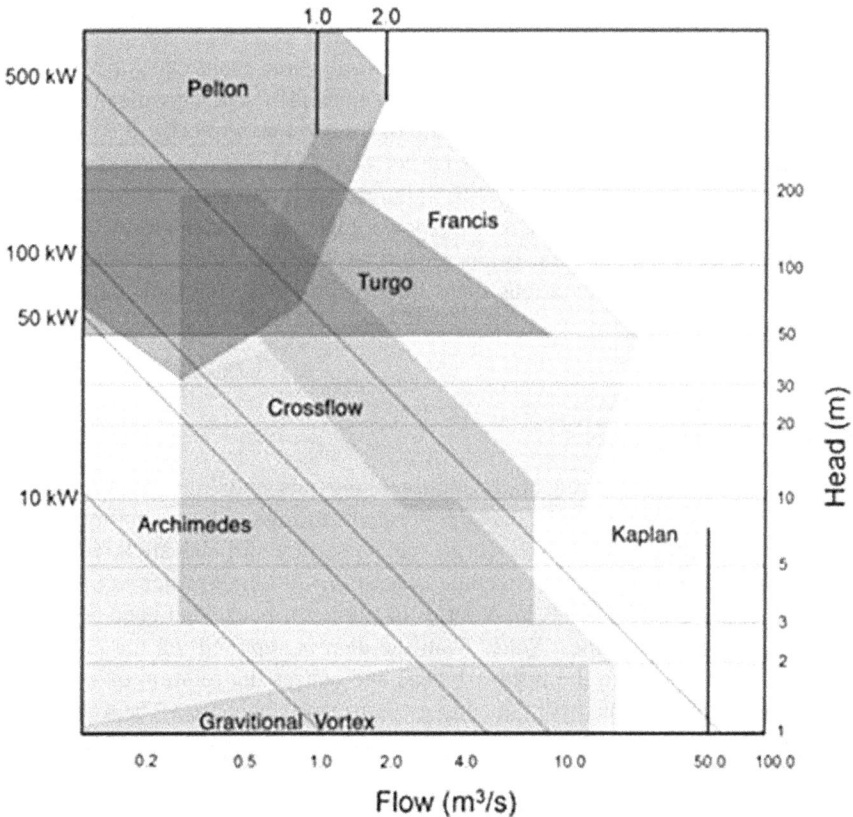

FIGURE 10.1   Selection criteria of various hydro-turbines [9].

TABLE 10.1
Head Conditions of Different Turbines

S. No.	Turbine Type	Turbines	Head (m)		
			High (>50 m)	Medium (10–50 m)	Low (<10 m)
1	Impulse	Pelton	√		
2	turbines	Turgo	√	√	
3		Cross-flow		√	√
4		Multijet Pelton		√	
5		Undershot water wheel			√
6	Reaction	Francis (spiral casing)		√	
7	turbines	Propeller			√
8		Kaplan			√
9		Francis (Open fume)			√
10	Gravity	Overshot water wheel			√
11	turbines	Archimedes screw			√

according to head and flow rates. Table 10.1 illustrates the head conditions of different impulse turbines.

Basically, the hydro-turbines are classified into three main categories, namely impulse turbines, reaction turbines, gravity turbines [10]. The classification of turbines is important to distinguish the failure of the turbine. Typically, the cavitation, abrasion, corrosion, fatigue, and material defect are the major modes of failure in impulse and reaction turbines [11, 12]. Failures due to fatigue in the material and material defects can also rely on how the power plant operates; however, the cavitation, erosion, abrasion, and corrosion depend on the water source conditions. The detailed discussion on the various types of hydro-turbines and their failures is as follows:

10.2.1 Impulse Turbines

Impulse turbines are driven under high-velocity jets and high head condition. However, these turbines can be installed for a wide range of head conditions varying from low to high. The Pelton, Turgo, and cross-flow turbines are three primary types of impulse turbines. A schematic diagram of working principle of impulse turbines is given in Figure 10.2a. A series of runner or buckets aligned around the shaft rim in Pelton turbines. Water from the dam is supplied via the high-speed nozzles which strikes to the turbine blades and causes the rotation of shaft thus, transforming the possible hydraulic energy to mechanical energy. The transformed mechanical energy is converted into the electrical energy with the help of a generator. Pelton turbines suffer from cavitation in the needles, nozzles, and the runner buckets. Typical wear on the surface of Pelton turbine buckets and nozzle is shown in Figure 10.3.

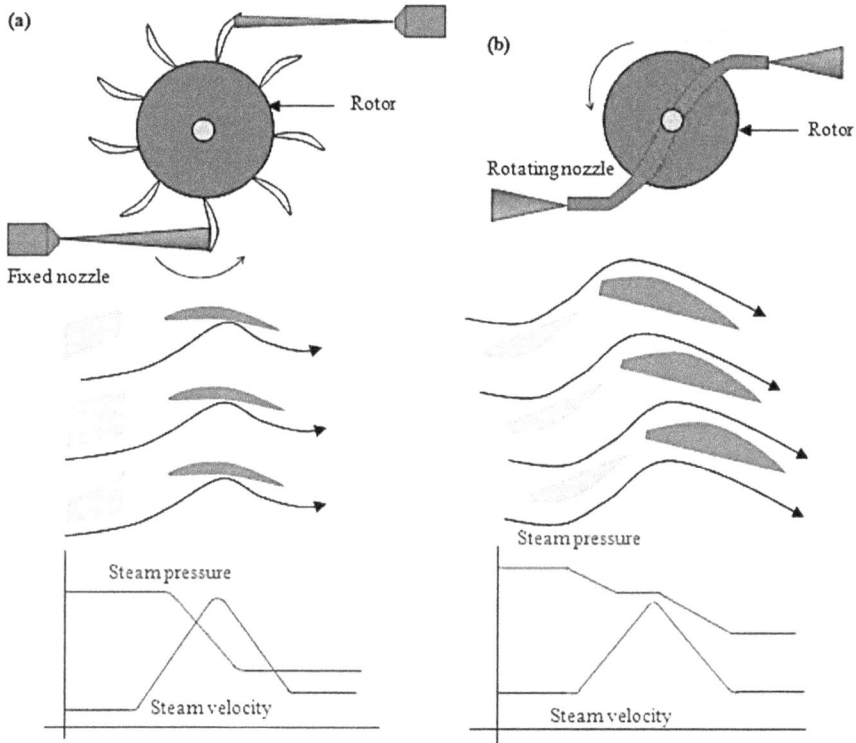

FIGURE 10.2   Working principle of (a) impulse and (b) reaction turbines.

10.2.2   REACTION TURBINES

Reaction turbines are driven under medium head conditions. The reaction turbine rotors are completely submerged in water and are enclosed in a pressure vessel. The runner blades are profiled to impose lifting forces across them, just like on the aircraft wings that rotate the runner more quickly than with a jet. A schematic diagram of working principle of reaction turbines is given in Figure 10.2b. Literature reports that the runners and draft tube cones of the Francis and Kaplan turbine undergo to cavitation turbine [10]. Figure 10.4 shows the major worn-out components of Francis turbine due to sediment erosion. Sediment erosion occurs in the various components of Francis turbine like blades, guide vanes, runners, seals, liners, passages, etc. [4]. This can lead to structural failures, leakages, deformation of runner, and blades, and reduced power outputs [5].

10.3   WEAR MECHANISMS DUE TO EROSION

As discuss earlier, an impulse turbine most probably fails as a result of erosion and cavitation whereas the reaction turbine will mostly fail because of erosion and abrasion [4, 10, 21]. In the process of corrosion, the electrochemical reaction causes the

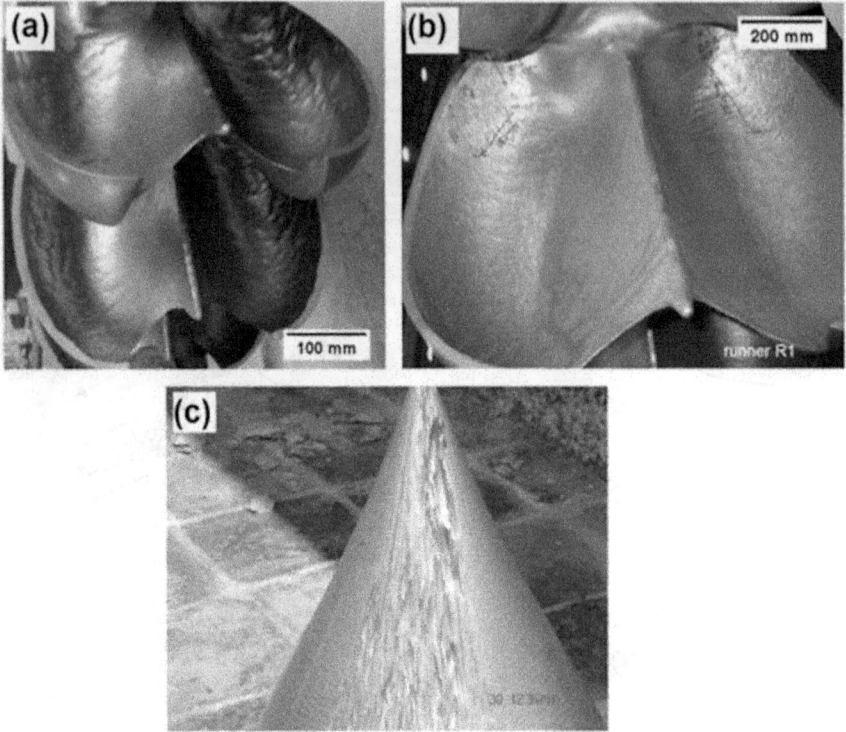

FIGURE 10.3 Typical wear on the surface of Pelton turbine (a) buckets of Toss Mini Hydel Project, Kullu, H.P., India [13], (b) Baspa II Hydropower Plant, Kinnaur, H.P., India [13, 14], and (c) nozzle of Chilime Hydropower Plant, Nepal [15].

material to deteriorate from the target surface. The two major mechanisms involved in the corrosion process are the 'pitting action' and 'crater formation' [22]. Figure 10.5 shows the primary mechanisms of abrasion, i.e., 'chip formation' and 'detachment', and the secondary mechanisms, namely 'crack formation' and 'fracture' [19]. The erosion of cavitation is caused by slurry bubble. Cavitation erosion depends not only on material toughness, ruggedness, and surface irregularities. On the other hand, erosion is a very complex process and was not yet completely understood.

Erosion is a complicated process, which involves several variables, including particle characteristics, target material properties, and method parameters. Finne [23] reported that material failure mode relies on initial behavior in material hardness. Due to the cutting action, erosion damage typically takes place in brittle materials while plastic deformation typically occurs for ductile materials. Levy [24] observed that ductile materials underwent higher erosion at the initial process, which become lower over a span of time. Mechanisms for wearing macroparticles for erosion have been described by Bellman and Levy [25]. The two mechanisms, namely shallow craters and platelets were described due to the action of macroscopic particles. In ductile substances, the various mechanisms for erosion such as deformation, micro-plowing, micro-cracking, lip-formation, platelets, and small craters occurs [26]. They also claimed that first a breakdown of the fragile materials contributes to the rapid impingement of particles,

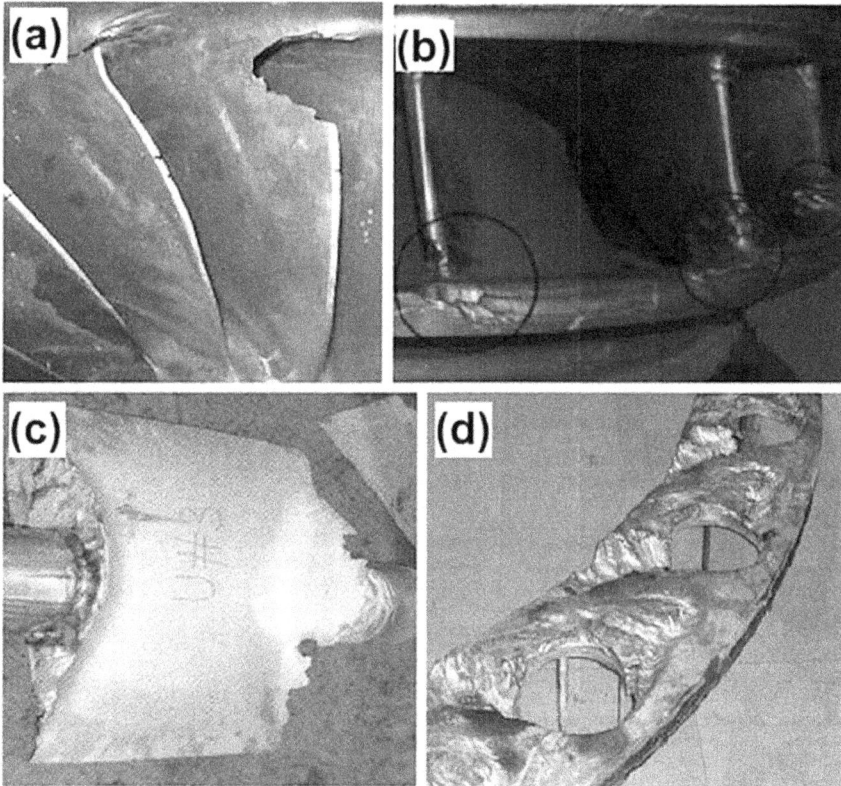

FIGURE 10.4   (a) Eroded runner after 1 year of operation at Jhimruk Hydropower plant, Nepal [16, 17], (b) surface erosion runner inlet at Cahua Hydropower plant, Peru [18, 19], (c) damaged guide vanes of Nathpa Jhakri Hydropower plant H.P., India [7, 17, 20], and (d) cheek plates at Nathpa Jhakri Hydropower plant H.P., India [7, 17, 20].

which ends with a breakdown of the surface of the object. Figure 10.6 illustrates the schematic diagram of erosion of ductile and fragile materials. The erosion in ductile and fragile materials can be differentiated from the primary wear mechanism. The erosion in ductile material starts with material deformation and result in grove formation with lips whereas erosion in fragile material starts with crack initiation and ends with fractured material in form of chips. In addition, Finne [23] has reported that if the particles accomplish the target at low angles of impact, the crater will build up. However, the pillage of the substance takes place as particles hit the target at high impact angles, and are eliminated by strong particle–target collisions.

It should be remembered that the above erosion processes are not absolute and may be combined at the same time. In order to overcome this tragedy, Sundararajan et al. [27] proposed a correlation in order to evaluate the erosion efficiency (η) of material, as shown below:

$$\eta = \frac{2VH}{M_P V_P^2} \tag{10.1}$$

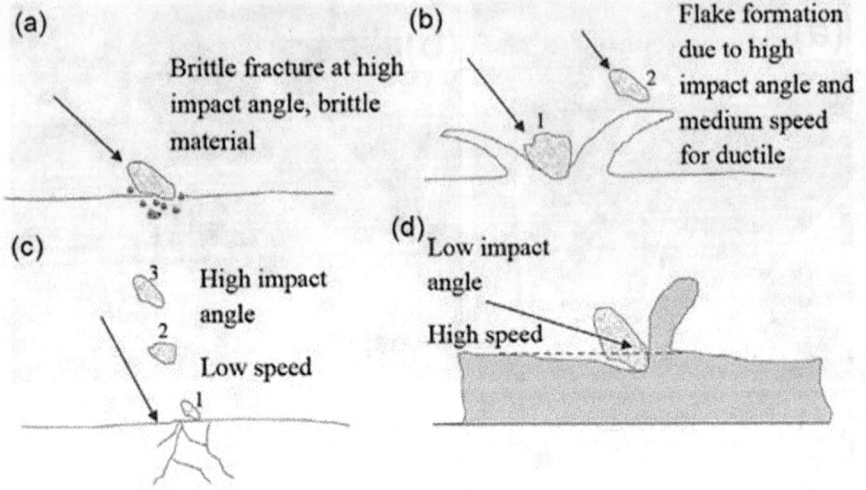

FIGURE 10.5   Various mechanisms of wear: (a) brittle fracture, (b) plastic deformation, (c) fatigue erosion, and (d) abrasive/cutting erosion [19].

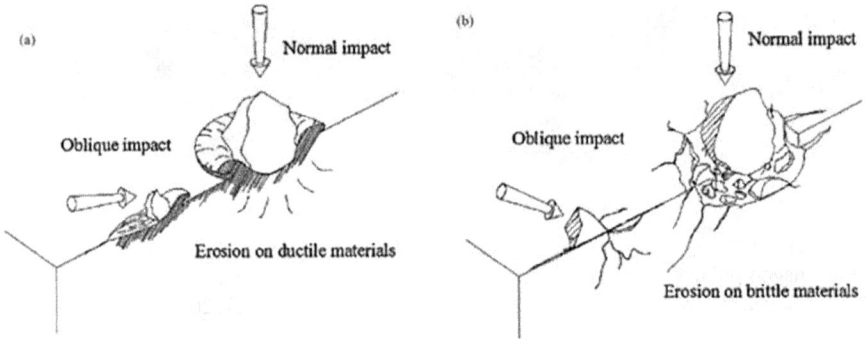

FIGURE 10.6   Schematic diagram of erosion in (a) ductile and (b) fragile materials [26].

where V = volume of removed material (m³), H_T = hardness of material (Hv), M_P = mass of particle (kg), and V_P = velocity of particle (m/s). They pointed out $\eta > 1$ represents the mechanism of fragility, i.e., brittle mechanism, while $\eta < 1$ is a mechanism of plastic erosion deformation, i.e., ductile mechanism.

The deformation (E_D) and cutting (E_C) mechanisms of erosion wear were defined by Clark and Wong [28] as written below:

$$E_r \begin{cases} E_C + E_D \\ \dfrac{\frac{1}{2}M_p\left(V^2{}_N\right)}{\varepsilon} + \dfrac{\frac{1}{2}M_p\left(V^2{}_T\sin 2\alpha\right)}{\varphi} \end{cases} \tag{10.2}$$

where the tangential and normal components of velocity of the particle are denoted by V_T and V_N, respectively. Symbols α, ε, and φ are the impingement angle, empirical constant of specific energy required for deformation erosion, and constant of kinetic energy required for cutting erosion, respectively.

The mechanism for determining the amount of energy required and spent for the material removal under oblique and normal impact was introduced by Grewal et al. [29]:

$$\xi = \frac{2V\sigma_{cr}\left(\dfrac{H_T}{K_T}\right)}{M_P V_P^2} \tag{10.3}$$

where K_T is the material toughness (MPa) and σ_{cr} is the critical stress for material (MPa). The value of σ_{cr} is equal to the ultimate tensile strength for brittle metals and alloys and the ultimate shear stress for ductile materials. When $\xi < 1$, plowing occurs, $\xi = 1$ micro-cutting occurs, and $\xi > 1$, brittle occurs.

10.4 DEPENDENCY OF WEAR IN HYDRO-TURBINES

Figure 10.7 shows the various factors that contribute in the sediment erosion. The sediment erosion is mainly dependent on the properties of sediment, flowing media and target materials, and different operating conditions in which the turbine works. The detailed study on various contributing parameters of sediment erosion.

10.4.1 PROPERTIES OF SEDIMENT PARTICLES

- *Particle Shape*
  Sediment particles have different shape, solidity, roundness that plays an important role in erosion phenomenon. Levy and Chik [30] reported that the wear generated by spherical particles was competitively lower than that by the angular particles. Bahadur and Badruddin [31] studied the influence of grain size and shape of particles (SiC, Al$_2$O$_3$, and Quartz) on solid particle erosion at high velocity. They performed the experiments using jet tester at high velocity

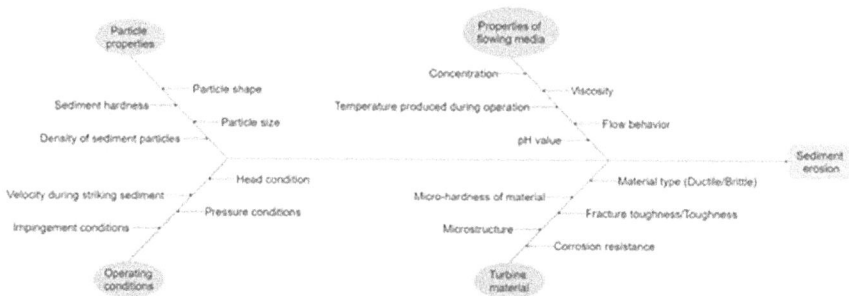

FIGURE 10.7 Various factors that contribute to the sediment erosion.

(40–65 m/s) and different particle sizes (10–500 μm). It was reported that the value of circularity factor (*CF*) for particles was found in between 0.70 and 0.97. Zou and Yu [32] had given a correlation to find the sphericity of particles, as expressed below:

$$\psi = \frac{A_s}{A_p} \qquad (10.4)$$

Hentschel and Page [33] reported another shape factor (*SF*) namely roundness factor, as given below:

$$RD = \frac{4 \times A_p}{\pi \times D^2_{maxfret}} \qquad (10.5)$$

Bouwman et al. [34] reported a correlation to calculate the *SF* value, by modifying the Riley (1941) correlation, as given below:

$$SF = \frac{4\pi A}{P^2} \qquad (10.6)$$

Heilbronner and Keulen [35] reported the Paris factor to investigate the shape of particles of fault rocks, as given below:

$$PF = 2\left(\frac{P_p - P_{ch}}{P_{ch_A}}\right) \qquad (10.7)$$

Desale et al. [36] studied the influence of *SF* value on ductile materials' erosion wear. They proposed a new correlation to calculate the *SF* value by modifying the Riley (1941) correlation, as written below:

$$SF = \left(SF_{Avg}\, SF_{max}\, SF_{min}\right)^{1/3} \qquad (10.8)$$

Woldman et al. [37] reported a dry wear study performed on rubber using sand and reported the range of shape factor for sand, i.e., 0.5–0.65. Walker and Hambe [38] performed an experimental study on erosion wear of white cast iron followed by image processing of particle shape. They performed the erosion wear experiments with the help of Coriolis tester, as presented the following correlation:

$$Er = K \times CF^m \qquad (10.9)$$

They found that as the *CF* value of sand (average = 0.73) decreases the erosion wear due to solid particles also increases. Literature reports that the particle shape majorly contributes to erosion wear mechanism. Spherically shaped particles resulted in plastic deformations and micro-cutting action which is due to rebounding nature of fly ash particles with target surface of the workpiece. Irregularly shaped blocky particles generate the plowing and smooth craters [38]. Angular-shaped particles generate the cracks and craters and during this process, chipping of erodent particles takes place [36].

- *Particle Size*
The erosion of materials is often affected by particle size substantially. Erosion is function of particle size and holds a power-law connection with particle size [39]. Elkholy [40] reported a relationship for erosion on the basis of velocity of solid–liquid (V), particle diameter (d), and concentration (C), as given by

$$E_w = k \times V^m \times d^n \times C^p \tag{10.10}$$

In the above equation, the symbols 'm', 'n', and 'p' are the exponents of velocity, particle diameter, and concentration, respectively. The exponent particle diameter (n) is normally between 0.3 and 2.0.

The greater particle size indicates that while hitting the target site, a wide region is damaged by the particle. The bigger particles often exhibit a high degree of kinetic energy and create more surface wear when in motion. The rise in particle size over 1000 μm has contributed to a continuous erosion ratio [41]. The solids flowing in river water are typically in a gritty shape and of different sizes. The particle size calculation is done by calculating the distribution of the particle size (PSD) or the size fraction set. By taking any international standard of sieves like ASTM, BSS, or IS, the size fraction distribution of particles can be calculated. The mean arithmetic diameter (d_i) can be calculated by using following equation [42]:

$$d_p = \sum_{i=1}^{n} fi \times \left(N_i \right) \tag{10.11}$$

where 'N_i' is B.S.S. number of particle size range and 'f_i' is the solid fraction. The erosion induced by the specific slurries decreases for the finer particle size fractions as compared to larger particle size fractions [42]. The different size fractions lead to change in particle–particle interactions and particle–target interaction phenomena [41].

- *Sediment Hardness*
Silt is very hard and its Vickers microhardness number lies in range 900–1000 VHN [3]. Verma [43] measured the microhardness value of silt, i.e., up to 5 Mohr which is sufficient for eroding the turbine components. Few researchers have developed correlations which can be used to calculate the erosion by using the relationship between particles hardness and target hardness. Wiederhorn and Hockey [44] suggested a statistical association of erosion with fracture toughness (k_T), specific radius (r_P), particulate density (ρ_P), and particulate strength (H_P):

$$E_T = k \times V_P^{2.8} \times r_P^{3.9} \times k_T^{-18} \times H_P^{0.48} \times \rho_P^{1.4} \tag{10.12a}$$

The empirical correlation stated by Wada and Watanabe [45] based on the hardness factor is as shown below:

$$E \prec \left(\frac{H_T}{H_P} \right)^k \tag{10.12b}$$

The symbol 'k' in the preceding equation is the exponent for the hardness ratio. The ratio of hardness can be expressed as the ratio of hardness of the material (H_T) to particles (H_P) [40]. Four different hardness materials ranging from 150 to 400 HB (Brinell hardness) were used by Elkholy [40]. They performed wear experiments and measured the weight loss of materials by adjusting the silica sand particle size while keeping the other parameters fixed. The hardness of silica sand was mentioned as 710 HB. They stated that the hardness ratio transformation point was as follows:

$$\frac{H_T}{H_P} = 1.9 \tag{10.12c}$$

The exponent 'k' changes, while the ratio of hardness rises or drops from above. The impact of particulate hardness on erosion of carbon steel (AISI 1020) was examined by Levy and Chik [30]. The erosion wear of AISI 1020 was raised up to a particle hardness of 700 kgf/mm², above which there was no further rise in erosion.

• *Density of Particles*
The surface density is also a consideration impacting the degree of erosion of the materials. The density of the material indicates larger solid particle content and also influences the target, which results in improved erosion performance, at constant values of certain parameters. Stack and Pungwiwat [46] experimentally used Al_2O_3 and SiC parts to conduct erosion experiments. Tests have shown that the Al_2O_3 is of a greater density than SiC particles, which has induced further mass loss of the target. Desale et al. [36] made a related form of similar finding. The mass density becomes a dominant factor at higher speeds. The impact energy of sediment relies explicitly on the mass density. Particles of higher mass density have high-speed kinetic energy. While these particles strike at the target material, the material degrades even more when they release their high cinematic energies.

10.4.2 PROPERTIES OF FLOWING MEDIA

The slurry viscosity, pH value, concentration, temperature, and rheological characteristics are included in the slurry properties. The detailed discussion is as follows:

• *Silt Concentration*
The majority of Himalayan rivers have very strong silt content (10 g/l or 10,000 ppm). During the monsoon season, it becomes much more large than other

seasons (up to 20,000 ppm) [3]. In general phenomenon of erosion, the silt concentration in water plays crucially. The silt is an unwanted material in hydropower plants, which comes in different amounts with the water (i.e., heterogeneous type of slurry). Material erosion generally increases as the concentrations of slurries are increased [42, 47–49]. The concentration exponent's value (p) is within the 0.9–1.3 range [29]. Dasgupta et al. [50] have established a study of erosion wear on steel samples by multisized sand slurry. They observed that as the solid content rose from 30% to 50%, the erosion rate was decreased. Nevertheless, the erosion rate decreases at low concentrations with the rise of the solid concentration as seen in the Figure 10.8 [51]. Other researchers recorded similar kind of findings [52, 53]. The explanation for this phenomenon is the rebounding particles' shielding effect. It was stated that, in general, the erosion rate increases to a critical value at the initial stages with high concentrations. The relations between striking and rebounding particles are important. The interactions between the particles of striking and bounding particles at solid concentrations (<1 wt.%) cause a loss of momentum and a change of direction. Therefore, the association with particulate matter decreases the contact between particle and target and therefore increases the erosion rate

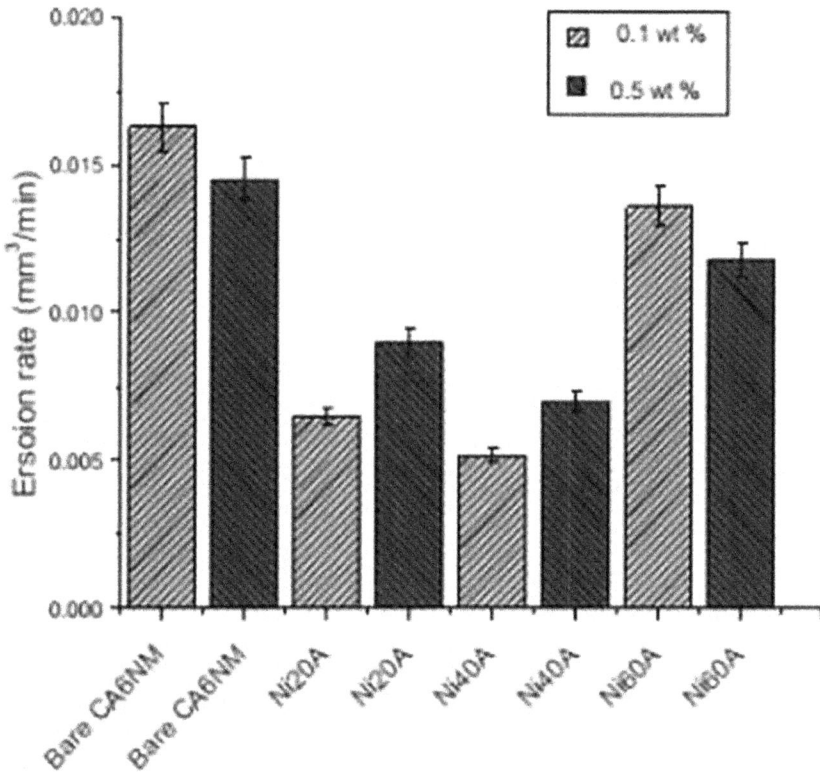

FIGURE 10.8 Erosion rate vs. solid concentration [51].

[54]. Particle rebounding relies on the properties of target materials. This effect is only restricted to a range of different parameters.

From the above, the concentration of slurry may be assumed to influence other parameters that have not yet been thoroughly examined. Erosion decreases with the increase in concentration percentage [55]. According to few researchers [50–54, 56, 57], the erosion rate decreases with rising slurry concentration. According to refs. [54, 56, 58], a change of concentration influences the both a phenomenon of particulate–target interactions and inter-particulate interactions. Finally, it can be said that further study is needed on the hidden aspects of solid concentration.

- *pH Value*
The pH of slurry governs the process of erosion and corrosion. While assessing erosion of materials, the pH content of slurry can play a significant part. Singh [56] reported the pH value of silt slurry of Naptha Jhakhri Dam, H.P., India, as 7.71–7.16 for concentration range of 30–60 wt.%. The erosion–corrosion of AISI 4330 steel in saline-sand slurry was investigated by Ballesteros et al. [59]. In comparison to saline-sand pH = 7, the volume loss from AISI 4330 was higher in the case of saline-sand's pH = 3. Karafyllias et al. [60] reported a relationship for the total volume loss (TVL) from materials:

$$TVL = M + C + S \tag{10.13}$$

In the above equation, the M, C, and S symbol reflects the loss of volume due to mechanical failure, mere electrochemical decay, and synergetic consequences of erosion and corrosion, respectively.

- *Flow Behavior*
Flow behavior of slurry depends upon the variety of factors like viscosity, particle size, particle shape, dimensionless numbers, and concentration. The rheology is the way to measure the flow characteristics of liquids containing solids. Specific parameters such as the shape, scale, density, and mass fraction are involved in the rheological feature of slurry [58]. In fact, the flow properties are often influenced by the concentration, as discussed in previous sections. The dimensionless numbers like Reynolds (Re) and Stokes (St) are influenced by the rise or drop in the concentration. Bong [61] found that slurry viscosity was increased by increasing slurry concentration, whereas Re decreased, as seen in the Figure 10.9.

The flows are known to be laminar for Re < 2100, transient for 2130 < Re < 4000 and turbulent for Re > 4000 in the case of flow through the pipelines. The erosion–corrosion performance of carbon steel was tested by Shehadeh et al. [63] using strong sand-water slurry varying from 3 to 9 g/l. It was noted that, the erosion–corrosion increases with increased rates and Re value (Figure 10.10). Concentration change often influence the movement of solid particles. The analysis of Stokes number (St) will help explain this phenomenon. The value of St increases with particle size and solid concentration, as Dabirian et al. [64]. However, Bartosik [65] investigated the significant effects on the shear stress among the particles and the target wall due to their strong concentration and particle size.

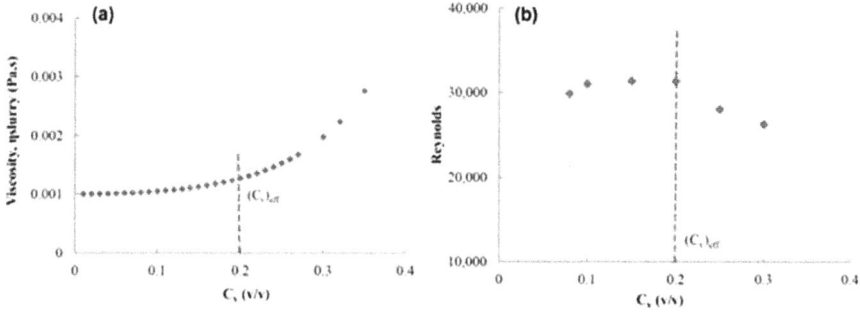

FIGURE 10.9   Effect of solid concentration on (a) viscosity and (b) Reynolds number [61, 62].

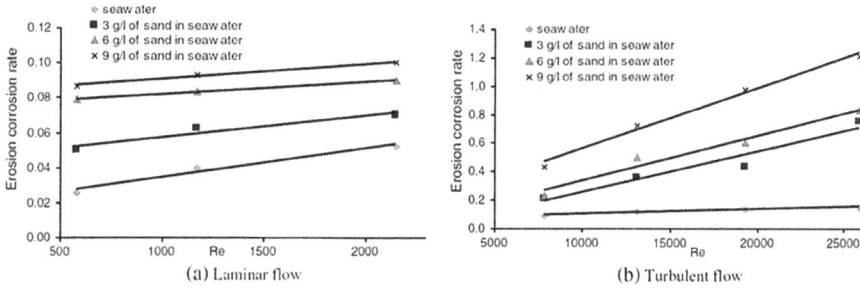

FIGURE 10.10   Erosion–corrosion rate (mg/m²/min) vs. Reynolds number [63].

10.4.3   OPERATING CONDITION

- *Velocity during Striking Sediment Particles*
  Erosion wear of the materials is strongly influenced by velocity of erodent particle. If particle of lesser mass impacts the target at very low velocity, the lower the kinetic energy of particle, resulting in small surface deformation. Even a smaller faction of erodent has adequate kinetic energy to induce permanent deformation of target surface. As the number of particles attain a high threshold energy with the increase in velocity, it results in a higher surface deformation rate [22, 66, 67]:

$$-\frac{dm}{dt} = kv^{n} \tag{10.14}$$

  where t is time period of investigation process, m is mass loss from the target surface, k is experimental constant, and n is the velocity exponent. The value of n is varies from 2 to 5 for different materials [68].

- *Impingement Conditions*
  The primary erosion mechanisms for the API X42 steel were also defined by
  Islam and Farhat [69] on the basis of the particulate impact angle and particle,
  as given below:
  - *Low Impact Angle:* At lower velocities, the metal is pressed forward and the
    ridges are shaped on the sides as the particles slid down on the surface. So,
    the 'plowing' under such conditions is the primary mechanism for erosion
    which is different from 'cutting'. It is noted during plowing that the material
    does not detached from the surface and 'lips' forms beside the erosion
    groove. On the other hand, at higher velocities, the cutting mechanism
    occurs primarily. In the cutting mechanism, the material detaches from the
    surface due to fracture of ridges and debris forms.
  - *High Impact Angle:* At lower velocities, the mechanisms of wear include
    plastic deformation and flattening. Plastic deformation and elimination of
    weak lips occurs due to the continuous effect of the particles' striking.
    The breaking of particles and the secondary metal cutting occurs at higher
    velocities and impact angles. When the erosive particle hits the embedded
    particles and deflects through the material, the target surface deteriorates.
    In this case, tiny amount of the substance is damaged, roughly of width
    2–3 µm.

10.4.4 MATERIAL TYPE

From the commencing of the 20th century, the market of materials is rapidly growing to a high level. A variety of materials are available in market which can withstand against different forces such as tensile force, compressive force, impact
loading, fatigue loading, etc. Researchers and designers have made a lot of efforts
in optimization in design changes with which the materials can withstand to these
properties. In surface failure processes, the phase microstructure, material properties like microhardness, toughness, elasticity, elongation, ductility, etc., plays an
important role.

Table 10.2 illustrates the various properties of turbine steels. Generally, high-
chrome steels are used for the fabrication of turbine components. The high-chrome
steels include 18Cr8Ni (austenitic steel), 13Cr1Ni (martensitic steel), and 13Cr4Ni
or 16Cr5Ni (martensitic–austenitic steel). Most of the modern turbines are made
from13Cr4Ni or 16Cr5Ni steels [70]. Swedish quality steel 13Cr6Ni was patented in
late 60s but the restriction of its use due to patent right favored the development of
13Cr4Ni, 16Cr5Ni, and 17Cr4Ni stainless steels, among which 13Cr4Ni and 16Cr5Ni
are widely used in the turbine industry. Austenitic steel 18Cr8Ni with less than 0.2%
Mo has good resistance to corrosion and cavitation, but their fatigue and creep resistance are poor [70]. This steel was used to fabricate the turbine runner also in Pathri
hydropower plant, Uttar Pradesh, and Mattur hydropower plant, Tamil Nadu, India
[71]. Martensitic steel 13Cr1Ni was developed as substitute for 18Cr8Ni, but corrosion, cavitation, and sand erosion resistance of this steel is poor because of low hardness, toughness, and less Ni content. Thapa and Brekke [72] reported that the

TABLE 10.2
Materials and Properties of Turbine Steels [70]

S. No.	Material	Tensile Strength (MPa)	Yield Strength (MPa)	Elongation (%)	Impact Strength (J)	Microstructure	Type
1	13Cr 1Ni 0.06C 0.4Mo	630	470	18	39	Martensite	Hard and brittle
2	13Cr 6Ni 0.06 0.4Mo	800	550	16	70	Martensite (70%) Austenite (30%)	
3	17Cr 4Ni 0.06C	880	650	12	59	Martensite+ Austenite+ Ferrite	Hard and ductile
4	13Cr 4Ni 0.04C 0.4Mo	823	686	23	81	Martensite	Hard
5	16Cr 5Ni 0.05C 1.5Mo	880	600	21	100	Martensite (65%) Austenite (35%)	Hard and ductile

13Cr4Ni and duplex steel exhibit better erosion resistance properties as compared to 16Cr5Ni turbine steel.

Microhardness is one of key properties which affects the erosion wear resistance of material [36, 73–77]. Microhardness of different material is listed in Table 10.2. The value of the hardness ratio of target hardness to particle hardness was stated by researchers [25, 78, 79]. A model of prediction of total erosion wear for different hardness materials (GPa, H_T) and for other parameters of the method, respectively, impact velocity, (m/s, V_P), and particle diameter (μm, d_P) was reported by Oka et al. [55]:

$$E_T = k \times V_P^m \times d_P^n \times H_T^{K_1} \times \sin \alpha^{n_1} \left(1 + H_T \left(1 - \sin \alpha\right)\right)^n 2 \qquad (10.15)$$

where m and n are exponents of speed and particle diameter, respectively. The symbols k and K_1 are the experimental constant and exponent of material hardness, respectively; n_1 and n_2 are constants which depend on the angle of impact of the particle and the target.

Levy and Hickey [80] recorded a higher wear rate of materials with higher hardness than materials with less hardness. However, the ductility of the substance is responsible for the plastic deformation through the transfer of kinetic energy of particles, lowering deterioration on the target surface. The erosion of standard SS 316L stainless steel, cold-rolled and case surface-hardened, was observed by Divakar et al. [81], as shown in Figure 10.11.

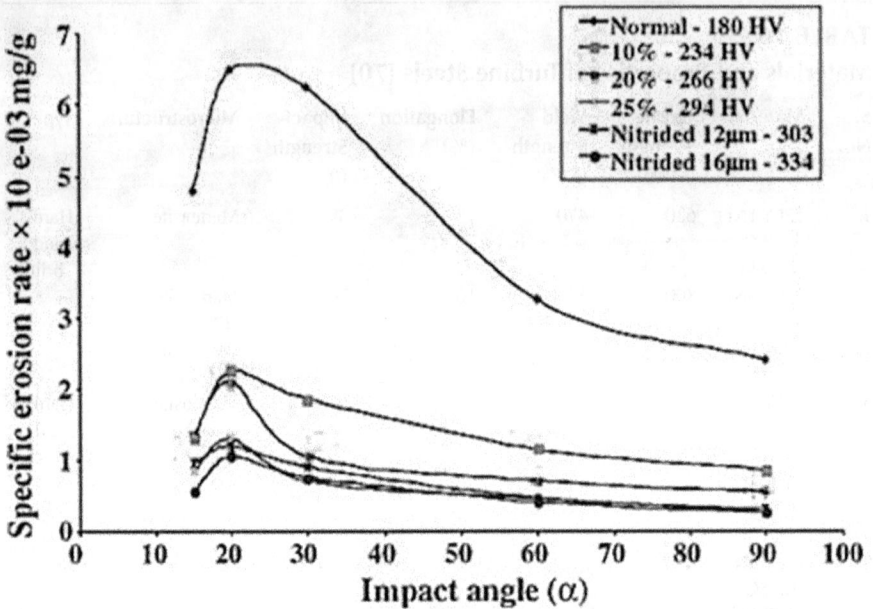

Legend:
- Normal - 180 HV
- 10% - 234 HV
- 20% - 266 HV
- 25% - 294 HV
- Nitrided 12μm - 303
- Nitrided 16μm - 334

FIGURE 10.11 Erosion rate of different materials at room temperature ($V = 32$ m/s, $d_p = 102$ μm) [81].

10.5 EXPERIMENTAL METHODS OF WEAR MEASUREMENT

Various types of testers were used to predict the wear in turbines [10]. Different types of test rigs, pilot plant test loops, and laboratory scale testers were developed by investigators to evaluate the wear in turbines. In initial researches, the field experiments were conducted to determine the erosion wear matching the actual working condition, but there are many drawbacks like time duration, material, and uncontrollable testing variable with high cost of experimentation [35, 82, 83]. Using the small-scale laboratory testing, highly accurate erosion data for the large parametric variations can be obtained. For the evaluation of the wear performance of target material in solid–liquid flow, the use of test rig is highly recognized. In addition, a major drawback is in using the test rig is the requirement of fresh slurry for erosion wear rate experimentation because the shape edges of slurry suspended particles is degraded during the experiments. Most commonly used wear measurement equipment are slurry jet erosion (SJE) wear test rig [29], slurry pot tester [84], Coriolis erosion tester [85], whirling arm test rig [86], and prototype test rigs [87].

Toward the end of 1985, the development of pot tester was started. Schumacher [88] developed a pot tester having 3 hubs with the capacity of holding 24 specimens, i.e., 8 specimen per hub. This tester worked similar to modern pot testers for the variation of process parameters but still had limitations in controlling the impingement angle and solid concentration. After 1990s, the investigators rapidly designed different types of pot testers. Clark [89, 90] designed a cylindrical pot tester in early 1990s. The main components of this tester are shown in Figure 10.12a. This tester was unique in terms

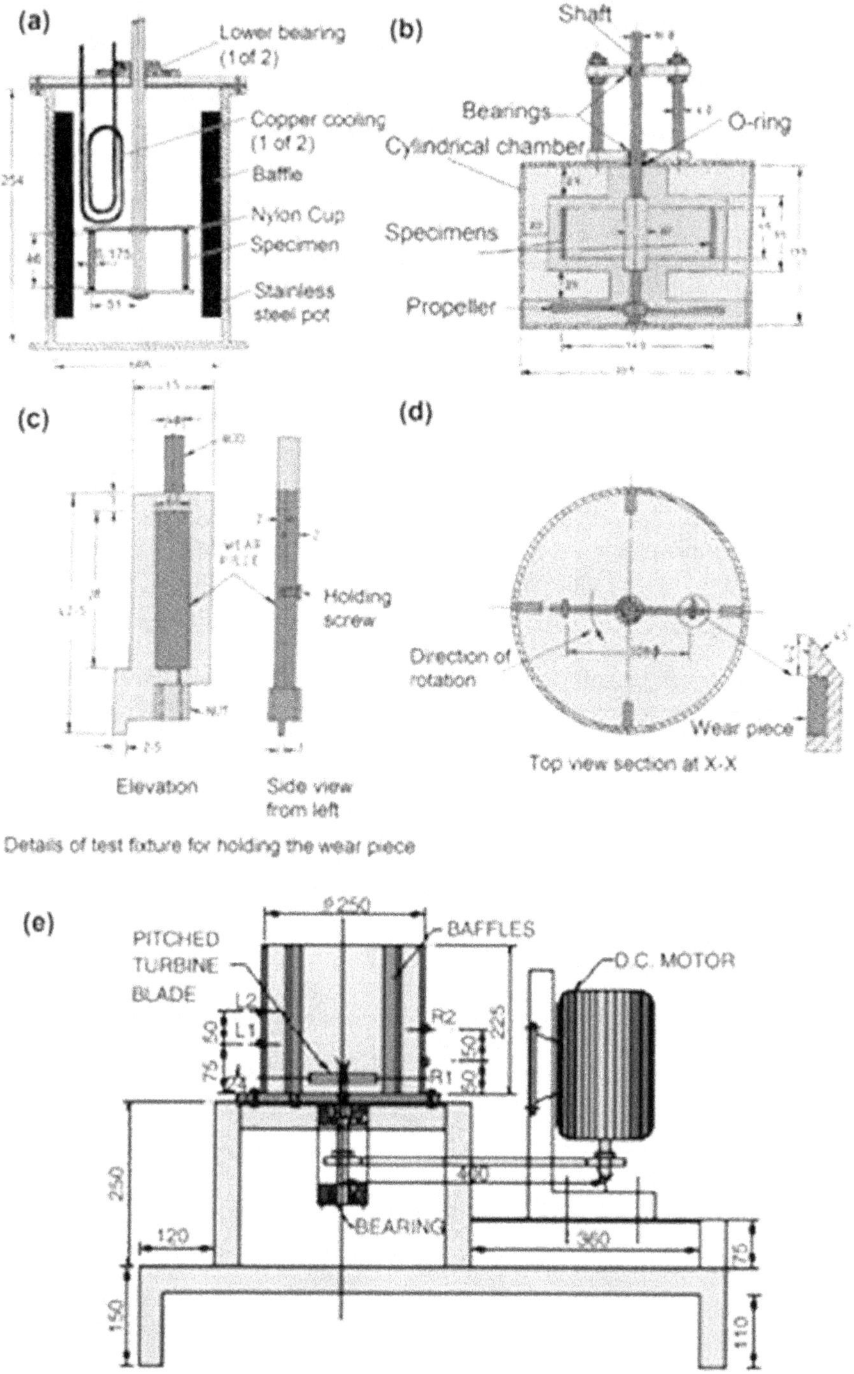

FIGURE 10.12   (a) Cylindrical erosion pot tester [89, 90], and (b–d) erosion pot and its components [73], and (e) slurry pot tester rig tester [91].

of avoiding pressure build up by providing ventilation and corrosion resistance to the specimen through using nylon spacers. The major limitation of this tester was that the shape of the specimen was limited to being cylindrical and its holder was limited to hold only 2 specimens. Gupta et al. [42] also developed the propeller-type pot tester, which is shown in Figure 10.12b. This tester overcame the previous testers in terms of variation of parameters like velocity, particle impact angle, and solid concentration. Gandhi et al. [73] modified the components of pot tester, which is shown in Figure 10.12c, d. They designed a fixture shown in Figure 10.12d. This tester was unique in terms of producing accurate velocities and impact angles. After the commencing of the 21st century, many researchers started to do further modifications in pot testers. Desale et al. [91] also modified the slurry pot tester for the evaluation of erosion at high solid concentrations with an extensive range of impact angles (0°–90°). A schematic illustration of this pot tester is presented in Figure 10.12e.

Coriolis tester was developed in 1980s by Tuzson [92]. In this tester, the slurry particles were moved through an overhead tank and strikes to the central point of a bowl that rotated by means of a motor. Then, these particles reached to a fixture that hold specimen. The working principle of this tester was Coriolis acceleration. According to this principle, the solid particles exert an outward centrifugal force when they move over a radial passage at high velocities. The centrifugal force speeds up the particles in the direction of the target and stimulates the particle–target interactions. The speed up of particles through centrifugal force is due to an acceleration which is known as Coriolis acceleration [85, 92]. The major advantage of Coriolis tester was the control on flow velocity and acceleration. Clark et al. [85] modified the *Coriolis erosion tester* which was earlier developed by Tuzson [92]. They designed the Coriolis tester to evaluate the erosion wear for pumps and pipelines for the flow of dense slurries. The main components of this tester are shown in Figure 10.13.

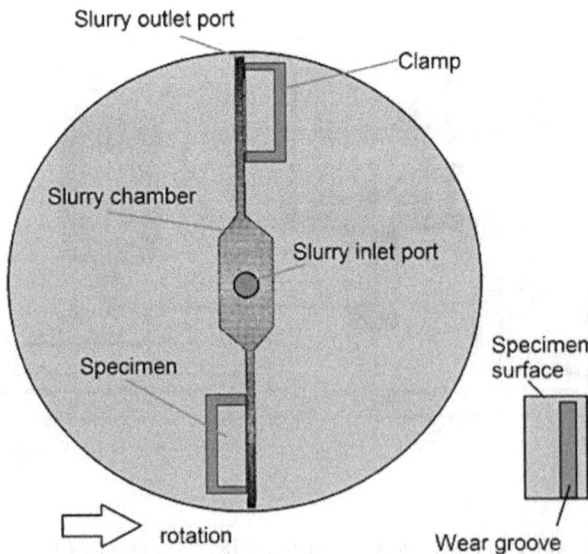

FIGURE 10.13   Coriolis erosion tester [85].

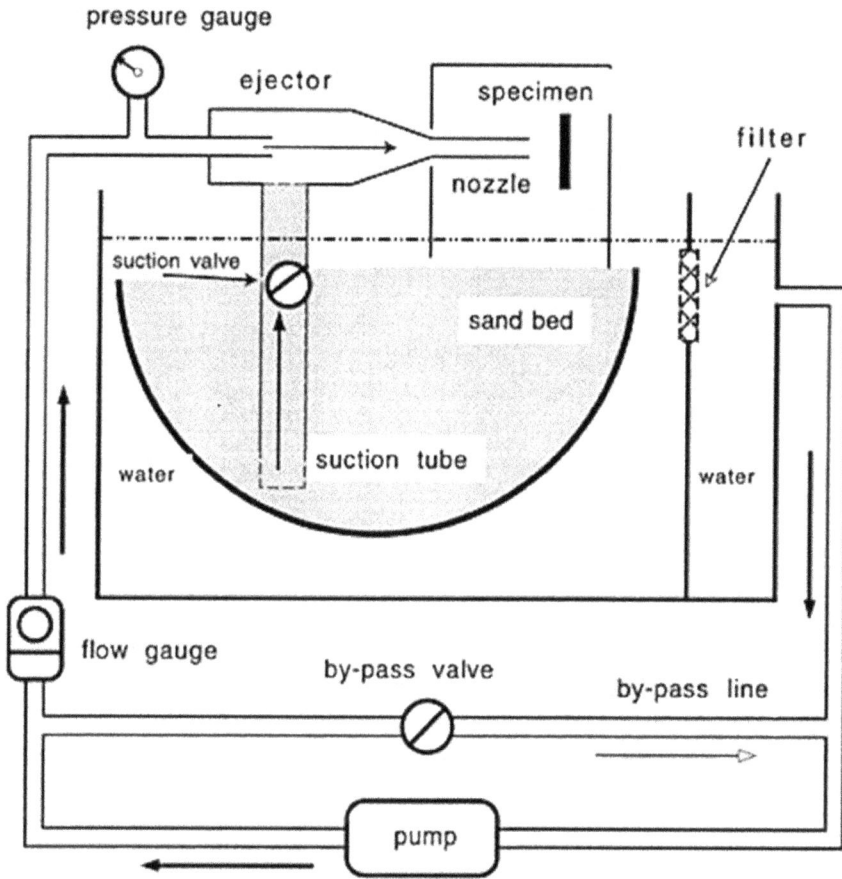

FIGURE 10.14   Schematic diagram of closed-loop erosion test rig [93].

SJE tester was also designed by various investigators. This tester utilizes the stream of jet of solid–liquid slurry to evaluate the erosion in terms of volume loss. A schematic diagram of *a closed-loop SJE tester* is shown in Figure 10.14. Matsumura et al. [94] developed a *nozzle jet tester* with capacity of holding 4 specimens. The uniqueness of this tester was lower power consumption, smaller-sized work pieces, lower erodent quantity, precise impact angle, and ease of control on concentration of particles. The major drawback of this tester was that the specimen remained submerged and the velocity remained unmeasured during experimentation.

Lin and Shao [95] modified the conventional *SJE tester* and advanced it for controlling the impingement angle and operation for the high concentration of particles. The drawback of this tester was toughness in maintaining the high velocity. Iwai et al. [96] designed a *micro-slurry-jet erosion tester* to estimate the intrinsic erosion properties for the effect of standoff distance on target surface, particle size and shape, solid concentration, time duration, and rotational speed. Walker and Robbie [82] also modified the conventional *SJE tester* by using an air-operated diaphragm pump for the circulation of the slurry and evaluated the erosion at high velocities. They used

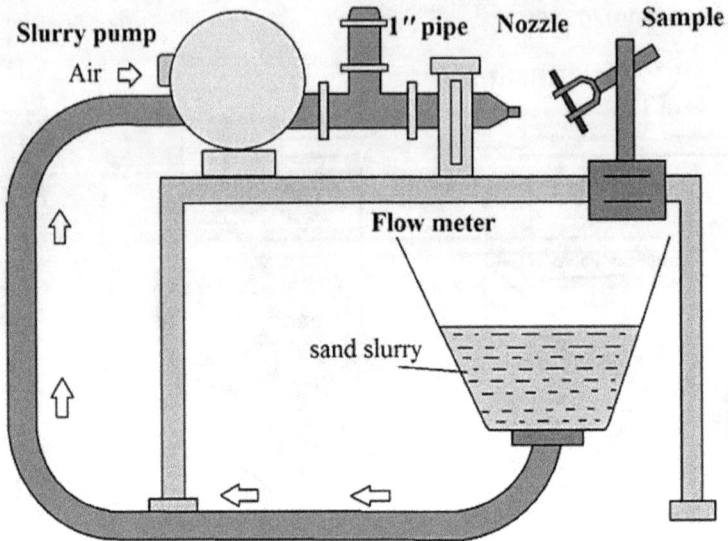

FIGURE 10.15 Schematic diagram of NRC slurry jet erosion tester [97].

the diaphragm pump to ensure the flow velocity through the nozzle. Xie et al. [97] developed a National Research Council (NRC) *SJE tester* and evaluated the erosion of different materials against the slurry at several impact angles. The components of NRC SJE are shown in Figure 10.15.

Whirling arm erosion tester is a collaborated version of tester which was developed by combining the slurry pot and erosion jet tester. This tester tests the erosion under the special type of condition. This tester contains 4 main components, namely (a) a motor that rotates the specimen at different velocities, (b) 2 arms with holders, (c) a slurry pot, and (d) a vacuum chamber that abolishes the aerodynamic effects from the slurry pot. In the early years, the *whirling arm erosion tester* was used to measure the erosion wear in dry condition only [98, 99]. The major drawback of this tester was the development of centrifugal stresses and extreme localized heating. Al-Bukhaiti et al. [86] developed a *whirling arm rig* to evaluate the erosion wear for a wide range of impact angles, as shown in Figure 10.16. The major advantage of this tester was that the specimen holder can be set-and-locked at any impact angle. The drawback of this tester was unable to develop of any erosion model.

10.6 PROTECTION TO WEAR APPLICATIONS USING THERMAL SPRAY COATINGS

Different types of surface modifications are employed to protect the offshore and hydraulic machinery. Thermal spray coating is basically a term used for the number of coating techniques employed on nonmetallic and metallic coatings. It is categorized into main three groups, according to the energy sources: electric arc, plasma arc spray, and flame spray. In this process, the coating can be deposited using powder, rod, or wire for required application. The high-speed molten droplet or particles are

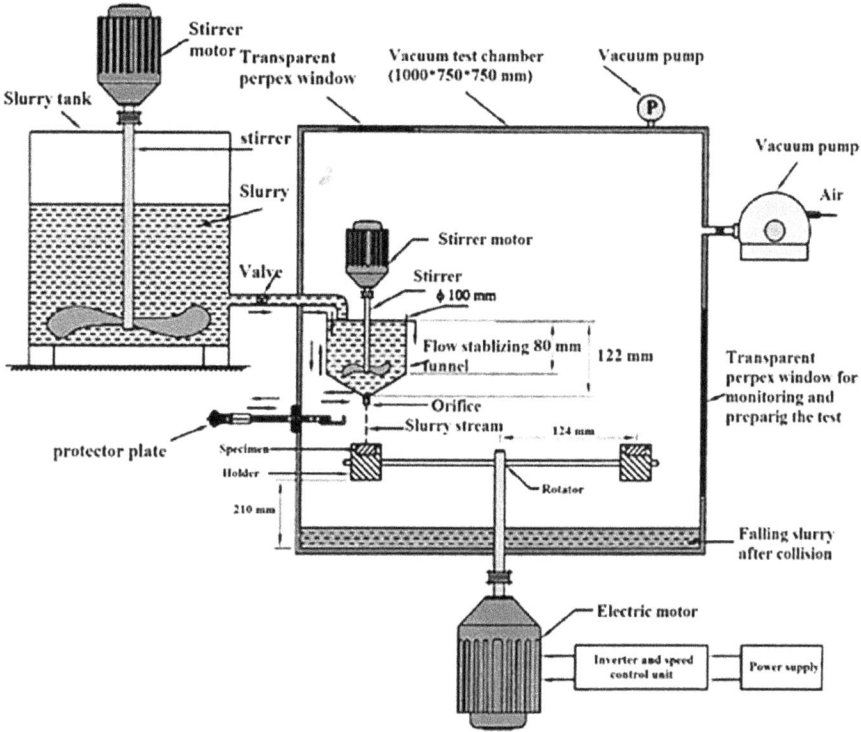

FIGURE 10.16   Schematic diagram of whirling arm rig [86].

propelled on the target surface by jet or fuel gases. The impact of molten particles results in the formation of thick layer of coating to the target surface and formation of homogenous lamellar structure. A major advantage of thermal spray coating technique is that a wide range of spraying powders are used for coating, which are not decomposed during the process. Hard substrates of high melting points can be deposited to fully treated surfaces without affecting the mechanical properties or without thermal shock. The disadvantage is limited-size coating, not suitable for deep coating.

There is a large historical background behind the development of thermal spray coatings. In initial years, M. U. Schoop started to develop oxyacetylene welding gun for the Sn (Ti) wire feed during 1882–1889 in Zurich (Switzerland) [100]. Later, the welding gun was used for the powder particles to deposit them on the target surface, at supersonic velocity. The coating layers of required thickness were formed by the gradational impacting of droplets. In 1908, electric arc spray was used by Schoop, which permits the use of Zn and steel by wire-arc metalizing, for using Zn as corrosion-resistant material on the base material [100]. After World War II, the plasma arc and thermal spray coatings significantly developed. A major advantage of the plasma arc spray process was considerably higher process temperature compared to the ignition spray jet. However, the development in HVOF process remained rapid till 1995. During this period, few of modifications were done in HVOF process which are hypersonic-velocity impact fusion (HVIF) and high-velocity-air fuel (HVAF) process. HVIF process was developed by Browning in 1992 [101]. This process also works on the same principle as HVOF but the static temperature of particles is kept below their melting points to avoid softening of the smallest particles. HVAF process was developed by Irving et al. [102] (introduced by Browning Thermal Systems, Inc.) in 1993. In this process, air is used instead of oxygen.

Coating deposition is generally performed at optimal process parameters available in popular literature. Applications of thermal spray coatings along with process parameters used for various industrial applications are shown in Figure 10.17. Cladding, nitriding, carburizing, and laser treatment are some of the other processes for improving erosion resistance of materials in high-temperature oxidation condition; however, they didn't show promising performance in offshore conditions [103]. Laser cladding has limitation as compared to thermal spray processes such as nonportable system, complication for larger parts, limitations of erosion- and corrosion-resistance properties, and inability to produce thin coatings and application to stainless steels (300 and 400 series), Ni/Co-based alloys (Inconel, Hastelloy, and Stellite) and carbides like tungsten carbide (WC) [104–106]. Plasma depositing arc-spraying process is most commonly used for the repair of equipment (piston rod, seals, pump plungers, shafts, and hubs). Surface-coating techniques like flame-spraying,

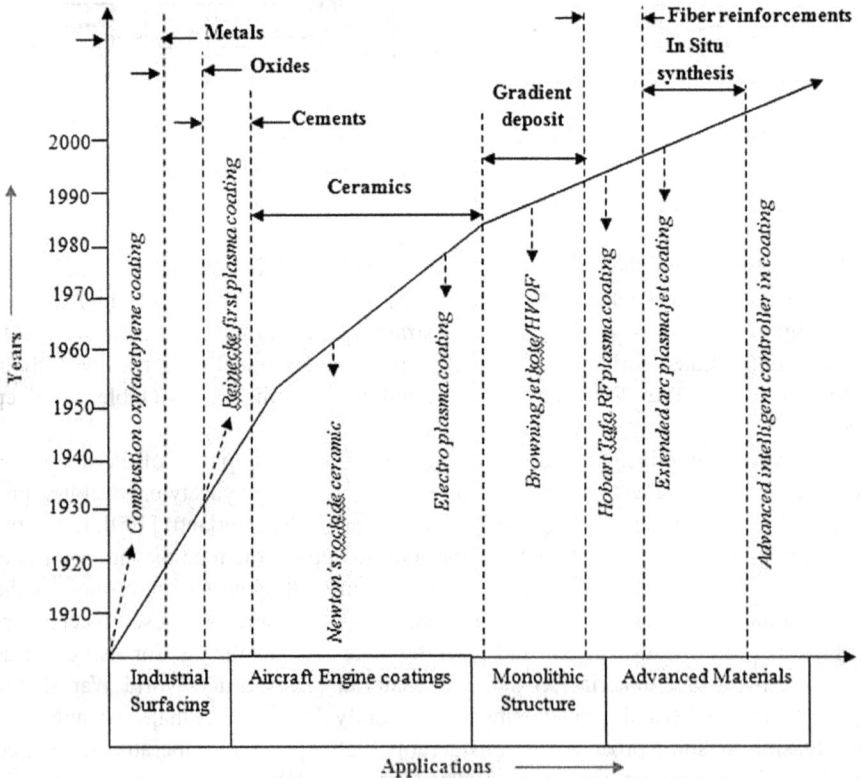

FIGURE 10.17 Applications of thermal spray coating in different industries.

detonation-gun (D-gun)-spraying, high-velocity oxy-fuel (HVOF)-spraying, plasma arc-spraying (PS), atmospheric plasma-spraying (APS), vacuum plasma-spraying (VPS), and laser-spraying (LS) technique are universally established protection methods to offshore and hydraulic machinery. In HVOF spray coating, the nozzle or gun is used to increase the energy of coating powder particles and operates at confined combustion. Typical HVOF coating is used to deposit the thermal-spraying powder at hypersonic velocities of fuel gases for the generation of extra energy. Properties and operation conditions of different thermal spray techniques are listed in Table 10.3.

10.7 VARIOUS TYPES OF THERMAL-SPRAYING POWDERS TO IMPROVE WEAR RESISTANCE PROPERTIES

Various types of materials are used for thermal spray coating which are classified in Table 10.4. Commercial thermal-spraying powders like Co-, WC-, or Ni-based alloys were used to improve the wear resistance of turbine materials exposed to erosion and corrosion. Generally, the selection of coating powder for the erosion wear applications is dependent on hardness and nature of the material. For the slurry erosion wear at lower impact angle of erodent particles with a target (ductile material), the material shows similar mechanism to abrasion and requires a higher hardness. In case of brittle material and impact angles closer to 90°, a high toughness coating is required to increase the performance of base material. A high fatigue resistance coating is required in case of cavitation or liquid impact. Thermal spray coating most commonly used for erosion wear include cermet coating like tungsten carbide–cobalt, chromium carbide (Cr_3C_2)–nickel chromium (particularly for temperatures above 540°C), fused self-fluxing alloys, non-ferrous alloys, aluminum bronze, monel, oxide ceramics like chromium oxide and alumina, and various alloys of iron, nickel, chromium, or cobalt. Ceramic and ceramic–metallic (cermets) are hard, sustain high temperatures, and are resistant to corrosion. Various coating powders used by researchers during the investigation are listed in Table 10.5. A detailed discussion about thermal-spraying powders is given as follows.

10.7.1 METAL-MATRIX COATINGS

There are several metal-matrix alloys which can be coated on different substrate materials by adopting the thermal spray process. Commonly used alloys for thermal spray coatings are Inconel, Hastelloy, Wallex, Stellite, Colmonoy, etc. Inconel is an austenitic nickel–chromium-based high-performance alloy which is oxidation- and corrosion-resistant. Inconel-625 and Inconel-718 are commercially used in a wide range of applications in erosion and corrosion environments. Hastelloy is a nickel–molybdenum alloy and known for its high resistance to corrosion. Hastelloy shows outstanding performance against highly oxidizing and reducing agents, and thus is suitable for severe corrosive environments [111]. Hastelloy is sprayed to valves and pipes used in the chemical processing and petrochemical industries. Chen et al. [112] studied the erosion corrosion behavior of the Hastelloy C22 coating. They found that an increase in flow velocity results in increase of total weight loss and corrosion

TABLE 10.3
Properties and Operation Conditions of Thermal Spray Techniques

S. No.	Spraying Technique	Particle Velocity (m/s)	Temperature (°C)	Adhesion (MPa)	Porosity (%)	Fuel/Power	Carrier Gas	Example
1	Flame-spraying (powder)	60–70	≈3100	6–10	7–12	Fuel gas (acetylene or propane)	Same fuel gas	Al, Zn for corrosion protection
2	Flame-spraying (wire)	120–140	≈3100		5	Fuel gas (acetylene or propane)	Same fuel gas	
3	Arc-spraying	100–170	≈7000	10–20	3–15	Electricity	Compressed air	
4	Plasma-spraying	150–600	≈15000	20–70	1–8	Electricity (W-cathode, Cu-anode)	Ar, He, N$_2$, or mixture	Ceramic coating
5	Detonation-gun-spraying	600–800	≈4200	60	<1	Acetylene + Oxygen	Ar, He, N$_2$, or mixture and N$_2$ for purging	
6	HVOF-spraying	600–1200	≈2750	>70	1–2	Oxy-propylene, Oxy-hydrogen and kerosene	Compressed air	Ceramic-metallic or metallic coating

TABLE 10.4
Materials Used for Thermal Spray Coating

S. No	Type/Phase	Material
1	Single-Phase	Alloys, polymer, ceramics, and intermetallic
2	Composite	WC/Co, Cr_3C_2/NiCr, NiCrAlY, and Al_2O_3
3	Layered	FGMs (functionally gradient materials)

current density; however, the passive current density decreases while velocity was increased above 4 m/s. The Hastelloy C22 coating showed lower corrosion resistance at 30° impingement angle due to dominant shear stress and highest value at 60° due to stronger protection of the passive film. Incoloy is a nickel–iron–chromium alloy which has high-temperature strength and used for high-temperature applications. Generally, Incoloy is non-suitable for severely corrosive environments; however, its few grades were designed with increased corrosion resistance [112]. Incoloy has resistance to seawater, brine, sour gas, and chloride, and hence can be sprayed in various applications used in oil and gas industries like propeller shafts, chemical processing equipment, gas turbines, hot vessels for food and water, aircraft, and tank trucks [113]. Yi et al. [113] studied the critical flow velocity (CFV) for erosion–corrosion of Inconel 600, Inconel 625, and Incoloy 825 alloys in saline-sand solution. They evaluated the CFV values of Inconel 600, Inconel 625, and Incoloy 825 as 13, 14 and 14 m/s, respectively. They also reported that Inconel 625 and Incoloy 825 showed higher CFV values due to their higher repassivation rates as compared to Inconel 600.

10.7.2 WC-BASED COATINGS

There are three universally established types of WC-based coatings, namely WC-10Co, WC-170Co, and WC-10Co-4Cr. WC-based coatings consist of hard carbides like WC and W_2C which enhance the hardness and brittleness of the ductile matrix. Tungsten carbides are manly present in two-phase cast eutectic (WC/W_2C) and macrocrystalline (WC) carbides [77, 114]. By thermal spray coating process, the W_2C and WC phases present in thermal-spraying powder enhances various mechanical properties, which improves the erosion performance of exposed surface [114]. WC-based coating shows significant effect on abrasion and erosion at room temperature. WC-based coating powder is deposited on different substrates either by HVOF or plasma-spraying technique. Plasma depositing arc-spraying method is used to deposit the thick layer of WC metal at different positions. HVOF-sprayed WC-based coatings are especially dense and adhered layer of hard-carbides as per the operation conditions. Decomposition of the WC powder can be controlled by the use of HVOF-spraying technique at high velocity with low temperature flame. Addition of Co and Cr power 4%–15% by weight increases the wear resistance of coating powder [77, 114]. Jia and Fischer [115] reported that carbide gain size is controlled in ductile

TABLE 10.5
Various Coating Powder Used by Researchers for HVOF-Spraying

Authors	Spray Technique	Coatings	Porosity	Average Microhardness (HV)	XRD Phase	Substrate Material
Machio et al. [107]	HVOF	WC-12Co WC-17Co WC-10VC-12Co WC-10VC-12Co	------	1101 1056 1045 942	------	Stainless steel
Goyal et al. [108]	HVOF	Cr_3C_2-NiCr WC-10Co-4Cr WC-10Co-4Cr + 5% (Nano-WC-12Co)	1.8	960 1677 1795	WC, W_2C, Co, and Co_3W_3C WC, W_2C, Co, and Co_3W_3C	CA6NM
Liu et al. [109]	HVAF	WC-10Co-4Cr + 10% (Nano-WC-12Co) WC-10Co-4Cr + 15% (Nano-WC-12Co)	------	1826 1873	WC, W_2C, Co, and Co_3W_3C WC, W_2C, Co, and Co_3W_3C	AISI 304
Goyal et al. [110]	HVOF	WC-12Co-4Cr	1.48	1160	WC, W_2C, Co, and Cr	CF8M
Grewal et al. [69]	HVOF	Ni-20Al_2O_3 Ni-40Al_2O_3 Ni-60Al_2O_3	1.3±0.2 1.8±0.16 2.5±0.25	563±90 714±78 1151±78	------	CA6NM

cobalt matrix which helps to increase the erosion performance of WC-Co coating effectively. WC-based coating enhanced the surface properties of base material for erosion wear resistance, because WC coating has the highest percentage (>90%) of fine grains having higher wear resistance. WC consists of a very small distance between two fine grains; due to this, the week grains can be sheltered. Erosion rate of WC-12Co and WC-10Co-4Cr HVOF coatings at different velocities and impact angles is shown in Figure 10.18. By comparison, it can be seen that WC-12Co HVOF coating shows higher erosion than WC-10Co-4Cr HVOF coating at 90° impact conditions [116]. WC particles detaches after the initial erosion that caused removal of binder that results in cracking and fracturing of coating [116], as seen from the SEM of erosion mechanisms (Figure 10.19).

10.7.3 CHROMIUM CARBIDE

Chromium carbide powder is deposited on the surface of metal substrates by different techniques like APS, LS, HVOF, etc. The Cr_3C_2 coating powder increases the percentage of chromium in the matrix of different powders which change the surface properties of base materials and corrosion resistance. The most versatile coating is the Cr_3C_2-NiCr coating, composed of Cr_3C_2 and NiCr particles [117]. In Cr_3C_2-NiCr coating, the Cr_7C_3 and $Cr_{23}C_6$ crystalline carbides form in very low intensity peaks in XRD patterns whereas Cr_3C_2 forms high intensity peaks [118]. By comparison, it can be seen that Cr_3C_2-25NiCr HVOF coating shows lower erosion than WC-12Co and WC-10Co-4Cr HVOF coating [116]. A typical SEM image of eroded Cr_3C_2-25NiCr is shown in Figure 10.19d. The cracking is the major erosion mechanism in Cr_3C_2-25NiCr HVOF coating [116]. Goyal et al. [119] reported that the HVOF Cr_2O_3-sprayed CA6NM steel have better erosion resistance than Cr_2O_3-sprayed CF8M steel.

10.7.4 NI/CR-BASED MATERIALS

Ni-Cr-based coatings are popular for the application of erosion and corrosion due to high-phase Ni and corrosion-resistant Cr content. The Cr added Ni (Pure metal), and

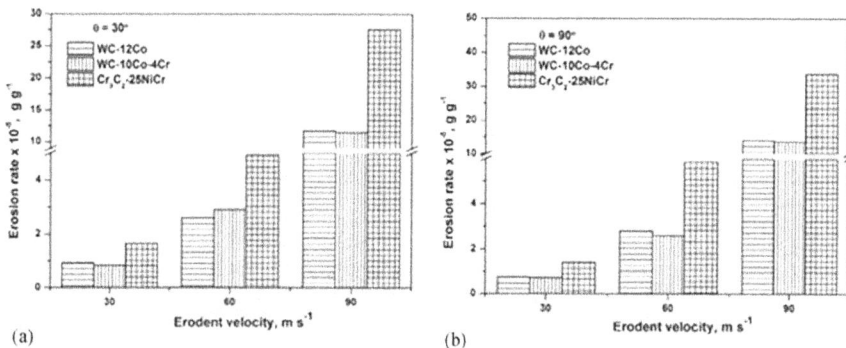

FIGURE 10.18 Erosion rate of WC-12Co, WC-10Co-4Cr, and Cr_3C_2-25NiCr coatings at impingement velocities of 30, 60, and 90 m/s at (a) 30° and (b) 90° [116].

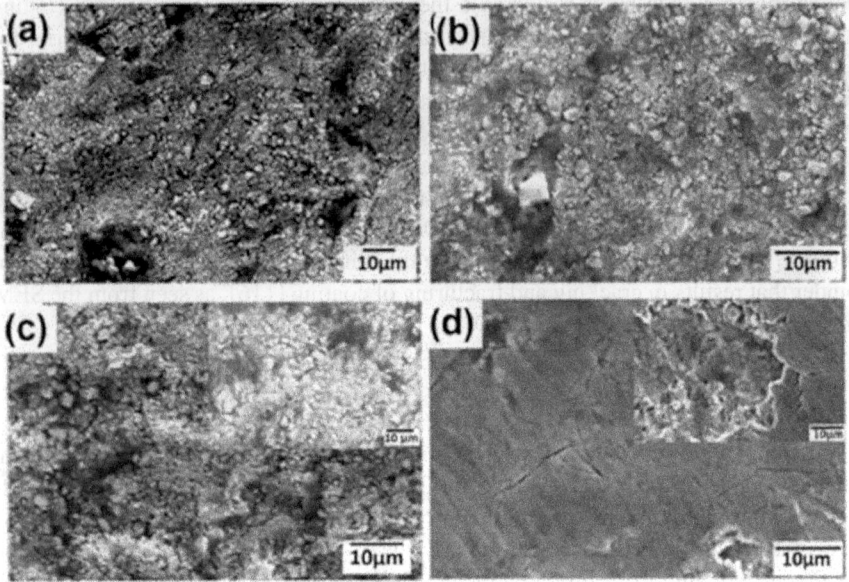

FIGURE 10.19 SEM images of eroded rate of (a) WC-12Co, 30 m/s, (b) WC-10Co-4Cr, 60 m/s, (c) WC-12Co, 90 m/s, and (d) Cr_3C_2-25NiCr, 90 m/s at 90° impact angle [116].

Cr_2O_3 added Ni (ceramics-based) coatings are well established in terms of their composition [84]. Singh et al. [49] studied the erosion performance of Ni-$20Cr_2O_3$, Ni-$20Al_2O_3$, and Al_2O_3-$20TiO_2$ HVOF coatings. Ni-$20Cr_2O_3$ and Ni-$20Al_2O_3$ coatings were superior over Al_2O_3-$20TiO_2$ coating; however, the Ni-$20Cr_2O_3$ coating has better erosion resistance than Ni-$20Al_2O_3$ coating. Grewal et al. [120] studied the erosion behavior of HVFS Ni-$20Al_2O_3$ coating and tested the optimum percentage of Al_2O_3 in Ni. The Ni-$40Al_2O_3$ coating showed the better performance as compared to the Ni-$20Al_2O_3$ and Ni-$60Al_2O_3$ coating. Erosion wear of various base materials was tested by depositing other Ni-based coatings like HVOF-sprayed NiCrBSiFe-WC(Co) [121], flame-sprayed NiCrBSi [122], flame-sprayed Ni-Cr-Fe-WC [123], HVOF-sprayed Ni-20Al [124], etc.

10.7.5 Co-Based Materials

In a study, the HVOF-sprayed Ni-based alloy (Colmonoy 88) shows lower erosion wear as compared to the HVOF-sprayed Co-based alloy (Stellite 6) coating [125], as illustrated in Figure 10.20a. From erosion mechanisms (Figure 10.20b, c), Colmonoy 88 was found brittle in nature as it shows fractures, craters, and pullouts. However, the Stellite 6 had a semi-brittle nature due to the presence of smears, craters, fractures, plowing, and lip formation mechanisms, as seen in Figure 10.20d, e. Szala et al. [11] tested the erosion performance of NiCrFeSiBC and CoCrWSiBC coatings. They found that Ni- and Co-based coatings had improved the erosion resistance of impeller material by 10%.

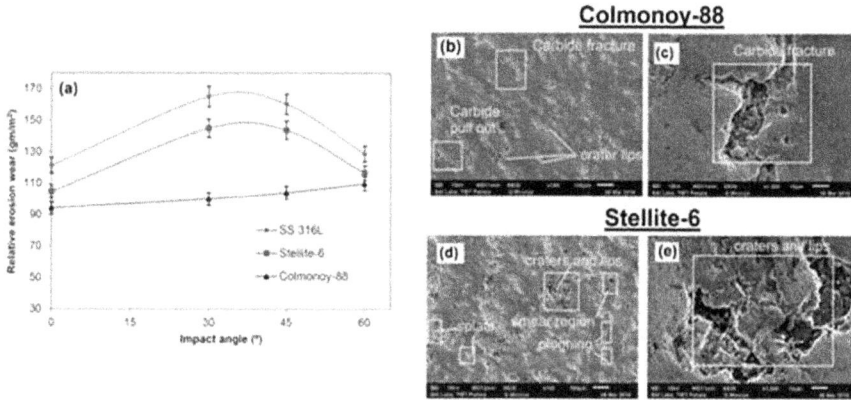

FIGURE 10.20 (a) Comparison of erosion wear between Colmonoy-88 and Stellite-6 HVOF alloy coating, and erosion mechanism on (b, c) Colmonoy-88 and (d, e) Stellite-6 at impact angle of $\alpha = 60°$ [125].

10.7.6 CERAMIC COATINGS

Recently, pure-ceramic and semi-ceramic coatings are also being used for erosion- and corrosion-resistant applications. The TiO_2, Al_2O_3, ZrO_2, Y_2O_3, and Cr_2O_3-based coatings are widely used by various researchers and inventors [126]. Ni-20Cr_2O_3 coating is an optimized composition of Ni Chromia coating. By comparison (Figure 10.21), the Ni-20Cr_2O_3 HVOF coating showed higher erosion compared to the WC-10Co-4Cr HVOF coating [84]. Goyal et al. [108] found that WC-10Co-4Cr-coated CF8M steel showed better erosion resistance as compared to the Al_2O_3-13TiO_2-coated CF8M steel (Figure 10.22). However, CF8M steel shows ductile erosion mechanisms, as seen in Figure 10.23a. On the other hand, WC-10Co-4Cr and Al_2O_3-13TiO_2-coated CF8M steels showed brittle erosion mechanisms, as seen in Figure 10.22b, c.

10.8 COATING PARAMETERS CONTRIBUTE IN CONTROLLING EROSION

Coating properties like hardness, toughness, fractural toughness, porosity, bulk modulus, elongation, coating type, bond strength, thickness, etc., play a crucial role in controlling the erosion wear. Table 10.6 illustrates the harness comparison of few hard coatings. From past results, it is observed that the harder coatings are more durable in erosion environment. The properties of thermal spray coatings mainly depend on the coating technique. Table 10.7 presents a comparison of coating parameters obtained from different flame-spraying processes [118]. The coating thickness and surface roughness are also important parameters in deciding the adherence, hardness, and other properties which are interlinked to erosion resistance. Generally, approximately 200 µm thickness of coating is preferred for better slurry erosion

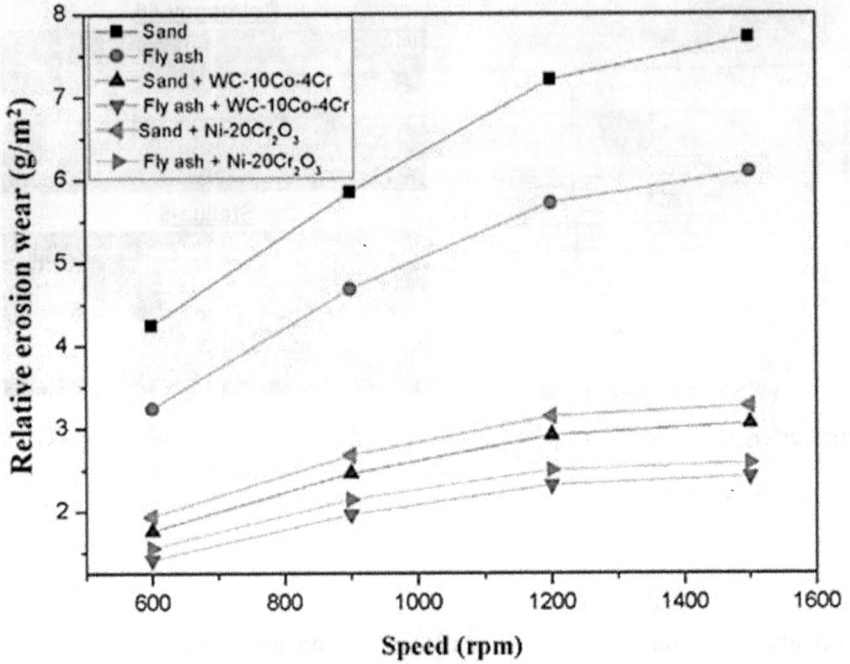

FIGURE 10.21 Comparison of erosion wear between Ni-20Cr$_2$O$_3$ and WC-10Co-4Cr HVOF coating [84].

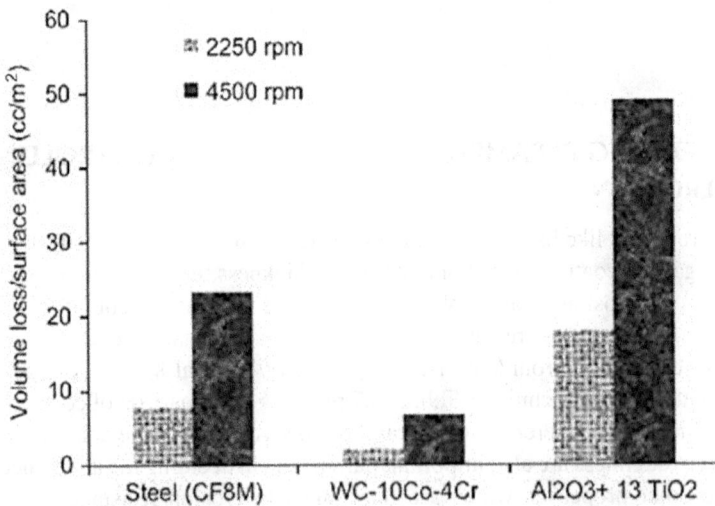

FIGURE 10.22 Volume loss comparison of bare, WC-10Co-4Cr, and Al$_2$O$_3$-13TiO$_2$-coated CF8M steel [108].

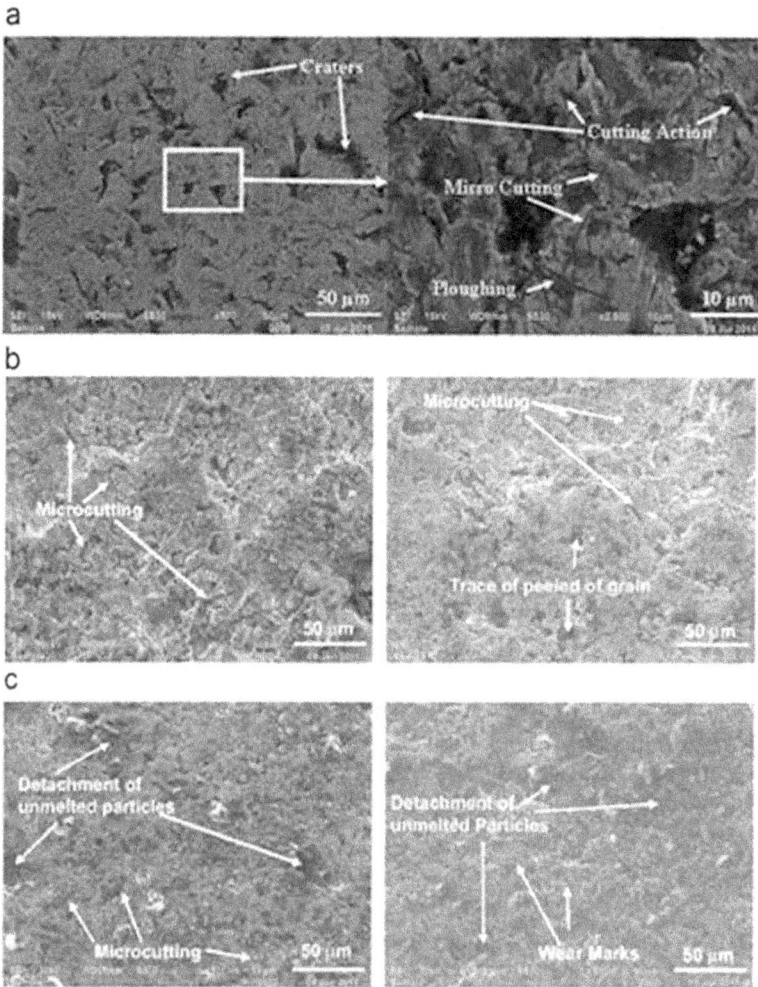

FIGURE 10.23 SEM images of eroded surface of the (a) bare CF8M steel, (b) WC-10Co-4Cr, and (c) Al₂O₃-13TiO₂-coated CF8M steel at 2250 rpm of pot tester's speed using 10,000 ppm slurry concentration and 100 mm average particle size [108].

TABLE 10.6
Typical Hardness Values HV$_{0.3}$ for the Main Commercial Compositions

S. No.	Powder	Vickers Hardness, HV$_{0.3}$	
		Ref. [118]	Refs. [119, 127, 128]
1	WC-12Co	1376	1250
2	WC-17Co	1358	1200
3	WC-10Co-4Cr	1274	1350
4	WC-(W, Cr)$_{2C}$.7Ni	1242	1300
5	Cr$_3$C$_2$-25NiCr	1080	1150

TABLE 10.7
Properties Achieved by Different Coating Processes [118]

Properties	Values				
	Flame Spray	Arc Spray	Atmospheric Plasma Spray	Detonation Gun Spray	HVOF Spray
Porosity (%)	10–20	10–20	1–7	0.55.0	~1%
Bond strength (MPa)	15–30	<20	>30	>70	>40
Thickness (μm)	100–2500	100–1500	50–500	~300	----

resistance [129]. Surface roughness of the coating should be as much lower as possible for erosion-based applications.

10.9 REMARKS

Thermal-spraying techniques are a universally used technology which is applied to various offshore applications. The life of components of hydraulic machineries can be improved either by changing the design of the component or by changing the selection of material or by changing the surface properties. So, thermal spray techniques can used in order to improve the surface properties of base materials used for the fabrication of different equipment and machinery. Various spraying techniques are available in literature to modify the surface of material, such as flame-spraying, arc-spraying, atmospheric plasma-spraying, vacuum-plasma-spraying, D-gun-spraying, laser-spraying, Underwater-spraying process, HVOF-spraying, and HVAF process. Coating powders are available in a wide range for the commercial applications in offshore industry. In this chapter, it is remarked that WC-based powders can be used to provide hard surfaces; however, these coatings don't perform well in high-temperature oxidation applications such as boiler. Ni-based alloy coatings are found to be promising for high-erosion-duty conditions in offshore industry. Cr-based and chromium carbide coatings are able to withstand erosion–corrosion conditions. Coating properties like hardness, toughness, fractural toughness, porosity, bulk modulus, elongation, coating type, bond strength, thickness, etc., play a crucial role in controlling the erosion wear. At last, it can be said that a good coating can be produced by choosing appropriate coating process and material, and optimizing the coating process as well as coating properties.

REFERENCES

1. 2019, *Ministry of Power Report*.
2. 2019, *International Energy Agency Report (IEA)*, Paris.
3. Awal, R., Chitrakar, S., Thapa, B. S., and Neopane, H. P., 2019, "Analysis A General Review on Methods of Sediment Sampling and Mineral Content Analysis," *J. Phys. Conf. Ser*, 1266(August), p. 012005.
4. Padhy, M. K., and Saini, R. P., 2009, "Effect of Size and Concentration of Silt Particles on Erosion of Pelton Turbine Buckets," *Energy*, 34(10), pp. 1477–1483.

5. Chitrakar, S., Dahlhaug, O. G., and Neopane, H. P., 2018, "Numerical Investigation of the Effect of Leakage Flow through Erosion-Induced Clearance Gaps of Guide Vanes on the Performance of Francis Turbines," *Eng. Appl. Comput. Fluid Mech.*, 12(1), pp. 662–678.

6. Kapali, A., Chitrakar, S., Shrestha, O., Neopane, H. P., and Thapa, B. S., 2019, "A Review on Experimental Study of Sediment Erosion in Hydraulic Turbines at Laboratory Conditions," *J. Phys. Conf. Ser.*, 1266(1), p. 012016.

7. Sangroula, D. P., 1970, "Hydropower Development and Its Sustainability With Respect to Sedimentation in Nepal," *J. Inst. Eng.*, 7(1), pp. 56–64.

8. Swarnakar, N. K., Rao, V. S., and Tripathi, S., 2008, "Innovative Use of Technology to Curb the Menace of Silt Erosion in Hydro Turbines," *Water Energy Abstr.*, 18(1), pp. 27–28.

9. Fraenkel, P., Paish, O., Bokalders, V., Harvey, A., Brown, A., and Edwards, R., 1991, "Micro-Hydro: A guide for Development workers. Power," *MDG Publ., Water Res. Library.*

10. Arne, K., 2014, *No Title*, Trondheim, Norway.

11. Szala, M., Hejwowski, T. J., and Lenart, I., 2014, "Cavitation Erosion Resistance of Ni-Co Based Coatings," *Adv. Sci. Technol. Res. J.*, 8(21), pp. 36–42.

12. Padhy, M. K., and Saini, R. P., 2012, "Study of Silt Erosion Mechanism in Pelton Turbine Buckets," *Energy*, 39(1), pp. 286–293.

13. Rai, A. K., Kumar, A., and Staubli, T., 2017, "Hydro-Abrasive Erosion in Pelton Buckets: Classification and Field Study," *Wear*, 392–393(April), pp. 8–20.

14. Karandikar, D. A., 2015, "*HVOF Coatings to Combat Hydro Abrasive Erosion*," *International Conference of the Hydropower Sustain Device*, Dehradun, India, pp. 171–179.

15. Bajracharya, T. R., Acharya, B., Joshi, C. B., Saini, R. P., and Dahlhaug, O. G., 2008, "Sand Erosion of Pelton Turbine Nozzles and Buckets: A Case Study of Chilime Hydropower Plant," *Wear*, 264(3–4), pp. 177–184.

16. Pradhan, P., Joshi, P., Biswakarma, M., and Stole, H., 2004, "*Sediment and Thermodynamic Efficiency Measurement at Jhimruk Hydropower Plant, Nepal in Monsoon 2003.*," *Proceedings of Ninth International Symposium on River Sedimentation*, China.

17. Singh, B., Thapa, B., and Gunnar, O., 2012, "Current Research in Hydraulic Turbines for Handling Sediments," *Energy*, 47(1), pp. 62–69.

18. Neopane, H.P., 2010, "*Sediment Erosion in Hydro Turbines*," Norwegian University of Science and Technology, Trondheim.

19. Chitrakar, S., Neopane, H. P., and Dahlhaug, O. G., 2016, "Study of the Simultaneous Effects of Secondary Flow and Sediment Erosion in Francis Turbines," *Renew. Energy*, 97, pp. 881–891.

20. Sharma, H. K., 2010, "Power Generation in Sediment Laden Rivers," *Int. J. Hydropower Dams*, 17 (6), pp. 112–116.

21. Neopane, H. P., Chitrakar, S., and Dahlhaug, O. G., 2019, "Role of Turbine Testing Lab for Overcoming the Challenges Related to Hydropower Development in Nepal," *IOP Conf. Ser. Earth Environ. Sci.*, 240(4), pp. 042012.

22. Rajaram, S. S., Harvey, T. J., and Wood, R. J. K., 2009, "Erosion-Corrosion Resistance of Engineering Materials in Various Test Conditions," *Wear*, 267(1–4), pp. 244–254.

23. Finne, I., 1960, "Erosion of Surfaces," *Wear*, 3, pp. 87–103.

24. Levy, A. V., 1996, *Solid Particle Erosion and Erosion-Corrosion of Materials*, ASM International, Ohio.

25. Bellman, R., and Levy, A., 1981, "Erosion Mechanism in Ductile Metals," *Wear*, 70(1), pp. 1–27.

26. Wang, Y. F., and Yang, Z. G., 2008, "Finite Element Model of Erosive Wear on Ductile and Brittle Materials," *Wear*, 265(5–6), pp. 871–878.

27. Sundararajan, G., Roy, M., and Venkataraman, B., 1990, "Erosion Efficiency-a New Parameter to Characterize the Dominant Erosion Micromechanism," *Wear*, 140(2), pp. 369–381.

28. Clark, H. M. I., and Wong, K. K., 1995, "Impact Angle, Particle Energy and Mass Loss in Erosion by Dilute Slurries," *Wear*, 186–187(PART 2), pp. 454–464.

29. Grewal, H. S., Agrawal, A., and Singh, H., 2013, "Design and Development of High-Velocity Slurry Erosion Test Rig Using CFD," *J. Mater. Eng. Perform.*, 22(1), pp. 152–161.

30. Levy, A. V., and Chik, P., 1983, "The Effects of Erodent Composition and Shape on the Erosion of Steel," *Wear*, 89(2), pp. 151–162.

31. Bahadur, S., and Badruddin, R., 1990, "Erodent Particle Characterization and the Effect of Particle Size and Shape on Erosion," *Wear*, 138(1–2), pp. 189–208.

32. Zou, R. P., and Yu, A. B., 1996, "Evaluation of the Packing Characteristics of Mono-Sized Non-Spherical Particles," *Powder Technol.*, 88(1), pp. 71–79.

33. Hentschel, M. L., and Page, N. W., 2003, "Selection of Descriptors for Particle Shape Characterization," *Part. Syst. Charact.*, 20(1), pp. 25–38.

34. Bouwman, A. M., Bosma, J. C., Vonk, P., Wesselingh, J. A., and Frijlink, H. W., 2004, "Which Shape Factor(s) Best Describe Granules?," *Powder Technol.*, 146(1–2), pp. 66–72.

35. Heilbronner, R., and Keulen, N., 2006, "Grain Size and Grain Shape Analysis of Fault Rocks," *Tectonophysics*, 427(1–4), pp. 199–216.

36. Desale, G. R., Gandhi, B. K., and Jain, S. C., 2006, "Effect of Erodent Properties on Erosion Wear of Ductile Type Materials," *Wear*, 261(7–8), pp. 914–921.

37. Woldman, M., van der Heide, E., Schipper, D. J., Tinga, T., and Masen, M. A., 2012, "Investigating the Influence of Sand Particle Properties on Abrasive Wear Behaviour," *Wear*, 294–295, pp. 419–426.

38. Walker, C. I., and Hambe, M., 2015, "Influence of Particle Shape on Slurry Wear of White Iron," *Wear*, 332–333, pp. 1021–1027.

39. Gandhi, B. K., Singh, S. N., and Seshadri, V., 2003, "A Study on the Effect of Surface Orientation on Erosion Wear of Flat Specimens Moving in a Solid-Liquid Suspension," *Wear*, 254, pp. 1233–1238.

40. Elkholy, A, 1983, "Prediction of Abrasion Wear for Slurry Pump Materials," *Wear*, 84(1), pp. 39–49.

41. Tilly, G. P., 1973, "A Two Stage Mechanism of Ductile Erosion," *Wear*, 23(1), pp. 87–96.

42. Gupta, R., Singh, S. N., and Seshadri, V., 1995, "Study on the Uneven Wear Rate in a Slurry Pipeline on the Basis of Mesurements in a Pot Tester," *Wear*, 184(4), pp. 169–178.

43. Verma, M., 1999, *"Silt Friendly Design of Turbine and Other Water Components,"* First *International Conference on Silting Problem in Hydropower Plants (CBIP)*, New Delhi, pp. 1–10.

44. Wiederhorn, S. M., and Hockey, B. J., 1983, "Effect of Material Parameters on the Erosion Resistance of Brittle Materials," *J. Mater. Sci.*, 18(3), pp. 766–780.

45. Wada, S., and Watanabe, N., 1987, "Solid Particle Erosion of Brittle Materials (Part 3)," *J. Ceram. Assoc. Japan*, 95(1102), pp. 573–578.

46. Stack, M. M., and Pungwiwat, N., 1999, "Slurry Erosion of Metallics, Polymers, and Ceramics: Particle Size Effects," *Mater. Sci. Technol.*, 15(3), pp. 337–344.

47. Desale, G. R., Gandhi, B. K., and Jain, S. C., 2011, "Development of Correlations for Predicting the Slurry Erosion of Ductile Materials," *J. Tribol.*, 133(3), pp. 1–10.

48. Prasad, B. K., Jha, A. K., Modi, O. P., and Yegneswaran, A. H., 2004, "Effect of Sand Concentration in the Medium and Travel Distance and Speed on the Slurry Wear Response of a Zinc-Based Alloy Alumina Particle Composite," *Tribol. Lett.*, 17(2), pp. 301–309.

49. Singh, J., 2019, "Analysis on Suitability of HVOF Sprayed Ni-20Al, Ni-20Cr and Al-20Ti Coatings in Coal-Ash Slurry Conditions Using Artificial Neural Network Model," *Ind. Lubr. Tribol.*, 71(7), pp. 972–982.

50. Dasgupta, R., Prasad, B. K., Jha, A. K., Modi, O. P., Das, S., and Yegneswaran, A. H., 1998, "Effects of Sand Concentration on Slurry Erosion of Steels," *Mater. Trans.*, 39(12), pp. 1185–1190.

51. Grewal, H. S., Arora, H. S., Agrawal, A., Singh, H., and Mukherjee, S., 2013, "Slurry Erosion of Thermal Spray Coatings: Effect of Sand Concentration," *Procedia Eng.*, 68(October 2015), pp. 484–490.

52. Chandel, S., Singh, S. N., and Seshadri, V., 2012, "Experimental Study of Erosion Wear in a Centrifugal Slurry Pump Using Coriolis Wear Test Rig," *Part. Sci. Technol.*, 30(2), pp. 179–195.

53. Hawthorne, H. M., 2002, "Some Coriolis Slurry Erosion Test Developments," *Tribol. Int.*, 35(10), pp. 625–630.

54. Kleis, I., and Kulu, P., 2008, "Influence of Particle Concentration," *Solid Particle Erosion*, Springer-Verlag London Limited, London, U.K. pp. 24–27.

55. Oka, Y. I., Okamura, K., and Yoshida, T., 2005, "Practical Estimation of Erosion Damage Caused by Solid Particle Impact: Part 1: Effects of Impact Parameters on a Predictive Equation," *Wear*, 259(1–6), pp. 95–101.

56. Singh, J., 2019, *"Investigation on Slurry Erosion of Different Pumping Materials and Coatings,"* Thapar Institute of Engineering and Technology, Patiala, India.

57. Turenne, S., Châtigny, Y., Simard, D., Caron, S., and Masounave, J., 1990, "The Effect of Abrasive Particle Size on the Slurry Erosion Resistance of Particulate-Reinforced Aluminium Alloy," *Wear*, 141(1), pp. 147–158.

58. Singh, J. P., Kumar, S., and Mohapatra, S. K., 2017, "Head Loss Investigations Inside 90° Pipe Bend for Conveying Of Fine Coal–Water Slurry Suspension," *Int. J. Coal Prep. Util.*, 00(00), pp. 1–17.

59. Peña Ballesteros, D. Y., Mendez Camacho, Y. E., and Barreto Hernandez, L. V., 2016, "Evaluation of the Synergistic Effect of Erosion-Corrosion on AISI 4330 Steel in Saline-Sand Multiphase Flow by Electrochemical and Gravimetric Techniques," *Int. J. Electrochem.*, 2016, pp. 1–7.

60. Karafyllias, G., Galloway, A., and Humphries, E., 2019, "The Effect of Low PH in Erosion-Corrosion Resistance of High Chromium Cast Irons and Stainless Steels," *Wear*, 420–421, pp. 79–86.

61. Bong, E., 2015, "Optimum solids concentration for solids suspension and solid–liquid mass transfer in agitated vessels," *Chem. Eng. Res. Des.*, 100, pp. 148–156.

62. Bong, E. Y., Eshtiaghi, N., Wu, J., and Parthasarathya, R., 2015, "Optimum Solids Concentration for Solids Suspension and Solid–Liquid Mass Transfer in Agitated Vessels," *Chem. Eng. Res. Des.*, 100, pp. 148–156.

63. Shehadeh, M., Anany, M., Saqr, K. M., and Hassan, I., 2014, "Experimental Investigation of Erosion-Corrosion Phenomena in a Steel Fitting Due to Plain and Slurry Seawater Flow," *Int. J. Mech. Mater. Eng.*, 9(1), pp. 1–8.

64. Dabirian, R., Mohan, R., Shoham, O., and Kouba, G., 2016, "Critical Sand Deposition Velocity for Gas-Liquid Stratified Flow in Horizontal Pipes," *J. Nat. Gas Sci. Eng.*, 33, pp. 527–537.

65. Bartosik, A., 2010, "Influence of Coarse-Dispersive Solid Phase on The 'Particles-Wall' Shear Stress in Turbulent Slurry Flow with High Solid Concentration," *Arch. Mech. Eng.*, 57(1), pp. 45–68.

66. Desale, G. R., Gandhi, B. K., and Jain, S. C., 2008, "Slurry Erosion of Ductile Materials under Normal Impact Condition," *Wear*, 264(3–4), pp. 322–330.

67. Islam, M. A., Alam, T., Farhat, Z. N., Mohamed, A., and Alfantazi, A., 2015, "Effect of Microstructure on the Erosion Behavior of Carbon Steel," *Wear*, 332–333, pp. 1080–1089.

68. Sharma, A., Kumar, A., and Tyagi, R., 2015, "Erosive Wear Analysis of Medium Carbon Dual Phase Steel under Dry Ambient Condition," *Wear*, 334–335, pp. 91–98.

69. Islam, M. A., and Farhat, Z. N., 2014, "Effect of Impact Angle and Velocity on Erosion of API X42 Pipeline Steel under High Abrasive Feed Rate," *Wear*, 311(1–2), pp. 180–190.

70. Jain, A., 1999, *"Silting Problems in Hydropower Projects: Indian Scenario," Proc. of 1st Int. Conf. on Silting Problems in Hydro Power Plants*, New Delhi, India, pp. 37–54.

71. Goel, D. B., and M.K. Sharma, 1996, *"Present State of Damages and Their Repair Welding in Indian Hydroelectric Projects," Proc. Silt Damages to Equipment in Hydro Power Stations and Remedial Measures*, New Delhi, India, pp. 137–152.

72. Thapa, B., and H. Brekke, 2004, *"Effect of Sand Particle Size and Surface Curvature in Erosion of Hydraulic Turbine," IAHR Symposium on Hydraulic Machinery and Systems*, Stockholm.

73. Gandhi, B. K., Singh, S. N., and Seshadri, V., 1999, "Study of the Parametric Dependence of Erosion Wear for the Parallel Flow of Solid-Liquid Mixtures," *Tribol. Int.*, 32(5), pp. 275–282.

74. Gandhi, B. K., Singh, S. N., and Seshadri, V., 2001, "Performance Characteristics of Centrifugal Slurry Pumps," *J. Fluid Eng.*, 123, pp. 271–280.

75. Clark, H. M., and Llewellyn, R. J., 2001, "Assessment of the Erosion Resistance of Steels Used for Slurry Handling and Transport in Mineral Processing Applications," *Wear*, 250–251(PART 1), pp. 32–44.

76. Clark, H. M. I., and Hartwich, R. B., 2001, "A Re-Examination of the 'particle Size Effect' in Slurry Erosion," *Wear*, 248(1–2), pp. 147–161.

77. Stack, M. M., and El-Badia, T. M. A., 2008, "Some Comments on Mapping the Combined Effects of Slurry Concentration , Impact Velocity and Electrochemical Potential on the Erosion – Corrosion of WC / Co – Cr Coatings," *Wear*, 264, pp. 826–837.

78. Evans, A. G., Gulden, M. E., and Rosenblatt, M., 1978, "Impact Damage in Brittle Materials in the Elastic-Plastic Response Regime," *Proc. R. Soc. Lond. A.*, 361, pp. 343–365.

79. Reddy, A. V., and Sundrarajan, G., 1986, "Erosion Behaviour of Ductile Materials with a Spherical Non-Friable Erodent," *Wear*, 111, pp. 313–323.

80. Levy, A., and Hickey, G., 2010, *Surface Degradation of Metals in Simulated Synthetic Fuels Plant Environments*, NACE Int., 82154, pp. 1–19.

81. Divakar, M., Agarwal, V. K., and Singh, S. N., 2005, "Effect of the Material Surface Hardness on the Erosion of AISI316," *Wear*, 259(1–6), pp. 110–117.

82. Walker, C. I., and Robbie, P., 2013, "Comparison of Some Laboratory Wear Tests and Field Wear in Slurry Pumps," *Wear*, 302(1–2), pp. 1026–1034.

83. Parent, L. L., and Li, D. Y., 2013, "Wear of Hydrotransport Lines in Athabasca Oil Sands," *Wear*, 301, pp. 477–482.

84. Singh, J., Kumar, S., and Mohapatra, S. K., 2017, "Tribological Analysis of WC–10Co–4Cr and Ni–20Cr2O3 Coating on Stainless Steel 304," *Wear*, 376–377, pp. 1105–1111.

85. Clark, H. M., Hawthorne, H. M., and Xie, Y., 1999, "Wear Rates and Specific Energies of Some Ceramic, Cermet and Metallic Coatings Determined in the Coriolis Erosion Tester," *Wear*, 233–235, pp. 319–327.

86. Al-Bukhaiti, M. A., Ahmed, S. M., Badran, F. M. F., and Emara, K. M., 2007, "Effect of Impingement Angle on Slurry Erosion Behaviour and Mechanisms of 1017 Steel and High-Chromium White Cast Iron," *Wear*, 262(9–10), pp. 1187–1198.

87. Thapa, B. S., Dahlhaug, O. G., and Thapa, B., 2015, "Sediment Erosion in Hydro Turbines and Its Effect on the Flow around Guide Vanes of Francis Turbine," *Renew. Sustain. Energy Rev.*, 49(October 2017), pp. 1100–1113.

88. Schumacher, W. J., 1987, *Ball Mill and Hub Test Methods for Slurry Erosion Evaluation of Materials*, ASTM Spec. Technol. Pub. Ohio, USA.

89. Clark, H. M., 2002, "Particle Velocity and Size Effects in Laboratory Slurry Erosion Measurements OR … Do You Know What Your Particles Are Doing ?," *Tribol. Int.*, 35, pp. 617–624.

90. Clark, M. C. I., 1993, "Specimen Diameter, Impact Velocity, Erosion Rate and Particle Density in a Slurry Pot Erosion Tester," *Wear*, 164–162, pp. 669–678.

91. Desale, G. R., Gandhi, B. K., and Jain, S. C., 2005, "Improvement in the Design of a Pot Tester to Simulate Erosion Wear Due to Solid-Liquid Mixture," *Wear*, 259(1–6), pp. 196–202.

92. Tuzson, J. J., 2015, "Laboratory Slurry Erosion Tests and Pump Wear Rate Calculations," *J. Fluids Eng.*, 106, pp. 135–140.

93. Fang, Q., Xu, H., Sidky, P. S., and Hocking, M. G., 1999, "Erosion of Ceramic Materials by a Sand/Water Slurry Jet," *Wear*, 224(2), pp. 183–193.

94. M. Matsumura, Oka, Y., and Yamawaki, M. 1987, "*Slurry Erosion-Corrosion of Commercially Pure Iron in Fountain-Jet Testing Facility-Mechanism of Erosion (Corrosion under Comparable Intensity of Erosion and Corrosion),*" *Proc. 7th Int. Conf. on Erosion by Liquid and Solid Impact, University of Cambridge*, Cambridge, Cambridgeshire, UK, Cambridge, p. 40.

95. Lin, F. Y., and Shao, H. S., 1991, "Effect of Impact Velocity on Slurry Erosion and a New Design of a Slurry Erosion Tester," *Wear*, 143, pp. 231–240.

96. Sun, B., Fan, J., Wen, D., and Chen, Y., 2015, "An Experimental Study of Slurry Erosion Involving Tensile Stress for Pressure Pipe Manifold," *Tribiology Int.*, 82, pp. 280–286.

97. Xie, Y., Jiang, J. J., Tufa, K. Y., and Yick, S., 2015, "Wear Resistance of Materials Used for Slurry Transport," *Wear*, pp. 1–7.

98. Tilly, G. P., and Sage, W., 1970, "The Interaction of Particle and Material Behaviour in Erosion Processes," *Wear*, 16(November 1969), pp. 447–465.

99. Madsen, B. W., 1985, "A Study of Parameters Using a New Constant-Wear-Rate Slurry Test," In: K. C. Ludema (Ed.), *Proceedings of the International Conference on Wear of Material*, American Society of Mechanical Engineers, New York, pp. 345–854.

100. Siegmann, S., and Abert, C., 2013, "100 Years of Thermal Spray: About the Inventor Max Ulrich Schoop," *Surf. Coatings Technol.*, 220, pp. 3–13.

101. Browning, J. B., 1992, "*Hypersonic Velocity Impact Fusion Spraying,*" *Proceedings of the International Thermal Spray Conference & Exposition*, Orlando, Florida, USA, p. 123.

102. Irving, R., Knight, R., and Smith, R. W., 1993, "The HVOF Process-the Hottest Topic in the Thermal Spray Industry," *Weld. J.*, pp. 25–30.

103. Babu, A., Arora, H. S., Singh, H., and Grewal, H. S., 2019, "Microwave Synthesized Composite Claddings with Enhanced Cavitation Erosion Resistance," *Wear*, 422–423, pp. 242–251.

104. Singh, S., Garg, P., and Goyal, D., 2017, "Comparison of Laser Cladding And Thermal Spraying Techniques Used For Wear Control," *Res. Cell An Int. J. Eng. Sci.*, 25(63019), pp. 222–229.

105. Meghwal, A., Anupam, A., Murty, B. S., and Berndt, C. C., 2020, *Thermal Spray High-Entropy Alloy Coatings : A Review*, Springer US.

106. Mostajeran, A., Shoja-Razavi, R., Hadi, M., Erfanmanesh, M., Barekat, M., and Savaghebi Firouzabadi, M., 2020, "Evaluation of the Mechanical Properties of WC-FeAl Composite Coating Fabricated by Laser Cladding Method," *Int. J. Refract. Met. Hard Mater.*, 88, p. 105199.

107. Machio, C. N., Akdogan, G., Witcomb, M. J., and Luyckx, S., 2005, "Performance of WC-VC-Co Thermal Spray Coatings in Abrasion and Slurry Erosion Tests," *Wear*, 258(1–4 SPEC. ISS), pp. 434–442.

108. Kumar, D., Singh, H., Kumar, H., and Sahni, V., 2012, "Slurry Erosion Behaviour of HVOF Sprayed WC–10Co–4Cr and Al2O3+13TiO2 Coatings on a Turbine Steel," *Wear*, 289, pp. 46–57.

109. Liu, S. L., Zheng, X. P., and Geng, G. Q., 2010, "Influence of Nano-WC-12Co Powder Addition in WC-10Co-4Cr AC-HVAF Sprayed Coatings on Wear and Erosion Behaviour," *Wear*, 269(5–6), pp. 362–367.

110. Goyal, D. K., Singh, H., Kumar, H., and Sahni, V., 2012, "Slurry Erosive Wear Evaluation of HVOF-Spray Cr 2 O 3 Coating on Some Turbine Steels," *J. Therm. Spray Technol.*, 21(September), pp. 838–851.

111. Karimi, A., Soltani, R., Ghambari, M., and Fallahdoost, H., 2017, "High Temperature Oxidation Resistance of Plasma Sprayed and Surface Treated YSZ Coating on Hastelloy X," *Surf. Coatings Technol.*, 321(May), pp. 378–385.

112. Chen, L., Bai, S. L., Ge, Y. Y., and Wang, Q. Y., 2018, "Erosion-Corrosion Behavior and Electrochemical Performance of Hastelloy C22 Coatings under Impingement," *Appl. Surf. Sci.*, 456(June), pp. 985–998.

113. Yi, J. Z., Hu, H. X., Wang, Z. B., and Zheng, Y. G., 2020, "On the Critical Flow Velocity for Erosion-Corrosion of Ni-Based Alloys in a Saline-Sand Solution," *Wear*, 458–459(January), pp. 13–21.

114. Thakur, L., Arora, N., Jayaganthan, R., and Sood, R., 2011, "An Investigation on Erosion Behavior of HVOF Sprayed WC-CoCr Coatings," *Appl. Surf. Sci.*, 258(3), pp. 1225–1234.

115. Jia, K., and Fischer, T. E., 1997, "Sliding Wear of Conventional and Nanostructured Cemented Carbides," *Wear*, 203–204, pp. 310–318.

116. Vashishtha, N., Khatirkar, R. K., and Sapate, S. G., 2017, "Tribological Behaviour of HVOF Sprayed WC-12Co, WC-10Co-4Cr and Cr3C2–25NiCr Coatings," *Tribiology Int.*, 105(June 2016), pp. 55–68.

117. Pelton, J. F., and Jr Koffskey, J. M., "Coating Composition, Method of Application, and Product Thereof," US Patent, 3, 150,p. 938. filed: 9.6.1960, granted: 29.9.1964 (also: GB 929,205)

118. Berger, L. M., 2015, "Application of Hardmetals as Thermal Spray Coatings," *Int. J. Refract. Met. Hard Mater.*, 49(1), pp. 350–364.

119. Goyal, D. K., Singh, H., Kumar, H., and Sahni, V., 2014, "Erosive Wear Study of HVOF Spray Cr 3 C 2 – NiCr Coated CA6NM Turbine Steel," *J. Tribol.*, 136(October), pp. 1–11.

120. Grewal, H. S., Agrawal, A., and Singh, H., 2013, "Slurry Erosion Performance of Ni-Al2O3 Based Composite Coatings," *Tribol. Int.*, 66, pp. 296–306.

121. Singh, J., Kumar, S., and Mohapatra, S. K., 2019, "Erosion Wear Performance of Ni-Cr-O and NiCrBSiFe-WC(Co) Composite Coatings Deposited by HVOF Technique," *Ind. Lubr. Tribol.*, 71(4), pp. 610–619.

122. Karimi, M. R., Salimijazi, H. R., and Golozar, M. A., 2016, "Effects of Remelting Processes on Porosity of NiCrBSi Flame Sprayed Coatings," *Surf. Eng.*, 32(3), pp. 238–243.

123. Arji, R., Dwivedi, D. K., and Gupta, S. R., 2009, "Some Studies on Slurry Erosion of Flame Sprayed Ni-Cr-Si-B Coating," *Ind. Lubr. Tribol.*, 61(1), pp. 4–10.

124. Saladi, S., Ramana, P. V., and Tailor, P. B., 2018, "Evaluation of Microstructural Features of HVOF Sprayed Ni–20Al Coatings," *Trans. Indian Inst. Met.*, 71(10), pp. 2387–2394.

125. Singh, J., Kumar, S., and Mohapatra, S. K., 2020, "Erosion Tribo-Performance of HVOF Deposited Stellite-6 and Colmonoy-88 Micron Layers on SS-316L," *Tribol. Int.*, 147(June 2018), p. 105262.

126. Singh, J., Kumar, S., and Mohapatra, S. K., 2019, "Tribological Performance of Yttrium (III) and Zirconium (IV) Ceramics Reinforced WC–10Co4Cr Cermet Powder HVOF Thermally Sprayed on X2CrNiMo-17-12-2 Steel," *Ceram. Int.*, 45(17), pp. 23126–23142.

127. Berger, L.-M., Zieris, R., and Saaro, S., 2005, "*Oxidation of HVOF-Sprayed Hardmetal Coatings*," *Proc Int Therm Spray Conf ITSC*, DVS Verlag, Düsseldorf, Basel, Switzerland, p. 8.

128. Oechsle, M., 2006, "*Carbide Containing Spray Powders and HVOF-Coatings*," *Conf Proc 7. HVOF-Kolloquium, 9–10 November 2006*, emeinschaft Thermisches Spritzen EV, Unterschleißheim, Erding, Germany, pp. 57–62.

129. Prashar, G., Vasudev, H., and Thakur, L., 2020, "Performance of Different Coating Materials against Slurry Erosion Failure in Hydrodynamic Turbines: A Review," *Eng. Fail. Anal.*, 115, p. 104622.

11 Bioinert and Bioactive Coatings for Human Body Implants

Ali Sabea Hammood, Rob Brittain, and Hitesh Vasudev

CONTENTS

11.1 INTRODUCTION

An aging population has increased the demand for implants to restore functions for those affected by disease or injury. By 2030, hip and knee replacements are expected to increase by 174% and 673%, respectively [1, 2]. This steep growth

DOI: 10.1201/9781003213185-11

has required biomaterials research to improve to match demand and increase the standard of living of those that will require implants.

The human body is a hostile environment and any implanted material will trigger a foreign body reaction (FBR) [3]. Previous generations of biomaterials that have been implanted into the body have failed due to corrosion, wear or simply could not match the mechanical requirements of the body function they are replacing often resulted in serious clinical complications [4]. Unfortunately, many materials that are able to match the difficult mechanical demands are metallic materials, namely stainless steel or titanium-based alloys. These materials have similar matching material properties to that of bone, but suffer from corrosion or wear limitations. To overcome these limitations, biomaterial coatings have been developed. These biomaterial coatings characteristically have excellent biocompatibility, wear- and corrosion resistance and are non-toxic. The material coatings generally come in two varieties, which either do not exhibit a response from the immune system (bioinert) or work with the surrounding tissue to ensure osteointegration (bioactive), but both while protecting the underlying metallic device. These coatings lack the mechanical strength and fracture toughness to be used alone but as a coating can be tailored to meet the demands of the environment [5].

This chapter will describe the interactions of bioinert and bioactive biomaterial coatings in the human body, along surface modification of these coatings on metallic-based implants. Thermal spraying has been identified as one of the most common application methods with the effect certain parameters have on corrosion- and wear resistance for the resultant films [6].

11.2 HUMAN BODY IMPLANT

11.2.1 INTRODUCTION

Biomaterials are synthetic or natural materials intended to function appropriately in a biological environment in which they are used to direct, supplement or replace the functions of living tissues of the human body [7].

- **Metallic Biomaterials As Implants**
  Biomaterials are synthetic or natural materials intended to function appropriately in a biological environment in which they are used to direct, supplement or replace the functions of living tissues of the human body [7].

  Biomaterials might be comprised of various types of materials, for example, metals, polymers, ceramics and composites. About (70–80%) of implants are from metallic materials because of their high mechanical properties, low cost and simplicity of production. Metallic biomaterials are used in a majority of devices for load-bearing parts, including artificial joints, osteosynthesis devices and spinal fixation and the number of applications increasing every year [8]. Figure 11.1 shows a variety of applications of metallic implants that are used in vivo for orthopedic surgery and dentistry.

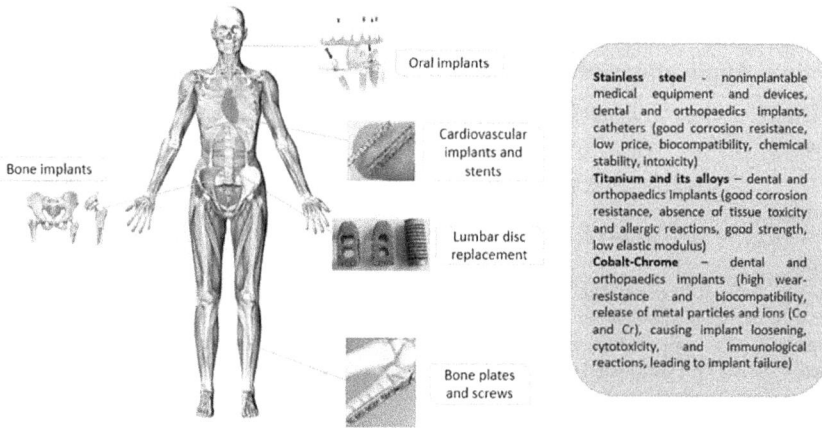

FIGURE 11.1 Application of metallic implants [9].

11.2.2 CRITERIA SELECTION OF MATERIALS TO IMPLANT

To ensure a biomaterial is suitable it must meet some essential requirements such as [10–12]:

1. Biocompatibility: One of the most significant features referring to how the material reacts with its environment.
2. Non-toxicity: The implant must not possess toxic influences because the toxic ions or other harmful products may be lead to cancer, allergies, deformities or inflammation.
3. Corrosion resistance: The implant must be able to resist corrosion. This characteristic is related to biocompatibility and toxicity.
4. Low modulus of elasticity: The materials that utilized in biomedical implant are usually possess a high modulus of elasticity approximately 5–10 times that of human bone, leading to stress shielding that may be cause death to surrounding bone cell with time.
5. Strength, ductility and fracture toughness: Researchers need to search about the implants with low dimensions that at the same time have generally high strength and ductility. In addition, they require to have high of fracture toughness to avoid crack and fracture on the application of high and cyclic loading through limbs motion.

11.2.3 STAINLESS STEEL

Stainless steel is the first metal material used in the biomedical field because of its lower cost, easy accessibility, excellent machinability, acceptable biocompatibility and great strength compared with others metallic implant materials [13].

Stainless-steel metals are broadly classified into four main categories: Martensitic, Ferritic, austenitic and duplex (austenitic plus Ferritic phase). Table 11.1 and Figure 11.2 show examples for biomedical implants made from stainless steel.

TABLE 11.1

Medical Applications of Categories of Stainless Steels

Material Type	Application Grade	Examples
Martensitic	Dental and surgical instruments	Bone curettes, chisels and gouges, dental burs, dental chisels, curettes and explorers
Ferritic	Very limited surgical instruments	Solid handles for instruments, guide pins and fasteners
Austenitic	A large number of non-implantable medical and many short-term implants like total hip replacements	Dental impression trays, coronary stents, orthopedic implants and fracture fixation devices
Duplex	Not yet applied in the biomedical field	Not yet applied in the biomedical field

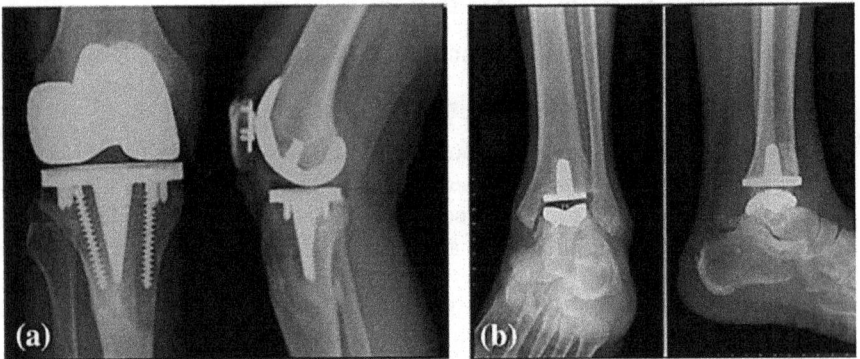

FIGURE 11.2   Examples of stainless steel implant: (a) knee, (b) ankle [14].

- **Standard Duplex Stainless Steel (2205 DSS)**

  Duplex stainless steels (DSS), SAF 2205 consists of two phases: an island-like austenite (FCC) and a continuous ferrite matrix (BCC) phase. In general, DSS possess properties better than both ferrite and austenite with better corrosion resistance and mechanical properties. The DSS possess (Ni) content lower than conventional 316L SS giving duplex an advantaged as Ni ions can lead to allergic reaction. Therefore, DSS can be an alternative material to austenitic stainless for biomedical application, with 2205 DSS properties like density = 7.8 g/cm^3, Young's modulus = 200Gpa, tensile strength = 655 MPa, coefficient of thermal expansion (10^{-6}/K) from 20°C to 500°C = 14, pitting resistance equivalent number (PREN) = 35–36, corrosion rate in 1% hydrochloric acid (mpy) = 0.1, corrosion rate in artificial saliva = 0.3, corrosion rate in 20% formic acid (mpy) = 1.3 [15, 16].

11.2.4   IMPLANT FAILURE

There are several reasons for the implant failure including [17, 18]:

1. Mismatch in modulus of elasticity between the bone and implant
2. Low fracture toughness
3. Corrosion and wear debris
4. Bacterial infection, etc.

Figure 11.3 explains the multitude of potential causes that lead to the failure of implants.

11.2.5   CORROSION MECHANISM OF METALLIC IMPLANTS

There are two main parameters linked with cause and mechanism of metallic implants corrosion:

1. Thermodynamic driving forces that cause the corrosion process involve oxidation and reduction reactions. Generally, two types of energy sources must be considered in the process of corrosion: (i) chemical driving force (ΔG) determine whether or not occurrence of corrosion, (ii) electrical (charge separation) or so-called electrical double layer (EDL) that originate electrical voltage in the interface between the metal with medium as shown in Figure 11.4 [19].

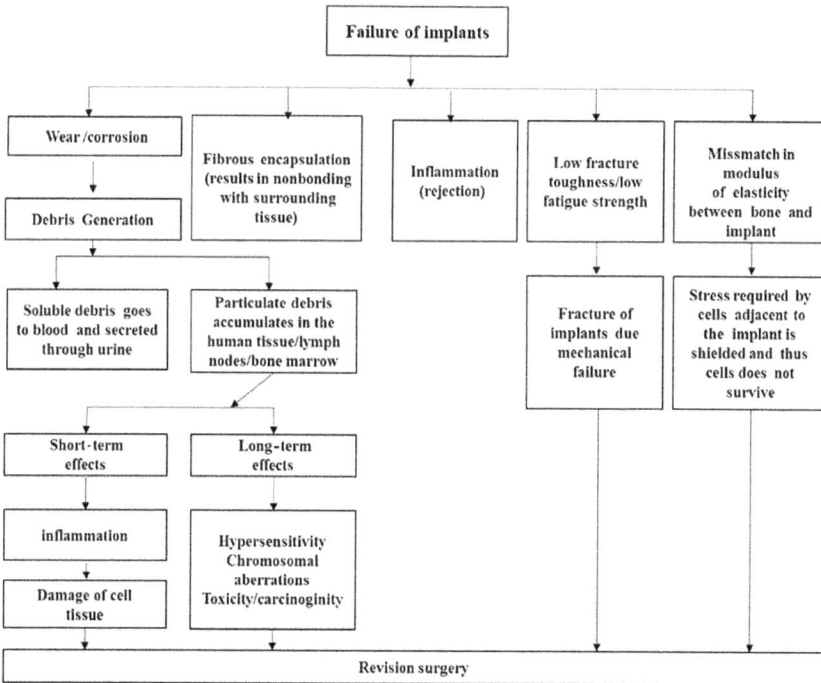

FIGURE 11.3   Different reasons of implant failure [19].

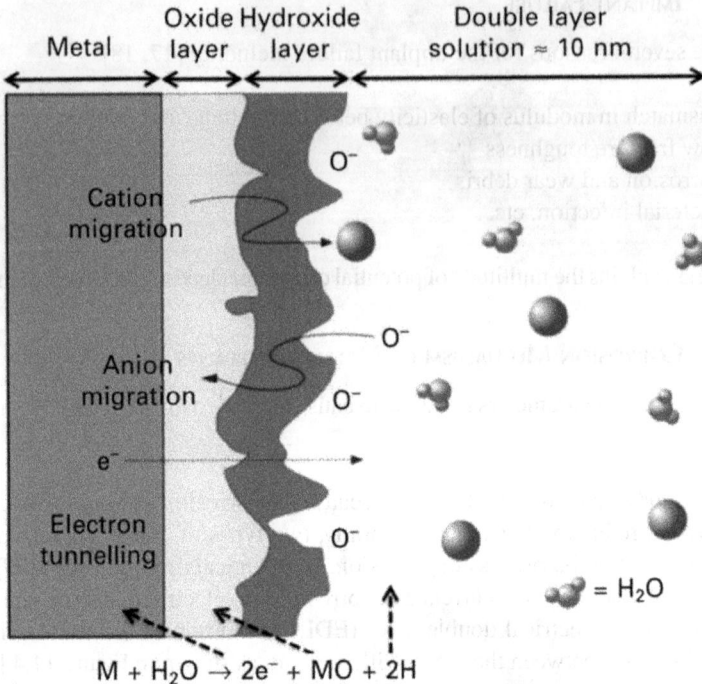

FIGURE 11.4 Schematic of the interface of a metallic implant surface in contact with a biological domain [19].

$$\Delta G = -nF\Delta E \tag{11.1}$$

where n is the number of valence ions, F is the constant of Faraday (95,000 C mol^{-1}), ΔE is the potential across interface between the metal with medium.

2. Physical barrier to the attack (oxide protective film) which limit the kinetics of corrosion. Generally, kinetic barriers to the process of corrosion prevent either the move of anions from medium to metal or the move of ions from metal to medium or the move of electrons in the interface between the metal with medium [19].

11.3 SURFACE MODIFICATION

11.3.1 INTRODUCTION

Researchers have concentrated on enhancing different surface engineering tools to improve the corrosion behavior and resistance to tribo-corrosion of the metallic implants in order to provide bioactive surfaces with good mechanical properties [20].

In addition to corrosion, stainless steel has low bioactivity, low biocompatibility and high elastic modulus making it unsuitable in arthroplasty implants. The best solutions to improve the corrosion behavior and blood cytocompatibility of metallic implants is deposition of a coating on the metal surfaces, which should also prevent the release of metal ions in the body [21, 22].

11.3.2 HYDROXYAPATITE

Calcium phosphates in various forms have been studied for coatings for implants. In this group Hydroxyapatite (HAp) was one of the first coating materials for metallic implants due to the similarity between it and bone. HAp was introduced to improve the fastening of an implant and the surrounding bone, with the first recorded clinical application being for femoral stems in hip replacements [23–25]. The benefits of these coatings was immediately recognized and proved able to stimulate osteointegration, enhancing the fixation of implanted devices. This enhancement is the result of the reaction of HAp to the environment stimulating the "expression of bone-related mRNAs and proteins in osteoblasts in vitro" [26]. Soon after implantation ions of calcium and phosphate are released, with proteins for the surround environment depositing on the surface forming a carbonate HAp scaffold layer for osteoblasts. This layer is reabsorbed by osteoclasts and superseded by bone formation [27]. HAp-based bioglasses and composite coatings have improved on the poor mechanical properties and enhanced the corrosion resistance [28, 29]. The mechanical, physio-chemical and biological properties of HAp are: density $(g/cm^3) = 3.156$, Young's modulus $(GPa) = 7\,0–120$, tensile strength $(MPa) = 40–100$, fracture toughness $(MPa\ m^{1/2}) => 1.0$, thermal expansion coefficient $(10^{-6} \times W/m/K) = 13.3$, thermal conductivity $(W/m/K) = 0.72$, biocompatibility, osteoconductivity and bioactivity = high, lattice structure = monoclinic and hexagonal, lattice parameters a = 9.84214, b = 2a, c = 6.8814, γ = 120° (monoclinic) and a = b = 9.4302, c =6.8911 γ = 120° (hexagonal) [21, 30–32].

11.3.3 CHITOSAN

Chitosan (CS) is one of the most common hydrophilic cationic polysaccharide consisting of N-acetyl-D-glucosamine and D-glucosamine, which is usually prepared through extracted and deacetylated from chitin [33]. Figure 11.5 shows the structures of chitin and chitosan.

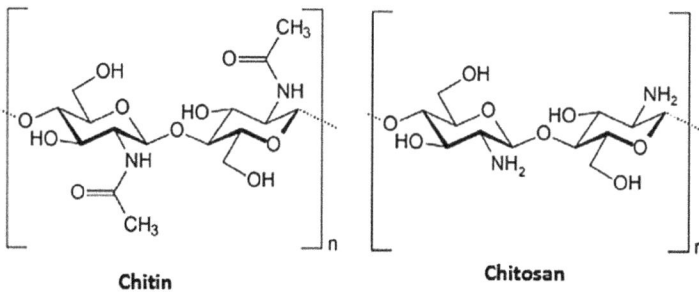

FIGURE 11.5 Chemical structures of chitin and chitosan [34].

CS is non-toxic, biodegradable and biocompatible. It has been widely investigated as a component of polyelectrolyte complex (PEC) due to its biocompatible and pH-dependent polycationic nature [35–38]. In addition, CS possesses the following properties [39]:

1. It has antibacterial and anti-acid effect.
2. It has antimicrobial action to avoid many microorganisms such as bacteria, fungi and algae.
3. It considers a preventive material to caries of dental.
4. It has the potential to improve drug absorption and stabilization of drug components to increase drug targeting.

11.3.4 ELECTROPHORETIC DEPOSITION

- Introduction
  Electrophoretic deposition (EPD) is a particular colloidal process that employs the mechanism of electrophoresis to the migration of charged particles suspended in a medium under applied an electric field [39]. There are three steps in the EPD process [40]:

  1. Formation of a stable suspension of particles.
  2. Migration of the particles toward the depositing electrode through the application of an electric field.
  3. Deposition of the particles on the electrode surface. A schematic of these steps in the EPD process is presented in Figure 11.6.

  There are two types of EPD depending on which electrode is chosen to deposit as shown in Figure 11.7 [1]. The first type is cathodic EPD, and the second type is anodic EPD [41, 42].

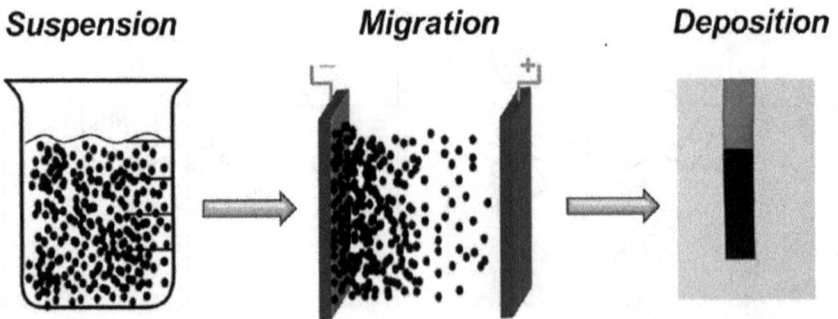

Suspension   **Migration**   **Deposition**

FIGURE 11.6   Schematic illustration of the EPD steps [40].

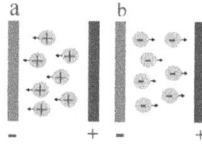

FIGURE 11.7   Schematic of the EPD types: (a) cathodic EPD; (b) anodic EPD.

- **Mechanism of Electrophoretic Deposition**

  Most materials obtain surface electric charge when they are brought into contact with an aqueous solution. The origin of the surface charge includes [43]:

  1. Dissociation of surface groups on the particles (depending on the pH of the suspension).
  2. Adsorption of ionized surfactants.
  3. Isomorphous substitution.
  4. Charged crystal surface fracturing.

  Literature shows that the EPD mechanism may be classified into four mechanisms [44]:

  a) Flocculation by particle accumulation mechanism.
  b) Electrochemical particle coagulation mechanism.
  c) Particle charge neutralization mechanism.
  d) Electrical double-layer distortion and thinning mechanism.

  A schematic of these mechanisms are shown in Figure 11.8.

- **Electrical Double Layer**

  Among the four mechanism of EPD, EDL is the most plausible for the EPD of bioceramics such as HAp because it considers the migration of a positively charged particle, toward the cathode without an increase in electrolyte concentration near the electrode [38]. There are three basic stages occurring in EDL [45]:

  a. When positively charged particles move toward the cathode electrode under applied an electric field, the double-layer envelope is distorted to become thinner in front of and wider behind the particle.
  b. The ions that carry similar charge with charged particle will migrate at the same direction of particles and interacts with the counter ions to reduce the double-layer envelope behind the particle.
  c. The next incoming particle can approach the particle to induce coagulation and deposition.

- **Derjaguin, Landau, Verwey and Overbeek Theory (DLVO)**

  The most comprehensive theory which describes the inter particle colloidal interaction is DLVO [46].

  Simply, DLVO theory describes the suspension stability by using two independent force concepts: Van der Waals and electrostatic forces. The first is attractive van der Waals force and the other is the repulsive electrostatic force that originates from the presence of EDL at the surfaces of particles [47].

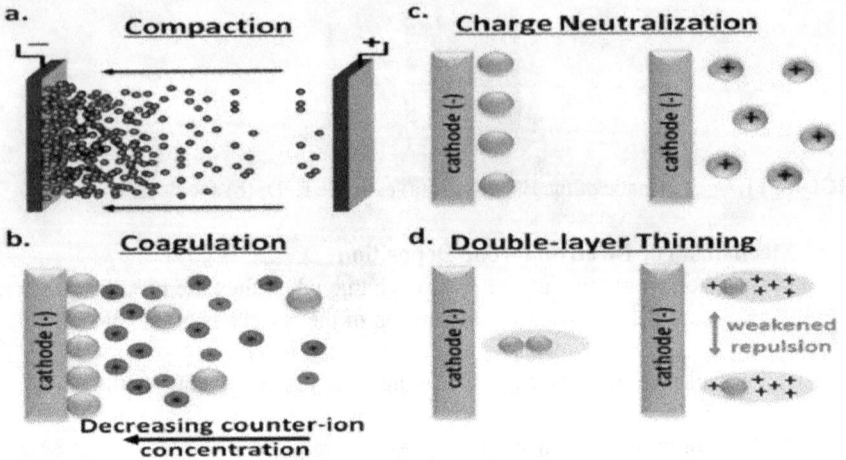

FIGURE 11.8 Mechanisms of depositions during EPD [40].

This theory explains that the system of colloidal stability is determined by the total reaction energy produced by the sum of attractive Van der Waals force (VA) and repulsive electrostatic force (VR) versus the interparticle distance during the following equation [48]:

$$VTotal = VR + VA \qquad (11.2)$$

The total potential energy of this interaction at a certain interparticle distance exhibits different forms based on the ionic strength of the system (see Figure 11.9). At appropriate ionic strength values, electrostatic stabilization of the colloids happens. On the other hand, in the presence of hydrophilic polymers, another important factor should be considered named steric stabilization [1]. Figure 11.10 shows a schematic of both steric and electrostatic process.

The electrostatic or the steric forces lead to particles repulsion of the same charge. Therefore, the electrostatic repulsive or the steric forces should be higher than Van der Waals force that lead to particles agglomeration [49, 50].

11.3.5 Factors Effecting Electrophoretic Deposition

The factors that can affect EPD technique can be classified into two groups [51–53]:

(I) The first group includes the parameters that related to EPD technique such as applied voltage, deposition time, distance between electrodes, as well as material of the electrodes.

(II) The second group includes the parameters that related to the suspension, for example, zeta potential, particle size, dielectric constant, dispersant concentration, viscosity and type of solution in the suspension.

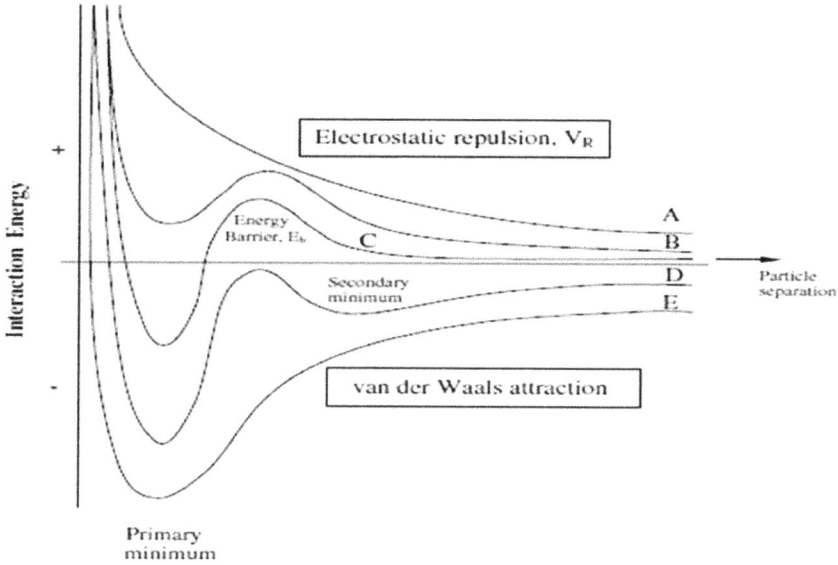

FIGURE 11.9 Schematic diagram showing the total interaction energy as a function of inter-particle distance between two particles [1].

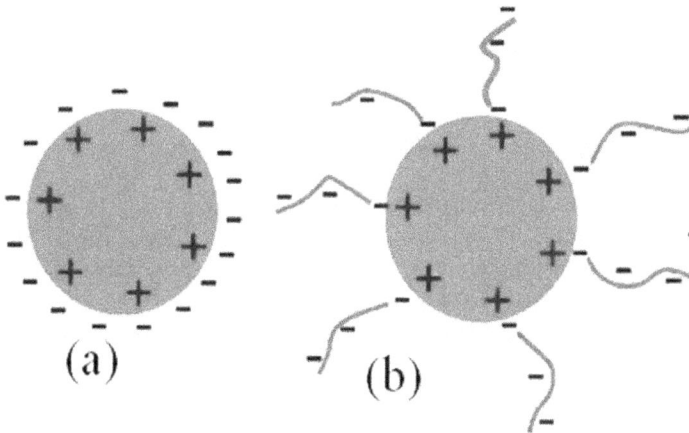

FIGURE 11.10 Graphical illustration of (a) electrostatic and (b) steric stabilization [1].

- Parameters related to the Process of EPD
  - o **Applied Voltage Effect**
    In general, increasing in the applied voltage during EPD technique leads to increasing coating layer thickness and cause heterogeneity in coating layer due to electrolysis in water, causing bubble formation [54].

o **Deposition Time Effect**

Previous studies demonstrated that increasing EPD deposition time leads to the reduction of deposition rate due to insulating film of particles that form on the surface of electrodes that reduce the electric field [55].

o **Deposition Electrodes**

The properties of the electrodes used for EPD also greatly affect the quality of the deposited thin films (e.g. surface finish, chemical composition, roughness and spacing). For example, electrodes that are spaced too far apart or having very low electrical conductivities will cause the electric field intensity to decrease between the electrodes which results in lower electrophoretic mobility and yields similar effects to that of lowering the applied voltage [56].

o **Temperature, Ultrasonic Bath and Stirrer**

The temperature of suspension is a significant factor in the coating during EPD where, previous studies showed that increasing the temperature leads to an increase in Brownian motion of particles and this lead to an increased impact between particles. The stability of suspension is very important in EPD technique. Therefore, researchers utilize various methods for the dispersion of particles in their solution, such as shaking and stirring. In addition, a simple method recently used to disperse particles is the ultrasonic bath [57, 58].

• Parameters Related to the Suspension

o **Particle Size**

It is essential in the EPD technique that the particles remain totally dispersed in the suspension to form uniform coatings. In case of larger particles, the particles tend to sediment due to gravity leading to thick coatings. On the other hand, the very small particles lead to aggregation and this causes heterogeneous coatings [59].

o **Dielectric Constant**

Dielectric constant of the liquid must be between [12–23, 60, 61] for the deposition to occur. If the dielectric constant is too low, the deposition will fail due to insufficient dissociative power. However, if the dielectric constant is too high, the size of the double layer region will reduce, leading to reduction of electrophoretic mobility of particles[62].

o **Viscosity of Solution**

The viscosity of a solution plays key roles in the electrophoretic mobility of particles and low viscosity is preferred for successful the EPD technique. The concentration of dispersed particles in the suspension has effects on the viscosity, and should therefore be considered [58, 62]. Some of the physical properties of common solvents include viscosity $[Cp] = 10^{-3}$ $[N.s.m^{-2}]$ and relative dielectric constant. The viscosity for the methanol, ethanol, n-propanol and acetone are 0.557, 1.0885, 1.9365 and 0.3087, respectively. The relative dielectric constant for ethanol, n-propanol and acetone are 32.63, 24.55, 20.33 and 20.7, respectively [52].

o **Zeta Potential**

Zeta potential of particles is an important parameter in the EPD technique. It is necessary to obtain a high and homogeneous surface charge of particle in the suspension. Zeta potential plays a role in the stability of the particles in the suspension, direction of particles motion, particle mobility and finally green density of the film [63].

Simply, the potential difference between the dispersion solution moving with the particle and the electrolyte solution that not moving with the particle is called (zeta potential) as shown in Figure 11.11. Particles with low zeta potential values tend to "clog" and form aggregates with each other, therefore resulting in the precipitation and eventual degradation of the suspension medium [64, 65].However, particles with high zeta potential have sufficient electrostatic repulsion to prevent this.

The zeta potential changes with pH value for the ceramic powders, particularly oxidizing in the aqueous suspensions if H^+ and OH^- are potential determining ions. The zeta potential is positive when the pH is at low value and negative if the pH is at high value [66].

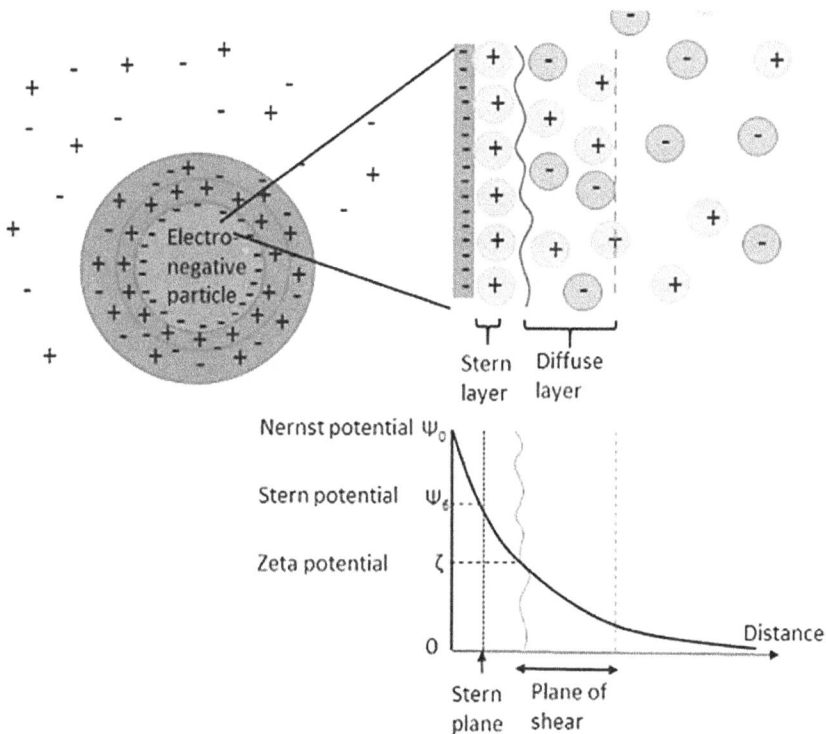

FIGURE 11.11   A diagram that explains the concepts of the stern layer, diffuse region and zeta possible [1].

11.4   BIOINERT COATINGS

11.4.1   What Are Bioinert Coatings

Bioinert materials do not interact, react or initiate a biological response when introduced into the body. Simply put, these materials do not have a negative impact on surrounding tissues, but will also not display a positive impact on surrounding tissue either. But it is common for a fibrous capsule to form around the coated implant. Bioinert coatings form a thin film between the underlying substrate material, the host tissue, and body fluids. There are many materials that are suitable for thin film coatings that are bioinert, with a variety of methods for deposition. The biocompatibility of a coating in this environment will depend on a variety of different parameters [67]:

- Hydrophobicity/hydrophilicity
- Wettability
- Surface charge
- Polarity
- Surface energetics
- Mobility of the surface molecules and smoothness

Therefore, great care must be taken when deciding which material is most suitable for the application.

11.4.2   Deposition of Biocoatings

Although numerous successful bioinert coatings have been achieved by various research groups over the years, bioceramic coatings have arguably had the greatest impact. Inert bioceramics have found themselves particularly useful as a coating for their tribological properties, acting as a protective coating, reducing the release of wear debris and metal ions from the surface of articulating joints.

The deposition of these coatings can be achieved by both physical and chemical methods, which include deposition techniques. The physical techniques like physical vapor deposition (PVD) and plasma spray. The chemical methods are chemical vapor deposition (CVD), sol-gel, electrodeposition and electrophoretic [68].

- **Physical Deposition of Bioinert Coatings**
  Physical deposition methods do not rely on chemically reactive processes for the formation of a thin coating; instead, they rely on thermal, electrical and kinetic energies to transfer a thin film onto a substrate. One of the most commonly used techniques is thermal spraying.

Thermal spraying of a biomaterial encompasses the heating of a stock material which is then sprayed (propelled) onto a surface forming the thin film. The stock materials can be metal and non-metal with powders, rods, wires or suspensions being used and deposited in a molten or semi-molten state. The heating is achieved by electrical (plasma/arc) or combustion. Thermal spraying allows thick films of bioinert materials to be deposited without the need for a high vacuum. Depending on the method of thermal spraying, it is possible for the particles to be propelled up to speeds of up to 1400 m/s [69]. Due to the high propeller speeds, these particles possess high kinetic energies, and thus upon surface impact, elastic and plastic deformation may occur. The solidified "splats" vary in diameter from a few to hundreds of microns with a thickness range of 1–20 microns [70]. Due to the versatility of thermal spraying, a wide range of biomedical coatings can be deposited.

Bioinert composite coatings of ZrO_2/SiO_2 are dense abrasive resistance coatings, with high hardness that can be applied by gas tunnel plasma spraying. This coating is achieved by mixing zirconia and silica powders into the feed with the spraying parameters given in Table 11.2, with adhesion aided by preheating of the substrate to 600 K.

The two gas flow rates (150 and 120 l/min) display different phase structures, with 120 l/min being composed of amorphous silica and zirconia with a small proportion of zirconium silicide, whereas 150 l/min exhibits a phase structure with amorphous silica, zirconium silicide and tetragonal silicide. These phases indicate that the highest flow rates are needed to decompose the ZrO_2 and SiO_2 particles. The wear- and corrosion resistance, along with adhesion improved with the higher flowrate, along with a denser structure, making them a suitable candidate for coatings in articulating joints such as hip or knee replacements, where softer HAp coatings are not suitable.

In some biomedical implants, an initial bioinert coating is first applied, followed by a bioactive coating to address adhesive issues from directly coating an implant with HAp [72].

These bioinert materials have also been shown to be beneficial for improving mechanical properties and adhesion when co-deposited with a bioactive material creating a biocomposite. Khor et al. [73] produced a HAp/YSZ/Ti-6Al-4V composite by plasma spraying a blended slurry powder feedstock of these materials. The resulting coating displayed greatly increased bond strength, microhardness and Young's modulus due to the microstructure of the resultant coating and also prevented the oxidation of the titanium alloy during the plasma spraying process and yttria-stabilized zirconia formation of a non-transformable tetragonal phase.

TABLE 11.2
Parameter for GTPS of ZrO_2/SiO_2 Coatings [71]

Powder	ZrO_2-50% SiO_2
Arc gun current (A)	50
Arc vortex current (A)	450
Gas flow rate (l/min)	120 and 150
Carrier gas flow rate (l/min)	7
Powder size (ZrO_2, SiO_2) (μm)	+45–75, +10–25
Spraying distance (mm)	50
Substrate traverse speed (mm/s)	40

11.5 BIOACTIVE COATINGS

11.5.1 WHAT ARE BIOACTIVE COATINGS

Bioactive coatings differ from bioinert coatings in that they interact with the environment, creating strong bonds between the device and the surround tissue. Most of the implants used in biomedical devices are metallic materials with surfaces that do not promote osteointegration. Developments have been made to create coatings that not only don't react negatively (bioinert), but are also osteoconductive. This allows the bone to strongly adhere to the surface to the implant while maintaining its integrity. Early implanted materials poor biocompatibility was an issue resulting in osteolysis [74, 75].

The growth of bioactive materials was achieved in the 1980s with bioglasses forming an apatite by-product which resembles that found in bones although the mechanical properties (particularly brittleness) made them unsuitable for large bone defects. There is often a trade off with bioactive materials, where trying to achieve the best bioactivity often results in poor mechanical properties.

Bioglasses can be used as a precursor for glass ceramic biomaterials; these have much better mechanical properties, similar to those of bones [76]. Overtime, it became clear that the best bioceramics would need to become porous to facilitate bone regeneration, with the pores acting like scaffolds. The design of these scaffolds are a key factor in the bone growth process of bioactive coatings, are required to be more than simply passive and should act as a template for bone growth. To that end, calcium phosphate-based coatings (in various forms) have shown to have some of the best bioactive properties, while being easily deposited, and thus are most widely studied group of coatings for implant materials and are considered earlier in this chapter. These bioactive coatings can be applied by a variety of different chemical and physical processes, with thermal spraying of bioactive coatings described.

11.5.2 THERMAL SPRAYING OF BIOACTIVE COATINGS

The application of bioactive coatings as stated can be achieved by various methods but thermal spraying (in various forms) is one of the more common methods. High-velocity oxy-fuel (HVOF) is one method used to apply HAp-based bioactive coatings. A simple schematic of the process is shown in Figure 11.12. The disadvantages of this method are mostly due to the different thermal expansions that exist between the coating and

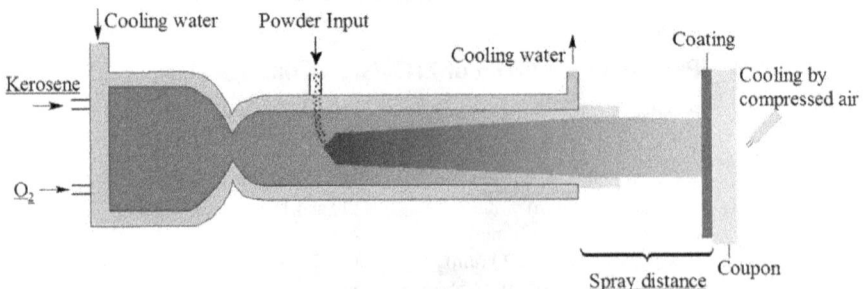

FIGURE 11.12   HVOF process for the application of HAp-based coatings [77].

the substrate during the deposition procedure, so care should be taken to reduce crack formation inherent to this method [77]. HVOF allows a variety of HAp-bioglass composite materials to be created which produce highly bioactive coating with osteoconductive properties with material properties greater than HAp alone.

Atmospheric plasma spraying (APS) is the most widely used method for the application of bioglasses, this can be achieved without calcium phosphate-based materials [78]. In order to use APS for bioglasses, there is a need to optimize the parameters to ensure the amorphous phase remains intact, as the plasma temperatures can reach up to 25,000°C, which could alter the resultant coating to a crystalline phase, altering the mechanical and chemical properties [78, 79]. Primarily, the substrate temperature and cooling rate have the biggest impact on the resulting phase of the bioglass, with fast cooling rates favoring an amorphous phase [80]. The standoff distance and the morphology of feeder material have also been shown to impact the crystallinity of the resultant coating phase, which is explained by the particles having higher thermal energy due to spending longer periods in the plasma [81]. The resultant bioglasses formed by APS typically have adhesion values of 4–61MPa, which is improved by the application of a bond coat [6].

A recent development of thermal spraying of bioactive glass coatings is the ability to promote osteogenesis and angiogenesis by the introduction of careful amounts of Co, Ce, Cu, Sr and Ag during the deposition process [82]. Although this recent development proves hopeful, incorrect proportions will lead to a cytotoxic response.

11.6   FUTURE DIRECTIONS AND CONCLUSIONS

11.6.1   FUTURE DIRECTIONS

A. Kocijan et al. in 2010 [83]: An electrochemical study of the passive film generated in artificial saliva (AS) and a simulated physiological solution on the 2205 DSS and AISI 316L stainless steel was performed. The results showed an improved resistance to corrosion of the 2205 DSS when compared with 316L SS due to an increasing growth of pits for the AISI 316L, which was also confirmed by the atomic force microscopy imaging. Therefore, the present study indicates a possible exploitation of 2205 DSS in orthopedic and orthodontic applications.

M. Conradi et al. in 2011 [84] studied the corrosion behavior of 2205 DSS and austenitic 316L SS in two physiological fluids, including AS and Hank's solution (PS). A study of productive oxide film formed on the 2205 DSS surface and 316SS was presented to determine the possibility of replacement of 316L by 2205 DSS in biomedical applications. They concluded that the resistance to corrosion of 2205 DSS was higher than 316L SS under the chosen experimental cases.

A. Kocijan et al. in 2012 [85] studied the DSS and austenitic SS in simulated body fluid (SBF) environments, investigated using X-ray photoelectron spectroscopy and electrochemical test to determine the possibility of alteration of 316 L SS by 2205 DSS in many medical applications. The results of the electrochemical tests demonstrate that the range passivity increased in 2205 DSS as compared with 316L SS. This proved the convenience of 2205 DSS for biomedical applications.

M. Farrokhi-Rad in 2018 [86] studied the comparison between spherical HAp (SHAp) and fiber HAp (FHAp) coated on 316L SS by EPD. It was concluded that the corrosion resistance of SHA deposition into SBF condition was higher than FHAp because of its less porous structure where the corrosive medium is able to penetrate the coating and attack the surface of substrate.

S. A. X. Stango et al. in 2018 [87] concentrated on modifying the surface of 316 L SS by Nd-YAG laser and coated it with HAp by EPD. The electrochemical behavior was studied in an SBF environment by using potentiodynamic polarization and EIS studies. As a result, the laser-textured coated samples possess higher resistance to corrosion than bare substrate because of formation of higher adhesion between the coatings and the substrate.

A. S. Hammoodinin 2020 [88] studied the EPD of HAp deposition on 2304 DSS with its improvement of microstructure; surface morphology and corrosion behavior in SBF and concluded that HAp coating on the 2304 DSS at 30 V for 2 min had the lowest corrosion rate value (0.277 mpy), i.e., high corrosion resistance in SBF, and thus, it is more corrosion-resistant than other uncoated samples.

11.6.2 Conclusions

From the literature reviews above studied by researchers, regarding the effect of EPD on the corrosion performance of 2205 DSS as well as the effect of the HAp-CS on corrosion performance of 2205 DSS with 316L SS, the following remarks can be listed:

1. It is necessary to utilize identical charged particles with solutions and dispersants for the formation coatings of different ceramic materials and obtain optimum control to the thickness of the layer.
2. HAp coating at smaller voltages is not suitable due to large porosity contents and thin coating layer, and coating at higher voltages is not suitable due to formed micro-holes by high hydrogen evolution, leading to increase in corrosion rate.
3. The coating time plays an important role in the thickness of the layer. The EPD deposition time is suitable from some seconds up to some minutes, depending on the coated material type.
4. Selection of applied voltages and EPD deposition time depend on the least value at which the deposition occurs and highest value which would not lead to spalling of coating film.
5. HAp coatings from methanolic suspension are not suitable and possess lower corrosion resistance than the bare substrate due to its highly cracked microstructure.
6. The CS enhances the stabilized HAp suspensions and allowed for cathodic EPD at low pH.
7. 2205 DSS exhibited excellent corrosion resistance compared to 316L SS in SBF environment, which proved the suitability of 2205 DSS in biomedical applications.

REFERENCES

1. Besra L and Liu M, "A review on fundamentals and applications of electrophoretic deposition (EPD)," *Progress in Materials Science*, vol. 52, no. 1, pp. 1–61, 2007.

2. Pishbin FDevelopment and Characterisation of Bioactive Coatings Based on Biopolymer and Bioactive Glass Obtained by Electrochemical Means, 2013. https://ethos.bl.uk/OrderDetails.do?uin=uk.bl.ethos.576010

3. Anoon IA. Advanced coating produced by electrophoretic deposition PhD Thesis. University of Technology, Iraq.

4. Amrollahi P*Development of Thermoelectric Thick-films via Electrophoretic Deposition*.2014. https://shareok.org/handle/11244/45118.

5. Sarkar A, Hah D. Electrophoretic deposition of carbon nanotubes on silicon substrates. *J Electron Mater.* 2012; 41 (11): 3130–3138.

6. Vallet-Regi M. *Bio-Ceramics with Clinical Applications*. John Wiley & Sons, Chichester, 2014.

7. Romanò CL, Romanò D, Meani E, Logoluso N, Drago L. Two-stage revision surgery with preformed spacers and cementless implants for septic hip arthritis: a prospective, non-randomized cohort study. *BMC Infect Dis.* 2011; 11(1): 129.

8. Prashar G, Vasudev H. Thermal Sprayed Composite Coatings for Biomedical Implants: A Brief Review. *J Thermal Spray Eng.* 2020; 2(1): 50–55.

9. Zhang BGX, Myers DE, Wallace GG, Brandt M, Choong PFM. Bioactive coatings for orthopaedic implants—recent trends in development of implant coatings. *Int J Mol Sci.* 2014; 15(7): 11878–11921.

10. Mantripragada VP, Lecka-Czernik B, Ebraheim NA, Jayasuriya AC. An overview of recent advances in designing orthopedic and craniofacial implants. *J Biomed Mater Res Part A.* 2013; 101 (11): 3349–3364.

11. Al-Amin M, Abdul Rani AM, Abdu Aliyu AA, Bryant MG, Danish M, Ahmad A. Bio-ceramic coatings adhesion and roughness of biomaterials through PM-EDM: A comprehensive review. *Mater Manuf Process.* 2020 Aug 17; 35 (11): 1157–1180.

12. Henao J, Poblano-Salas C, Monsalve M, Corona-Castuera J, Barceinas-Sanchez O. Bioactive glass coatings manufactured by thermal spray: A status report. *J Mater Res Technol.* 2019; 8(5): 4965–4984.

13. Mahajan A, Sidhu SS. Surface modification of metallic biomaterials for enhanced functionality: A review. *Mater Technol.* 2018; 33(2): 93–105.

14. Resnik M, Benčina M, Levičnik E, Rawat N, Iglič A, Junkar I. Strategies for improving antimicrobial properties of stainless steel. *Materials (Basel).* 2020; 13 (13): 2944.

15. Schweitzer PA. *Metallic Materials: Physical, Mechanical, and Corrosion Properties.* Vol. 19. CRC press; 2003.

16. Association IM. *Practical Guidelines for the Fabrication of Duplex Stainless Steels.* London, UK. 2009; 1–64.

17. Outokumpu SA, *Handbook of Stainless Steel.* AvestaResarch Centre, Outokumpu, Finland. 2013;

18. Farrer JCM. *The Alloy Tree: A Guide to Low-Alloy Steels, Stainless Steels, and Nickel-base Alloys.* CRC Press; 2004.

19. Chen Q and Thouas GA. Metallic implant biomaterials. *Mater Sci Eng R: Rep*, 2015; 87: 1–57.

20. Gunn R. *Duplex Stainless Steels: Microstructure, Properties and Applications.* Elsevier; 1997.

21. Bhattacharya A. *Stress Corrosion Cracking of Duplex Stainless Steels in Caustic Solutions.* Georgia Institute of Technology; 2008.

22. McGuire MF. *Stainless Steels for Design Engineers.* Asm International; 2008.

23. Geetha M, Singh AK, Asokamani R, and Gogia AK, Ti based biomaterials, the ultimate choice for orthopaedic implants: A review, *Prog. Mater. Sci.* 2009; 54 (3): 397–425.

24. Alvarez-Armas I, Degallaix-Moreuil S. *Duplex Stainless Steels*. John Wiley & Sons; 2013.

25. Dowson D, Neville A. *12 - Tribology and corrosion in hip joint replacements: materials and engineering*A2 - Revell, P.A. BT - Joint Replacement Technology (Second Edition)*. Woodhead Publishing; 2014. p. 401–442.

26. Heidersbach R. *Metallurgy and Corrosion Control in Oil and Gas Production*. John Wiley & Sons; 2018.

27. Duplex and high-strength stainless steels - Outokumpu Forta datasheet. 2016.

28. Talbot DEJ, Talbot JDR. *Corrosion Science and Technology*. CRC press; 2018.

29. De Groot K, Geesink R, Klein C, Serekian P. Plasma sprayed coatings of hydroxylapatite. *J Biomed Mater Res*. 1987; 21(12): 1375–1381.

30. Furlong RJ, Osborn JF. Fixation of hip prostheses by hydroxyapatite ceramic coatings. *J Bone Joint Surg Br*. 1991; 73(5): 741–745.

31. Adell R, Lekholm U, Rockler B, Brånemark P-I. A 15-year study of osseointegrated implants in the treatment of the edentulous jaw. *Int J Oral Surg*. 1981; 10(6): 387–416.

32. Knabe C, Howlett CR, Klar F, Zreiqat H. The effect of different titanium and hydroxyapatite-coated dental implant surfaces on phenotypic expression of human bone-derived cells. *J Biomed Mater Res Part A An Off J Soc Biomater Japanese Soc Biomater Aust Soc Biomater Korean Soc Biomater*. 2004; 71(1): 98–107.

33. Daculsi G, Laboux O, Malard O, Weiss P. Current state of the art of biphasic calcium phosphate bioceramics. *J Mater Sci Mater Med*. 2003; 14(3): 195–200.

34. Sebdani MM, Fathi MH. Novel hydroxyapatite–forsterite–bioglass nanocomposite coatings with improved mechanical properties. *J Alloys Compd*. 2011; 509(5): 2273–2276.

35. Say Y, Aksakal B. Enhanced corrosion properties of biological NiTi alloy by hydroxyapatite and bioglass based biocomposite coatings. *J Mater Res Technol*. 2020; 9(2): 1742–1749.

36. Asri RIM, Harun WSW, Hassan MA, Ghani SAC, Buyong Z. A review of hydroxyapatite-based coating techniques: Sol–gel and electrochemical depositions on biocompatible metals. *J Mech Behav Biomed Mater*. 2016; 57: 95–108.

37. Djokić SS. *Electrodeposition and Surface Finishing: Fundamentals and Applications*. Springer New York; 2014. (Modern Aspects of Electrochemistry).

38. Oakes L. *Controlling Nanomaterial Assembly to Improve Material Performance in Energy Storage Electrodes*. Vanderbilt University; 2016.

39. Choi JI. *Electrophoretic Deposition of Highly Efficient Phosphors for White Solid State Lighting using near UV-Emitting LEDs*; 2014.

40. Younes I and Rinaudo M, Chitin and chitosan preparation from marine sources. Structure, properties and applications, *Marine Drugs* 2015; 13 (3): 1133–1174.

41. Javadian S, Kakemam J. Intermicellar interaction in surfactant solutions: A review study. *J Mol Liq*. 2017; 242: 115–128.

42. Moussa DT, El-Naas MH, Nasser M, Al-Marri MJ. A comprehensive review of electrocoagulation for water treatment: Potentials and challenges. *J Environ Manage*. 2017; 186: 24–41.

43. Dickerson JH, Boccaccini AR *Electrophoretic Deposition of Nanomaterials*. Springer New York; 2011. (Nanostructure Science and Technology).

44. Arias LEC *Electrophoretic Deposition of Organic/Inorganic Composite Coatings on Metallic Substrates for Bone Replacement Applications: Mechanisms and Development of New Bioactive Materials Based on Polysaccharides*. Friedrich-Alexander-Universität Erlangen-Nürnberg (FAU); 2015.

45. Corni I, Ryan MP, Boccaccini AR. Electrophoretic deposition: From traditional ceramics to nanotechnology. *J Eur Ceram Soc*. 2008; 28(7): 1353–1367.

46. Trung MSVQ. *Electrophoretic Deposition of Semiconducting Polymer Metal/Oxide Nanocomposites and Characterization of the Resulting Films*. 2006.

47. Jugowiec D, Łukaszczyk A, Cieniek Ł, Kowalski K, Rumian Ł, Pietryga K, et al. Influence of the electrophoretic deposition route on the microstructure and properties of nano-hydroxyapatite/chitosan coatings on the Ti-13Nb-13Zr alloy. *Surf Coatings Technol.* 2017; 324: 64–79.
48. Goudarzi M, Batmanghelich F, Afshar A, Dolati A, Mortazavi G. Development of electrophoretically deposited hydroxyapatite coatings on anodized nanotubular TiO₂ structures: corrosion and sintering temperature. *Appl Surf Sci.* 2014; 301: 250–257.
49. Sarkar P and Nicholson PS, Electrophoretic deposition (EPD): Mechanisms, kinetics, and application to ceramics, *J Am Ceram Soc.* 1996; 79, (8): 1987–2002.
50. Saxena A, Rout A.*The Study of Hydroxyapatite and Hydroxyapatite-Chitosan Composite Coatings on Stainless Steel by Electrophoretic Deposition Method.*2011. http://ethesis. nitrkl.ac.in/2560.
51. Geeson JM. *Electrophoretic Deposition of Graphene Enhanced Aluminum and Bismuth Trioxide Nanothermite Thin Films.* University of Missouri--Columbia; 2016.
52. Sa'adati H, Raissi B, Riahifar R, Yaghmaee MS.How preparation of suspensions affects the electrophoretic deposition phenomenon. *J Eur Ceram Soc.*2016;36(2): 299–305.
53. Cheeke JDN. *Fundamentals and Applications of Ultrasonic Waves.*CRC press; 2017.
54. Ammam M. Electrophoretic deposition under modulated electric fields: A review. *RSC Adv.*2012;2(20): 7633–7646. https://pubs.rsc.org/en/content/articlehtml/2012/ra/ c2ra01342h
55. Antoniac IV. *Handbook of Bioceramics and Biocomposites.* Springer; 2016.
56. El-Meliegy E, Van Noort R. *Glasses and Glass Ceramics for Medical Applications.*Springer Science & Business Media; 2011.
57. Poitout DG. Biomaterials used in orthopedics. In: *Biomechanics and Biomaterials In Orthopedics.* Springer; 2016. p. 13–19.
58. Dey A, Mukhopadhyay AK. *Microplasma Sprayed Hydroxyapatite Coatings.* CRC Press; 2015.
59. Basu B, Katti DS, Kumar A. *Advanced Biomaterials: Fundamentals, Processing, and Applications.* John Wiley & Sons; 2010.
60. Harold M. *COBB, The History of Stainless Steel.* ASM International; 2010.
61. Fritz J. The use of 2205 duplex stainless steel for pharmaceutical and biotechnology applications. *Int Molybdenum Assoc.* 2011; 3.
62. Grumezescu A. *Nanobiomaterials in Hard Tissue Engineering: Applications of Nanobiomaterials.* William Andrew; 2016.
63. Arshanapalli SA. *Fabrication of Hydroxyapatite Coated Magnesium Alloy for Orthopedic Bio-Degradable Metallic Implant Applications.* Wichita State University; 2013.
64. Hench LL. *Introduction to Bioceramics, An* (2nd Edition). World Scientific Publishing Company; 2013.
65. Gshalaev VS, Demirchan AC. *Hydroxyapatite: Synthesis, Properties, and Applications.* Nova Science Publishers; 2012. (Biomaterials--properties, production, and devices series).
66. Suopajärvi T *Functionalized Nanocelluloses in Wastewater Treatment Applications.* 2015.
67. Spencer P, Misra A. *Material-tissue Interfacial Phenomena: Contributions from Dental and Craniofacial Reconstructions.* Woodhead Publishing; 2016.
68. Nuss KMR, von Rechenberg B. Biocompatibility issues with modern implants in bone-a review for clinical orthopedics. *Open Orthop J.* 2008; 2: 66.
69. Prashar G, Vasudev H. Application of thermal spraying techniques used for the surface protection of boiler tubes in power plants: Thermal spraying to combat hot corrosion. In *Advanced Surface Coating Techniques for Modern Industrial Applications* 2021 (pp. 112–134). IGI Global.

70. Khor KA, Wang Y, Cheang P. Thermal spraying of functionally graded coatings for biomedical applications. *Surf Eng.* 1998; 14(2): 159–164.

71. Morks MF, Kobayashi A. Development of ZrO_2/SiO_2 bioinert ceramic coatings for biomedical application. *J Mech Behav Biomed Mater.* 2008; 1(2): 165–171.

72. Rocha RC, Galdino AG de S, da Silva SN, Machado MLP. Surface, microstructural, and adhesion strength investigations of a bioactive hydroxyapatite-titanium oxide ceramic coating applied to Ti-6Al-4V alloys by plasma thermal spraying. *Mater Res.* 2018; 21(4).

73. Khor KA, Gu YW, Pan D, Cheang P Microstructure and mechanical properties of plasma sprayed HA/YSZ/Ti–6Al–4V composite coatings. *Biomaterials.* 2004; 25(18): 4009–4017.

74. Sheikh Z, Brooks PJ, Barzilay O, Fine N, Glogauer M. Macrophages, foreign body giant cells and their response to implantable biomaterials. *Materials (Basel).* 2015; 8(9): 5671–5701.

75. Anderson JM, Rodriguez A, Chang DT. Foreign body reaction to biomaterials. In: *Seminars in Immunology.* Elsevier; 2008. p. 86–100.

76. Antoniac I. *Bioceramics and Biocomposites: From Research to Clinical Practice.* Wiley; 2019.

77. Evcin A, Bohur BG. Coating of different silica sources containing hydroxyapatite for Ti6Al4V metal substrate using HVOF technique. *Arab J Geosci.* 2019; 12(6): 220.

78. Filho OP, La Torre GP, Hench LL. Effect of crystallization on apatite-layer formation of bioactive glass 45S5. *J Biomed Mater Res An Off J Soc Biomater Japanese Soc Biomater.* 1996; 30(4): 509–514.

79. Pawlowski L. *The Science and Engineering of Thermal Spray Coatings.* John Wiley & Sons; 2008.

80. Monsalve M, Ageorges H, Lopez E, Vargas F, Bolivar F. Bioactivity and mechanical properties of plasma-sprayed coatings of bioglass powders. *Surf Coatings Technol.* 2013; 220: 60–66.

81. Helsen JA, Proost J, Schrooten J, Timmermans G, Brauns E, Vanderstraeten J. Glasses and bioglasses: Synthesis and coatings. *J Eur Ceram Soc.* 1997; 17(2–3): 147–152.

82. Hoppe A, Mouriño V, Boccaccini AR. Therapeutic inorganic ions in bioactive glasses to enhance bone formation and beyond. *Biomater Sci.* 2013; 1(3): 254–256.

83. Dupleksnega K The corrosion behaviour of austenitic and duplex stainless steels in artificial body fluids. *Mater Tehnol.* 2010; 44(1): 21–24.

84. Conradi M, Schön PM, Kocijan A, Jenko M, Vancso GJ. Surface analysis of localized corrosion of austenitic 316L and duplex 2205 stainless steels in simulated body solutions. *Mater Chem Phys.* 2011; 130(1–2): 708–713.

85. Kocijan A, Conradi M, Schön PM. Austenitic and duplex stainless steels in simulated physiological solution characterized by electrochemical and X-ray photoelectron spectroscopy studies. *J Biomed Mater Res Part B Appl Biomater.* 2012; 100(3): 799–807.

86. Farrokhi-Rad M. Electrophoretic deposition of fiber hydroxyapatite/titania nanocomposite coatings. *Ceram Int.* 2018; 44(1): 622–630.

87. Stango SAX, Karthick D, Swaroop S, Mudali UK, Vijayalakshmi U. Development of hydroxyapatite coatings on laser textured 316 LSS and Ti-6Al-4V and its electrochemical behavior in SBF solution for orthopedic applications. *Ceram Int.* 2018; 44(3): 3149–3160.

88. Hammood AS. Biomineralization of 2304 duplex stainless steel with surface modification by electrophoretic deposition. *J Appl Biomater Funct Mater.* 2020; 18: 2280800019896215.

89. Li DQ. *Electrokinetics in Microfluidics,* Academic; 2004.

12 Effect of Microwave Heating on the Mechanical and Tribological Properties of the Thermal-Sprayed Coatings

Amit Bansal, Hitesh Vasudev, and Lalit Thakur

CONTENTS

12.1 INTRODUCTION

Wear and corrosion are the primary causes of plastic surface deterioration. Abrasive wear and rust affect machine tool components such as grey cast iron beds and slides. As a result, these components can exhibit outstanding tribological and mechanical characteristics [1–3]. The surface properties of the metallic components such as cast iron, different grades of stainless steel (SS-304, SS-316, etc.) and advanced materials (titanium and its alloys) can be modified by selecting a suitable surface modification technique like carburizing, nitriding, cyaniding and coatings/claddings. The selection of technique depends upon the coating materials and applications where the material is actually exposed in harsh environmental conditions. Among all these coatings techniques, the thermal spray coating techniques are most preferred due to their ease of operation and versatile nature. Many researchers have confirmed the presence of micro-porosities and the formation of cracks in thermal spray coating processes (flame spraying, plasma spraying, D-gun spraying and high-velocity oxy-fuel spraying [HVOF]) [4–6]. The WC-CO-Cr and

DOI: 10.1201/9781003213185-12

Al_2O_3-40TiO_2-based coatings were performed by utilizing a D-gun-spraying technique for the improvement of abrasion resistance of the coated specimen. However, the higher velocity associated with D-gun method resulted in the decomposition of powder. This led to higher porosity content in the as-sprayed coatings and further reduction in the bond strength of the deposited coatings [7]. The occurrence of such defects on the surfaces of the coatings can even deteriorate the sub-surface region of the coatings. Among thermal spray techniques, HVOF is commonly used due to its inherent attributes such as relatively dense and hard coatings with considerably less porosity compared with other thermal spray techniques. The HVOF-based coatings are extensively applied to mitigate the abrasive, erosive and adhesive wear in a number of applications. The improvement in surface properties of the engineering components by utilizing a HVOF technique is the most acceptable solution for solving many industrial problems related to erosion, corrosion and wear of component material.

From the above discussion, it has been found that the HVOF coatings are most commonly applied among thermal spray processes due to its relatively better properties than other thermal spray techniques. Nevertheless, there are still defects associated with the HVOF coatings. These include poor mechanical interlocking between individual splats, presence of micro-porosity in considerable proportions and cracks in the deposited coatings. The oxidation of the coatings will occur when performing the HVOF operation in the atmospheric conditions. These resulted in pores in the coatings material. Further, the wear and corrosion resistance of the coatings will effected by the presence of these pores. The corrosive media will easily be entered in the surface/subsurface of these regions. Further, due to high velocity associated with HVOF-spraying process, the all-powder particles were not completely melted. Due to the presence of these unmelted/partially melted powder particles, a lot of residual stresses are produced inside the coated specimens, which deteriorates the wear and corrosion properties of the coatings. However, by utilizing an inert atmosphere in thermal spraying, a relatively high-quality coating can be produced, but this route of producing coatings will be highly expensive. From the above discussion, it has been found that there are always some defects associated with the as-sprayed thermal-sprayed coatings. However, the mechanical and tribological properties of the thermal-sprayed coatings can be tailored by choosing a suitable surface modification technique like laser processing (laser glazing), microwave glazing and electron beam processing [8–11]. The laser-treated Stellite-6 and Ni-20 Cr coatings showed reduced porosity in the as-sprayed coatings [5]. The laser heating applied on plasma-sprayed WC-Co coatings resulted in healing of the microstructure and increased the microhardness of as-sprayed coatings [12]. The annealing performed on plasma-sprayed bio-medical coatings (HA reinforced with Al_2O_3) at 700°C resulted in an increase in mechanical properties of the as-sprayed coatings [13]. The heat treatment operation was performed on the HVOF-sprayed WC-15NiCr coatings deposited on the mild steel substrate materials, in the inert atmosphere at 750°C. The heat-treated coatings resulted in the improved abrasive wear resistance due to healed microstructures developed in the coating [14]. The effect of heat treatment (hot isostatic pressing at 920°C for 2 hrs at 103 MPa in an argon atmosphere) on the tribo-mechanical output of HVOF-sprayed WC-12Co coatings was investigated in another report. The crystallization of amorphous coating phases was discovered in the post-treated coatings. By comparing the formed coatings to the as-deposited coatings, the crystalline phase increased the elastic modulus and hardness [15].

12.2 MICROWAVE POST-PROCESSING

Amongst all post-processing techniques employed for thermal spray processes, the laser glazing is the most popular technique applied as post-processing technique for enhancement of tribological and mechanical properties of as-sprayed thermal-sprayed coatings. However, laser glazing is not a cost-effective process. Further, the limitations associated with laser processing and electron beam processing is the high maintenance cost and high processing cost. Microwave processing of materials has recently gained attention as a result of intrinsic characteristics such as uniform heating and volumetric heating phenomena associated with microwave processing of materials. Further, the thermal gradient during microwave heating is not steep, which results in less residual stresses inside the materials. The microwave heating can be effectively utilized for post-processing of the thermal-sprayed coatings. The thermal-sprayed coatings processed by using this novel microwave route has improved tribological and mechanical properties. Sharma et al. [9] first reported the effect of microwave glazing on the thermal-sprayed coatings. Sharma et al. [9] reported the enhancement in the application envelope of atmospheric plasma-sprayed alumina–titania-based ceramic composite coatings (approximately 0.9 mm) through post-processing route carried out by microwave hybrid heating (MHH) technique. The experiments were performed in a domestic microwave oven (model: NC 5054D) for a duration of 30 min by utilizing charcoal as a susceptor material. The ceramic coating is mainly transparent to microwave energy at room temperature, but it starts absorbing microwave radiation at elevated temperature. The microwave-processed coatings have higher densities than the as-sprayed coatings due to the transformation of alpha alumina phase into gamma alumina phase. The gamma alumina phase is generally more ductile in nature. Therefore, it facilitates improved surface finish of as-sprayed coatings by covering most of the irregularities present in the coatings. There is a flow of dominant gamma alumina phase during microwave exposure which resulted in the glazing of atmospheric plasma-sprayed alumina–titania coatings. The authors reported that there is possible healing of the surface defects (cracks and pores) by material flow in the glazed surface which resulted in significant improvement in microhardness (~1550 HV) and surface finish (R_a ~ 1μm) of the glazed surface. The authors further illustrated that there are three zones developed in the microwave-processed alumina–titania coatings as shown in Figure 12.1. Zone A indicates the presence of thin amorphous layer having glazed appearance. Zone B indicates the formation of epitaxial growth in the as-sprayed alumina–titania coatings due to volumetric and internal heating phenomenon associated with MHH technique. This phenomenon resulted in high temperature with slow cooling rate at the middle of coatings, which facilitates the formation of epitaxial growth in the middle of coatings. Zone-C indicates the no-change zone due to better conduction of heat through it.

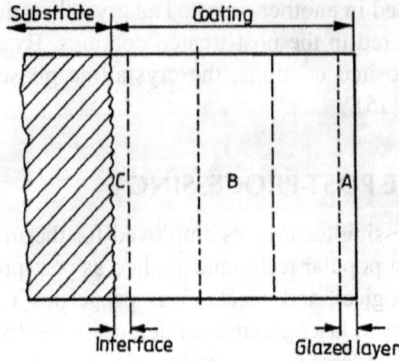

FIGURE 12.1 Schematic illustrations showing different zones developed along the cross section of microwave-irradiated plasma-sprayed coatings [9].

12.3 RECENT STUDIES ON MICROWAVE POST-PROCESSING OF THERMAL SPRAY COATINGS

Sharma and Krishnamurthy [16] also evaluated the erosive wear behaviour of microwave-glazed ceramic coatings by utilizing a high frequency impulse in an ultrasonic frequency generating facility. The boron carbide (B_4C) particles were used as an abrasive particle in the erosive wear test. They reported that the there is a reduction in porosity and increase in both (wear resistance and microhardness) values of the microwave-processed coatings was observed due to the homogenous and dense microstructure developed in the microwave irradiated plasma-sprayed coatings. There is an improvement in various properties like increase in the microhardness value of around 12.31%, wear resistance of 33.33% and surface smoothness of around 30% was observed in the microwave-processed coatings as compared to as-sprayed ceramics coatings. Further, there is a decrease in porosity of around 25.9% was observed in the microwave-processed coatings as compared to as-sprayed coatings. Zafar and Sharma [6] reported the improvement in mechanical and tribological properties of the microwave-processed flame-sprayed coatings (Ni-based) deposited on austenitic stainless steel (SS-304) grade. The Ni-based flame-sprayed coating (around 300 μm) was deposited on SS-304 substrate material by utilizing a 5PM thermal spray gun. The pressure of main gasses like acetylene and oxygen were maintained at 0.4 and 0.15 MPa, respectively, during thermal-spraying process. However, the flame-sprayed coatings exhibit poor adhesion with the substrate material. Therefore, a suitable post-processing technique like MHH (power = 1 kW; f = 2.45 GHz; exposure time = 900 s) were utilized for improving their mechanical and tribological properties. The microwave-processed coatings exhibit homogenized microstructure, increased densification and improved cohesive strength between the deposited splats across the complete thickness of coatings. An improvement in microhardness (18% increase) and fracture toughness (35% increase) was observed in the microwave-processed coatings. There is a decrease in both porosity (8% decrease) and surface roughness (44% less) was observed in the microwave-processed coatings. Furthermore, during abrasive wear test, it has been observed that the microwave-processed coatings exhibit lower material pull out than the flame-sprayed coating due to presence of homogenous and dense microstructure in the microwave-processed coatings.

Vasudev et al. [17] reported the improvement in mechanical and tribological properties of the microwave-processed HVOF coatings (Inconel-718-based) deposited on grey cast iron substrate material. Grey cast iron is one of the most commonly used material for making machine tool components like beds, etc. To improve the surface properties of the grey cast iron components, the HVOF coatings of Ni-based superalloy such as Inconel-718 (around 250 μm) were deposited on it (grey cast iron component) by utilizing a thermal spray gun. But, the presence of microporosity was an inherent feature associated with HVOF coatings. Therefore, post-processing performed through industrial microwave route (Make: Enerzi Microwave Systems) was adopted for improving the microstructural properties of the HVOF-sprayed Inconel-718 coatings. The microwave was operated at a power of 1 kW for a duration of 900 s. The authors reported that a homogenous microstructure was developed in the microwave-processed Inconel-718 coatings. During microwave exposures, there is microwave coupling of Alloy-718 coatings at high temperatures. This resulted in the internal heat generation and caused healing of pores because of melting of splats at elevated temperature.

The SEM micrographs of the top surface of the as-sprayed and post-processed Alloy-718 coating at different magnifications are shown in Figure 12.2. Figure 12.2 b, d and f represents the formation of melted splats after microwave heating.

Splats melt at high temperatures, covering much of the irregularities in the Inconel-718 coatings as they are sprayed. Scanning electron micrographs (SEM) of as-sprayed and microwave aided post-processed coatings are seen in Figure 12.3a and Figure 12.3b and c, respectively, for reference. The smooth microstructure with lower porosity was observed in the microwave-processed coatings as compared to as-sprayed coatings. The XRD results further endorsed the densification of coatings by producing lattice strains inside the microwave-processed Inconel-718 coatings. This lattice strains resulted in shifting in the major peaks of the microwave-processed coatings as compared to as-sprayed coatings. Overall, the microwave-processed coatings resulted in high fracture toughness and improved microhardness and abrasive wear resistance than the as-sprayed coatings.

Prasad et al. [18] investigated the tribological characteristics (abrasive wear behaviour) of the microwave-processed HVOF coatings (superalloy powder CoMoCrSi) deposited on titanium-based substrate (grade-15) material. The coating powders were processed by utilizing a high-energy ball milling technique to obtain a sufficient number of intermetallic phases in the coating powders. The abrasive wear resistance was measured by utilizing pin on disk apparatus as per ASTM G-99 standard. The microwave-processed coatings exhibited adhesion strength of around 72.65 MPa, which is significantly higher than the adhesion strength (61.31 MPa) of as-sprayed HVOF coatings. The wear rate of microwave-fused coatings is significantly lower than the as-sprayed coatings. The wear rate of microwave-assisted fused coatings at various temperatures like 200°C, 400°C and 600 °C is reduced by 6, 8 and 10 times, respectively, than the as-sprayed coatings. Moreover, the friction coefficient of microwave-assisted coatings at various temperatures like 200°C, 400°C and

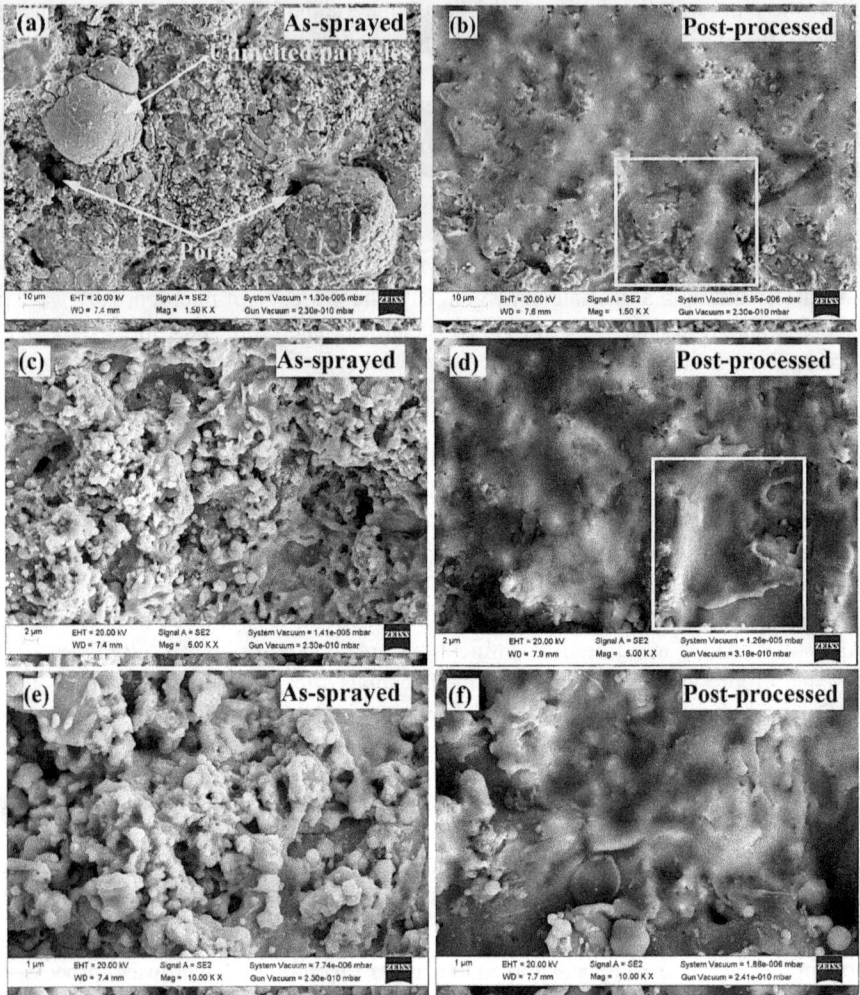

FIGURE 12.2 Typical top surface SEM micrographs of (a, c, e) as-sprayed-coated specimen at 1500×, 5000×, 10000×, respectively (b, d, f); microwave post-processed Alloy-718 coatings at 1500×, 5000×, 10000×, respectively [17].

600°C is reduced by 25%, 38.14% and 41.76%, respectively, than that of the as-sprayed coatings. Prasad et al. [19] investigated the tribological characteristics (dry sliding wear behaviour) of the microwave-processed flame-sprayed coatings (CoMoCrSi-Cr_3C_2 and CoMoCrSi) deposited on titanium-based substrate material. The coating powders (CoMoCrSi) were produced by utilizing a high-energy ball milling process and later the 30wt% Cr_3C_2 was added as reinforcement in it. The authors reported that there is an enhancement of the adhesion strength of the coatings after microwave exposure. The adhesion strength of the microwave-fused CoMoCrSi-Cr_3C_2- and CoMoCrSi-deposited coatings was observed to be 51.74 and 46.54 MPa,

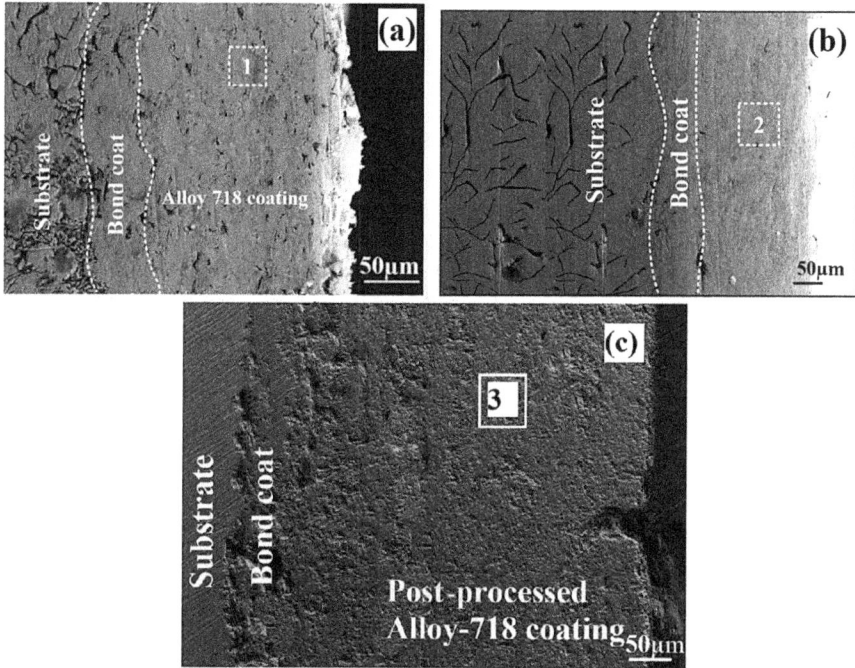

FIGURE 12.3 SEM microstructure along the cross section of (a) as-sprayed-coated specimens (b, c) post-processed-coated specimens [17].

respectively, whereas the adhesion strength of the as-sprayed $CoMoCrSi-Cr_3C_2$ and CoMoCrSi-deposited coatings was observed to be 31.02 and 28.12 MPa, respectively. Therefore, an increase of 65.46% and 66.79% in adhesion strength of the $CoMoCrSi-Cr_3C_2$- and CoMoCrSi-deposited coatings was observed after being subjected to microwave exposure. The microwave-fused coatings exhibit significantly improved hardness than as-sprayed coatings due to the presence of hard particles such as TiC in the microwave-fused coatings. During microwave exposure, the powders particles were melted and they interact with each other through diffusion process and it resulted in the formation of TiC in the microware-induced coatings. The overall $CoMoCrSi-Cr_3C_2$-based coatings performed better (in terms of mechanical and tribological behaviour) in both (as-sprayed and microwave-fused) conditions than the CoMoCrSi-based coatings due to presence of hard particle reinforcement like Cr_3C_2 in it. The microwave-fused coatings exhibit lower wear rate and volume loss at 10 and 20 N loads compared to the uncoated substrate materials and as-sprayed coatings. The wear rate increases with increase in temperatures for both (as-sprayed and microwave-fused) coatings. However, the microwave-fused coatings have lower wear rates than the as-sprayed coatings due to the formation of metallurgical bonding in the microware fused coatings. The presence of defects like cracks, unmelted particles and pores resulted in the lower wear rates of the as-deposited flame-sprayed coatings on titanium-based substrate material.

Prasad et al. [20] reported the improvement in tribological properties (high temperature sliding wear resistance) of the microwave-processed HVOF-sprayed CoMoCrSi-Cr_3C_2 coating on a titanium alloy substrate (grade 15). Both (as-sprayed and coated) the specimens were subjected to sliding wear test with various temperatures (200°C, 400°C and 600°C) and applied load of 10 and 20 N, respectively. The coatings fused through microwave heating resulted in less porosity, homogenous microstructure, metallurgical bonding and greater hardness as compared to the as-sprayed coated specimens. The microwave-fused coating resulted in improvement in wear resistance as compared to as-sprayed coatings due to predominantly metallurgical bonding in the microwave-treated coatings. The authors reported that there is a presence of tribo-oxide layer during sliding action in the microwave-fused coatings, which resulted in the improvement of the wear resistance of the microwave-fused coatings. Prasad et al. [21] investigated the tribological characteristics (dry sliding wear behaviour using a pin on disc apparatus at 200°C, 400°C and 600°C) of the microwave-processed HVOF-sprayed coatings (CoMoCrSi-WC+CrC+Ni and CoMoCrSi-WC+12Co) deposited on titanium-based substrate material. The coating powders (CoMoCrSi) were produced by utilizing high-energy ball milling process and later the reinforcement of CrC, WC and WC+12Co was done in it. The purpose of ball milling of powders is to improve the inter-metallic Laves phases by reducing the particle size of the coating powders. An improvement in metallurgical and mechanical properties (porosity, microhardness, surface roughness and adhesion strength) of the microwave-fused coatings was observed as compared to the as-sprayed coatings. The improvement in properties were due to the formation of homogenous microstructure in the microwave-fused coatings. There is an improved cohesive strength between the individual splats of coatings after microwave exposure which resulted in higher microhardness in the microwave-fused coatings. In microwave-fused coatings, there is a formation of various stable oxides like CoO, Co_3O_4, MoO_2 and Cr_2O_3. During wear testing at high temperatures, these oxides phases act as a lubricant which protect the underlying surface of the coating's material. Therefore, microwave-fused coatings exhibit lower wear rate as compared to as-sprayed coatings.

12.4 CONCLUSIONS

Microwave treating of thermal spray coatings is a relatively new development in the field of material processing. This technique is gaining popularity for treating thermal spray coatings for the enhancement of the tribological and mechanical properties of the thermal-sprayed coatings. The microwave-processed thermal-sprayed coatings exhibit lower surface toughness, less wear rate and improved microhardness and wear resistance than the as-sprayed coatings. There is a need for further explore this technique for better understanding the effects of individual magnetic and electric fields, microwave material interaction, effects of microwave heating on the microstructure of the deposited coatings to establish structure–property correlation, and physics of heat transfer, for the proper utilization of this technique. In the past decade, a lot of work has been done for utilizing this technique in treating thermal-sprayed coatings. More work is needed to further understand the mechanics involved, which could lead to increased use of this type of electromagnetic energy. In the future, the microwave approach will undoubtedly be a safer candidate for handling thermal-sprayed coatings.

REFERENCES

1. H. Vasudev, L. Thakur, H. Singh, and A. Bansal, A study on processing and hot corrosion behaviour of HVOF sprayed Inconel 718-nano Al_2O_3 coatings, *Materials Today Communications*, 2020b, 25, no. August, 101626.
2. H. Vasudev, L. Thakur, A. Bansal, H. Singh and S. Zafar, High temperature oxidation and erosion behaviour of HVOF sprayed Inconel-718 coating, *Surface and Coatings Technology*, 2019, 362,366–380.
3. H. Vasudev, L. Thakur and H. Singh, A review on tribo-corrosion of coatings in glass-manufacturing industry and performance of coating techniques against high temperature corrosion and wear, *i-manager's Journal on Material Science*, 2017,3,38–48.
4. M. Afzal, M. Ajmal, A.N. Khan, A. Hussain and R. Akhter, Surface modification of air plasma spraying WC–12% Co cermet coating by laser melting technique, *Optics & Laser Technology*, 2014, 56, 202–206
5. B.S. Sidhu, D. Puri and S. Prakash, Mechanical and metallurgical properties of plasma sprayed and laser remeltedNi–20Cr and stellite-6 coatings, *The Journal of Materials Processing Technology*, 2005a, 159, 347–355.
6. S. Zafar and A.K. Sharma, Microstructure and mechanical properties of microwave post-processed Ni coating, *Journal of Materials Engineering and Performance*, 2017, 26,1382–1390.
7. H. Vasudev, L. Thakurand H. Singh, A. Bansal, Mechanical and microstructural behaviour of wear resistant coatings on cast iron lathe machine beds and slides,*KovoveMaterialy, Metallic Materials*, 2018,56(1),55–63.
8. R. Shoja-Razavi, Laser surface treatment of stellite6 coating deposited by HVOF on 316L alloy, *The Journal of Materials Engineering and Performance*, 2016, 25, 2583–2595.
9. A.K. Sharma, S. Aravindhanand R. Krishnamurthy, Microwave glazing of alumina–titania ceramic composite coatings, *Materials Letter*, 2001a,50,295–301.
10. Q. Wang, L. Li, G. Yang, X. Zhao and Z. Ding, Influence of heat treatment on the microstructure and performance of high-velocity Oxy-fuel sprayed WC-12Co coatings, *Surface and Coatings Technology*, 2012, 206, 4000–4010.
11. J.K.N. Murthy, D.S. Rao and B. Venkataraman, Effect of grinding on the erosion behaviour of a WC–Co–Cr coating deposited by HVOF and detonation gun spray processes, *Wear*, 2001, 249, 592–600.
12. J. Mateos, J.M. Cuetos, E. Fernandez and R. Vijande, Tribological behaviour of plasma-sprayed WC coatings with and without laser remelting, *Wear*, 2000, 239,274–281.
13. A. Singh, G. Singh, andV. Chawla, Influence of post coating heat treatment on microstructural, mechanical and electrochemical corrosion behaviour of vacuum plasma sprayed reinforced hydroxyapatite coatings, *Journal of the Mechanical Behavior of Biomedical Materials*, 2018, 85, 20–36.
14. T. Ben Mahmud, T. I. Khan and M. A. Farrokhzad, Heat treatment effect on wear behaviour of HVOF-sprayed near-nanostructured coatings, *Surface Engineering*, 2017, 33, 1, 72–82.
15. N. Al Harbi and J. Stokes, *Optimizing HVOF spray process parameters and post-heat treatment for Micro/Nano WC – 12 % Co, mixed with Inconel-625 Powders: A critical review*, *32 International Manufacturing Conference*, Belfast, 2015, 1–12.

16. A. K. Sharma and R. Krishnamurthy,Sliding wear chracterisation of microwave glazed plasma- sprayed ceramic composites,*Proceedings of the Institution of Mechanical Engineers, Part J*, 2009,224,497–511.
17. H. Vasudev, P. Singh, L. Thakurand A. Bansal, Mechanical and microstructural characterization of microwave post processed Alloy-718 coating, *Materials Research Express*, 2020a, 6, 1265f5.
18. C. D. Prasad, S. Joladarashi, M. R. Ramesh, M. S. Srinath, and B.H. Channabasappa, Influence of microwave hybrid heating on the sliding wear behaviour of HVOF sprayed CoMoCrSi coating, *Materials Research Express*, 2018b,5, 086519.
19. C. D. Prasad, S. Joladarashi, M. R. Ramesh, M. S. Srinath, and B.H. Channabasappa, Microstructure and tribological behavior of flame sprayed and microwave fused CoMoCrSi/CoMoCrSi-Cr3C2 coatings,*Materials Research Express*, 2018a, 6(2), 026512.
20. C. D. Prasad, S. Joladarashi, M. R. Ramesh, M. S. Srinath, and B.H. Channabasappa, Effect of microwave heating on microstructure and elevated temperature adhesive wear behavior of HVOF deposited CoMoCrSi-Cr$_3$C$_2$ coating, *Surface and Coatings Technology*, 2019, 374, 291–304.
21. C. D. Prasad, S. Joladarashi, M. R. Ramesh, M. S. Srinath, and B.H. Channabasappa, Channabasappa, 2020, Comparison of microstructural and sliding wear resistance of HVOF coated and microwave treated CoMoCrSi-WC+CrC+Ni and CoMoCrSi-WC+12Co composite coatings deposited on titanium substrate, *Silicon*, 2020, 12, 3027–3045. doi: 10.1007/s12633-020-00398-1.

13 Plasma Spray Coating
A Weapon to Fight with Erosion and Corrosion Phenomena

Biswajit Swain, S. S. Mohapatra, and A. Behera

CONTENTS

13.1 TYPES OF SURFACE DEGRADATION

Degradation is known as the loss of applicable properties of materials which proceeds steadily because of exposure to work-related conditions. We can distinguish the aspects that enhance the deprivation of engineering materials that are, mechanical loading, irradiation, elevated temperature, and aggressive environment.

But temperature turns over the total volume of the machineries and brings some amount of volumetric fluctuations in the microstructure. The deprivation caused by the environmental influence is normally inadequate to near-surface region and includes processes like erosion, corrosion, oxidation, thermal degradation, etc.

13.1.1 EROSION

Erosion is a phenomenon where erosive agents associated with winds and streams remove rock, soil, or dissolved material from one area to another (Figure 13.1). This natural process is possible from the active motion of erosive agents, i.e., snow, water, ice (glaciers), wind, plants, animals, and humans, furthermore, depending on the type of erosive material. The erosion is categorized based on the medium of erosion such

DOI: 10.1201/9781003213185-13

317

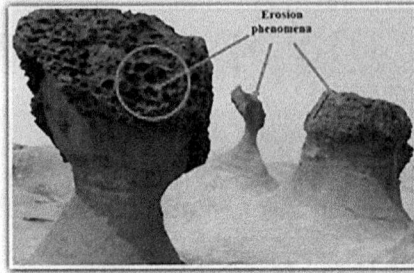

FIGURE 13.1   Erosion phenomena in nature.

as water erosion, snow erosion, wind erosion, zoogenic erosion, and many more. The erosive interaction of the materials leads to the breakdown of rock into the clastic residue. The breakdown is due to an erosion known as physical or mechanical erosion. On the other hand, chemical erosion occurs as the soil or rock is eroded from particles' removal and liquefication into a solvent. The continuous shifting and degradation of the naturally occurring landmasses due to natural erosion is measured from the effect of geomorphic drivers, such as natural flooding, wind abrasion, glacial plucking, abrasion, rainfall, and foundation wear in rivers. The effect of these phenomenon decides how fast a surface is battered. The rate of physical erosion is complex to describe as it depends on the sloping exteriors and properties associated with rainfall and thunderstorms. Therefore, it is expected that there is a possible link between the erosion rates and the amount of wind-swept material interacting with the surface. In polymers, the surface and bulk erosion are two quantities that measure the erosion and degradation of the polymer. However, in surface erosion, the outer surface of the polymer is degraded. The interior of the material does not degrade until the outer layers are eliminated through erosion.

13.1.2   CORROSION

Corrosion is the spontaneous electrochemical oxidation of metal in reaction with an oxidant. For example, rusting is the corrosion of iron that forms iron oxides (Figure 13.2). The corrosion causes damage as it creates oxides or salts of the metal

FIGURE 13.2   Corrosion in machine parts.

resulting in characteristic orange appearance. Corrosion in materials other than metals is termed as "degradation". The term is popular in the context of ceramics and polymers. Corrosion is a constant threat as it damages the valuable properties of the materials like strength, appearance, and permeability to fluids and gases, which significantly affect the applications' structural aspects.

From corrosion, it often appears that many physical alloys disintegrate as they interact with humidity in the air. The aforesaid phenomenon is resulted in corrosion. The phenomenon can occur near pits or crack and stretch across a wide area, homogeneously degrading the surface; however, corrosion occurs on uncovered surfaces as it depends on diffusion. Thus, methods like passivation and chromate coating decrease the material's exposed surface area, resulting in improved corrosion resistance.

Nevertheless, only uniform corrosion is predictable. The other forms of corrosion are unpredictable and especially hard to model. Here, the chemistry of corrosion is complex, but fundamentally, it is an electrochemical phenomenon. Therefore, there is the formation of an anode and a cathode. For example, during the corrosion of iron, oxidation takes place at a specific spot and it acts as an anode. Furthermore, the electron discharges at the anodic spot, which transfers over the metal. The electron travels to a spot where it reduces oxygen in the presence of h^+. The hydrogen ion is present due to the formation of carbonic acid from the dissolution of carbon dioxide from the air into the water from the moist air condition of the atmosphere the spot functions as a cathode.

13.1.3 OXIDATIVE DEGRADATION

The degradation due to oxidation is an autocatalytic process. In this attack of the hydrogen atoms, the molecules interact to produce hydroperoxides. Moreover, polymers' solidity is inversely proportional to the number of hydrogen atoms attached to the polymer chain's carbon atoms. It is also known that the heavy metal can supplement the degradation of metals like copper. Furthermore, the presence of hydrogen-donating compounds like peroxide decomposers can assist in further subduing the process of degradation. The mixtures of different stabilizers can together generate a combined effect that inhibits degradation. Microcrystalline wax can limit the effect of ozone on naturally occurring polymers such as natural rubbers and other elastomers in the same context.

Organic materials degrade in the presence of oxygen. Therefore, many products of oxidative degradation appear moulded. Moreover, elevated temperatures, heat, and catalysts, metals, and metal ions further increase the damage from the oxidation.

The separation of oxidation of resulting products is difficult in the bulk samples. Samples with low molecular weights are used to analyse and study the effect of degradation from the oxidation. A sample with lower molecular weight presents moderate ease in isolating the oxidation degradation for the study. Furthermore, the testing of lower molecular weights is more useful as most polymers have physical elements parallel with low molecular compounds.

To instil consistency in the overall physical and chemical properties of the material, the polymer is treated under heat and mechanical stress during the processing of the materials. During the entire lifecycle of the product, light, oxygen, heat, and moisture are permanently present throughout the whole cycle.

13.1.4 THERMAL DEGRADATION

Thermal degradation occurs at predetermined temperatures with chemical damage without the involvement of other compounds. Similarly, polymers degrade at high temperatures in the absence of air. However, this varies from the thermal oxidation, which can also occur at lower elevated temperatures.

The maximum temperature of the polymers primarily depends on the thermal degradations. Additionally, the injection moulding of these polymers depends on factors like the maximum temperature. Therefore, it is also a significant constraint on the polymers' manufacturing.

When the temperatures increase, the chains in polymers can be broken and interact with different chains to alter the polymer's properties. The reactions change the molecular weight of the polymer. Moreover, the reactions also affect the ductility, colour, cracking, embrittlement, and a general reduction in physical properties. In addition to these changes, optical properties also vary. These thermal breakdowns may result in volatile, toxic, flammable formation as the polymers form complex mixtures of compounds. Thermal degradation of polymers occurs in three distinct steps: initiation, propagation, and termination.

The beginning of degradation requires the breaking of polymer chain due to heat or light. Initiation of polymers generates a highly reactive and unstable polymer "free radical" (r) and an h-atom with an unpaired electron (h).

The propagation of thermal degradation includes a variation of reactions and the formation of peroxy radical. A hydrogen atom from another polymer chain reacts to form a hydroperoxide (rooh) and reforms a free radical (r). The hydroperoxide further divides into two new radicals (ro) + (oh), which permits diffusing other polymer molecules' reaction. The process then stimulates depending on annihilating the hydrogen from the polymer chain.

The termination of the process occurs with the formation of inert products with polymer chains linking up with free radicles the chain's termination by placing stabilizers in the polymer matrix.

Methods of thermal degradation are:

1. Depolymerization: When the temperature is varied, the edge of the polymer chain develops; this illustrates a low free radical with low activity. The polymer misses the monomer one by one.
2. Random chain session: These backbones will then crumble erratically. The chain can be breaking down at any location. Therefore, molecular weight diminishes rapidly.
3. Side group elimination: Polymer groups connected to the chains' support side have bonds weaker than the bonds connecting the chain. When the polymer is heated, the side groups bare off from the chain before breaking into minor pieces.

Thermogravimetric analysis is a technique used to study thermal behaviour where the sample is treated in a controlled manner at a predetermined atmosphere. The sample's mass is measured during the entire process; when a polymer sample undergoes reduction, the gaseous outcomes such as carbon monoxide, water vapour, and carbon dioxide are produced along with a reduction in mass.

13.2   EROSION

13.2.1   Types of Erosion

Solid Particle Erosion

Solid particle erosion is a dynamic process that results in removing materials from a surface due to its interaction with fast-moving solid elements constantly impinging on the surface. The erosion results in the wear of components, roughening, and degradation of the surface. The changes in the surface morphology significantly affect the mechanical and chemical properties of the components.

With the increased use of polymer composite materials in various applications under increasingly dynamic conditions, one of the significant threats faced by the composite is exposure to solid particle erosion. The phenomenon can be attributed to significant wearing in pipelines carrying sand slurries in petroleum refining, helicopter rotor blades, and pump impeller blades, also present in high-speed vehicles and also aircraft operating in desert environments. Therefore, solid particles' erosive behaviours on polymer composites' surface have gathered attention in the research.

Polymer composites display higher erosion resistance than unreinforced polymers. Nevertheless, polymer composite materials display poor erosion resistance when compared to metallic materials.

In the turbine blade, solid particle erosion is present from the foreign objects that enter the inlet of steam turbines. A large portion of these particles is produced in the boiler tubes. Hence, during the plant operation, the dimensions of the boiler tube's inner surface increase gradually due to the deposition of foreign particles. After achieving a critical thickness, the scale detaches from the inner tube wall in the presence of transient thermal stress: during the turbine in a steady-state operation, but it is more liable under transient operation such as start-up procedure because thermal stress is higher.

The presence of heavy erosion decreases the turbine's overall efficiency due to continuous material loss when compared to the turbine's normal wear or ageing. Accordingly, these factors influence the overall cost of production and maintenance of the turbine blades.

Wear is a complicated phenomenon involving several mechanical, chemical, and material factors which influence the complex mechanisms. These studies focus on the microstructural features of these obstacles—important parameters like surface temperatures and melting, strain rate effects. The characteristics and mechanisms are similar in oxidative wear. Additionally, any progress in understanding this phenomenon will further enhance the evolution of more wear-resistant materials and systems. The studies are conducted to build with solid particle erosion in mind.

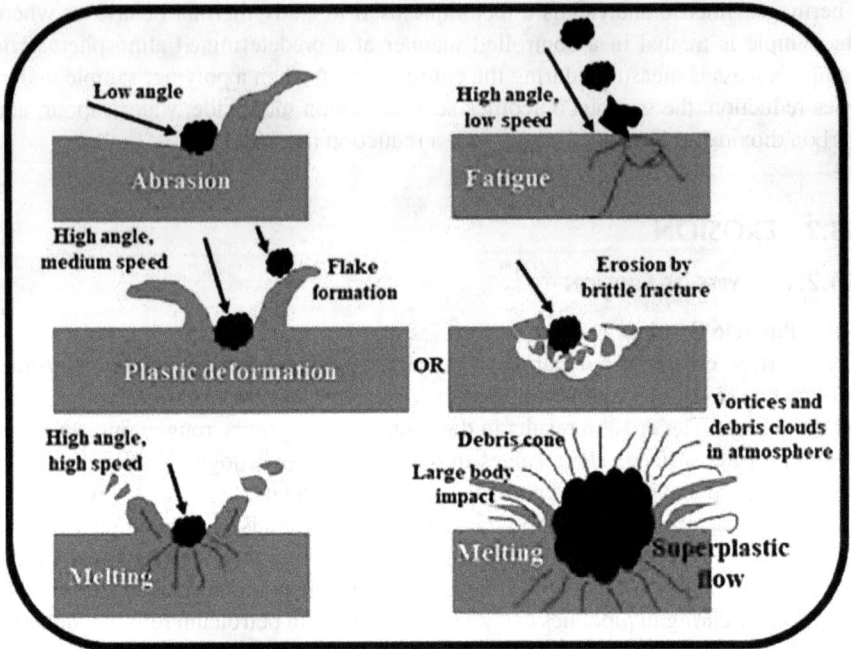

FIGURE 13.3   Solid particle erosion wear mechanism.

Some of the mechanisms of solid particle erosion wear are illustrated in Figure 13.3.

Slurry Erosion

Slurry erosion is a surface degradation phenomenon which is caused due to the striking of solid particles present in the slurry. Here, the degradation of these materials depends on various factors, such as flow conditions of fluid (velocity of fluid, particles concentration, particle striking angle, density of the fluid, chemical reactions within liquid, temperature of the liquid) and properties of erodent, such as size, shape, hardness, strength, and mechanical and endurance properties of the target material. Thus, factors inducing slurry erosion are enormous, and the loss of materials affects these properties. The impingement velocity and the mass of the single particle enhance the momentum. The concentration of particles also impacts the rate of erosion. The velocity of these particles influences the total energy of these particles. The aforesaid erosion phenomenon includes the formation of cavitation bubbles at solid particle surfaces.

Erosion rate influencing factors are:

1. Particle concentration in the slurry: More concentration results in more erosion loss. When estimated per unit of erodent mass, a higher concentration diminishes erosion damage.
2. Velocity of slurry and particles: The increase in velocity of solid particles results in higher erosion.

3. The angle of impingement of erodent on the wearing surface: Ductile materials causing more material loss in lower impact angle, whereas brittle materials causing more loss in higher impact angle.
4. Particle size: The rate of erosion increases with the increases in the size of the solid particles.
5. Particle shape: The roughness of the edges of the solid particles results in more wear on the material's surface.
6. Particle density: More excellent energy dissipation takes place with denser particles leads to more damage per particle impact.
7. Hardness of the erodent: Erodent with high hardness leads to more damage than softer erodent impingement.
8. Behaviour of the liquid: Liquid having both low density and viscosity results in higher erosion rate.
9. Characters of target material: Erosion resistance improves in materials with elevated hardness along with toughness.

Test Methods

It is common to use erodent (silicon carbide, alumina, quartz sand, or ores) where the particle size is below 1 mm in diameter. These particles are suspended in water and strikes the target materials during the test.

Slurry Jet Method

Here, pre-mixed fresh slurry is driven through a small-bore ceramic tube, giving rise to a jet (of known velocity) directed at an angle of 90°, against the target specimen.

The test is conducted by forming a circular-symmetric depression on the surface of the target. Therefore, the quantity of material loss is defined with weight loss. However, the data estimated from the device may be compromised by the absence of an unworn surface reference datum.

Flow and particle impact analysis of the slurry jet has also have endeavoured. Jet diffusion on the target surface, interference of the particles around the impact, and a spectrum of particle velocities in the jet are issues commonly present.

The slurry test can accommodate the variation in impact angles of the solid particles.

Cavitation Erosion

Cavitation erosion is a process of surface corrosion and surface material deterioration due to the formation of vapours or gas pockets inside the fluid flow. These pockets are made due to lower pressure below the liquid's saturation vapour pressure and the barrage's erosion on the surface.

Cavitation erosion occurs when the fluids' operating pressure drops beneath its vapour pressure, creating gas pockets resulting in bubble formation, and ultimately collapse. The condition can form a jam, and can prevent the flow of cooling fluids, further intensifying the problem. The locations where this is most likely to occur are: under the pressure of the pumping action, especially if operating near the net positive suction head; at the exit valve or regulator, essentially when running in a near-closed

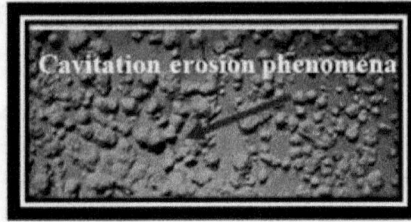

FIGURE 13.4   Cavitation erosion of surface.

site; pipe elbows and expansions. This kind of corrosion prevails where ultrapure
water is present as the fluid and damages the parts in a centrifugal pump. It will also
contribute to additional types of erosion–corrosion found in different pipe joints.
Reduction of the hydrodynamic pressure gradients, designing to evade pressure
drops below the liquid, and air access vapour pressure are some means to effectively
mitigate the corrosion at the design level. Solid coating and cathodic protection are
supplementary means to increase resistance.

Mechanisms

Cavitation-corrosion is a material degradation phenomenon where the surface is initially eroded due to the downfall of bubbles in a liquid which later corrodes. This
phenomenon is also known as cavitation erosion. However, a small difference is seen
with regards to damages caused during cavitation-corrosion where both corrosion
and cavitation are responsible for the damage to the material, but during cavitation-
erosion, only cavitation is taken into account because it is quite difficult to pin point
the corrosion effect in erosion. Hydraulic turbines, turbine blades, propellers, submerged planes, and pump impellers are primary locations for cavitation corrosion.
As we understand, when the fluid flowing has lesser pressure, the water initially
boils, resulting in the formation of bubbles at room temperature. Therefore, as the
piston moves inward, the volume contracts, and the pressure rises, resulting in the
collapse of the bubble. When the process is done at high speeds at various parts of the
jet engine such as turbine blades, propellers, and impellers, in a repetitive manner, it
contributes to the formation of bubbles and then causes cavitation corrosion as the
bubbles collapse. This occurs as the collapsing bubbles damage the protective layers
on the metal surface. Additionally, the unprotected metal surface is degraded with the
corrosion and a protective layer can be regenerated. Bubble formation and the collapsing of the bubbles resulting in plastic deformation and ultimately the degradation
of the surface of the metal. Moreover, the corrosion assists the generation and movement of additional cavitation. Thus, cavitation-corrosion is a combined action of the
mechanical and corrosion processes.

Damage from the cavitation can be mitigated by reducing the possibility of bubble
formation and collapse, which can be designed by tweaking the hydrodynamic
arrangement that modifies the smaller pressures. Generally, the pumps maintain an
upsurge of pressure to avoid bubble formation close to the points in the system at low
pressure. Additionally, the materials in the hydrodynamic systems are chosen in a
fashion to avoid any damage from corrosion and mechanical damage.

Explanation of Cavitation Erosion

Cavitation erosion ordinarily constitutes an attack on the exterior of material by gas or vapour bubbles, forming a sudden collapse from the fluctuation in pressure near the surface. There is a hydrodynamic generation of low pressure under the influence of several flow parameters, such as liquid viscosity, temperature, pressure, and nature of the flow. This surface starts weakening from the swift flow of bubbles repeatedly striking the surface—the pounding results in bending, accompanied by pitting.

Furthermore, cavitation erosion can degrade the surfaces of metals and non-metals. The cavitations generate undesirable levels of noise, and vibrations can reduce the lifetime of the samples. In submarines, the noise produced from the cavitation erosion damage increases the potential of enemy detection during wartime.

Corrosion

Corrosion is a consequence of the chemical interaction between a metal or metal alloy and its environment. It is spontaneous and destructive. The process restores the metal to its former compound state comparable to the ores from which the metals were extracted.

Forms of Corrosion
Uniform Corrosion

Usually, uniform corrosion is the most occurring corrosion, with uniform removal of metal from the surface. Hence, uniform corrosion requires a uniform composition and metallurgy, including an environment that remains consistent in enveloping the metal surface. However, these conditions may not be expected in many types of equipment, and consequently, during tests, a small portion of non-uniform corrosion is acceptable.

A widely used example of uniform corrosion is atmospheric corrosion and uniform corrosion of steel in an acid solution. Usually, an alloy corrodes uniformly at a low rate while in service. The low rate of degradation makes the corrosion highly predictable in comparison to its other types.

Galvanic Corrosion

Galvanic corrosion occurs when two dissimilar alloys are in contact with each other while in a corrosive environment—the galvanic corrosion results in one of them corroding while the other is protected. The protection or preferential corrosion of these alloys depends on their relative position in the galvanic series. The preferential corrosion also results in the nobler alloy's protection from corrosion, while the corroded alloy in the couple is termed to be more active in the galvanic series.

Some materials can endure in either/both states possible in the series. On the contrary, the metal's corrosion also depends on the potential reversals and unpredicted galvanic corrosion. This corrosion's preferential nature close to the dissimilar metals' intersection is a distinctive feature of galvanic corrosion. Moreover, the extent of corrosion decreases as the point of interest moves away from the junction. The reduction in corrosion is attributed to higher resistance due to a longer electrolyte path. The electrolyte conductivity also limits the current to a small area close to the intersection.

Crevice Corrosion

The crevice is a sheltered location created from contact with a different material. Here, corrosion occurs more frequently as the secondary material may be part of a fastener made of the identical or a different alloy, a coat of mud, sand, or a different insoluble solid. However, this corrosion tends to occur in metal-metal crevices preferentially. Additionally, gasket corrosion or corrosion from a deposit of insoluble solids occurs with crevice formation from a non-metallic material when it is in contact with a metal surface. In this case, the crevice is composed of differing alloys, or if the deposit is conductive, the rate of corrosion may compound multiple times with the addition of galvanic corrosion.

Moreover, crevice corrosion can occur from exposure to the atmosphere and the retention of water in the crevices, while the exterior surfaces can drain and dry. The crevice corrosion of stainless steels in aerated salt solutions is widely known. Corrosion products of Fe, Cr, and Ni from the steel accumulate in these crevices and produce highly acidic chloride solutions, which exponentially increases the rate of corrosion.

Pitting Corrosion

Pitting corrosion occurs when a protected surface forms exposed pockets of corrosion that appears localized in nature. The dimension of the pits may also vary significantly. In stainless steels and nickel alloys with chromium, the alloys primarily rely on a passive film for corrosion resistance. These films are particularly susceptible to pitting with the film's wearing at isolated portions, and pitting appears to have an analogous mechanism with crevice corrosion in stainless steels. The pits behave like a self-serving crevice that limits transport across the bulk solution and the acid chloride pit anode.

Environmentally Induced Cracking

Environmentally induced cracking is the phenomenon where ductile material fails due to uniform corrosion. There are three distinct types of failure that are part of EIC: stress corrosion cracking (SCC), corrosion fatigue cracking (CFC), and hydrogen-induced cracking (HIC). These failures occur under varying types of loading and environmental conditions. SCC occurs in specific environmental conditions when the alloys are under static tensile stress. Again, pure metals are less prone to SCC. The corrosion rate further decreases with the presence of a passive surface film under oxidizing conditions.

On the other hand, CFC occurs under cyclic loading and unloading in a corrosive environment. The presence of the corrosive environment increases the probability of corrosion which leads to failure. In this case, both alloys and pure metals are susceptible, and no specific environment is required.

HIC is generated with the diffusion of hydrogen into the alloy lattice. The diffusion occurs as the reaction which results in the evolution of hydrogen gas produces atomic hydrogen at the surface during corrosion, electroplating, cleaning and pickling, or cathodic protection. Furthermore, HIC is accelerated by cathodic polarization.

Intergranular Corrosion

Due to grain boundaries, the reactive components or passivating elements present in the metal matrix may segregate or deplete at the grain boundary. This phenomenon is termed intergranular corrosion, also known as an intergranular attack. Consequently, the grain boundaries' local regions are less corrosion-resistant, leading to preferential corrosion at the grain boundary. Sometimes, these mechanisms are severe enough to drop grains out of the surface. This corrosion mainly occurs in alloy systems.

A typical example of IGC happens in austenitic stainless steels as chromium depletes from the grain boundaries during heat treatment by metallurgical reaction with carbon. The resulting composition is highly vulnerable to IGC.

Dealloying and Dezincification

Dealloying, also known as selective leaching, is a form of corrosion that occurs in solution and results in the preferential removal of an element from the matrix. Dezincification is the dealloying in brass. The phenomenon of dealloying is also a commonly cited example of selectively leaching. In brass, the dealloying occurs as zinc present is very active, and it readily leaches out of brass, leaving the material in a porous form with degraded mechanical properties. The dezincification is easily recognizable as it forms a separate layer with red colour. This preferred corrosion happens in areas under films and places that are not effortlessly accessible, resulting in a catastrophic failure due to its unexpected and random nature.

Graphitic corrosion is a different example of selective leaching. It is the dealloying of iron from grey cast iron, resulting in a weak, porous inert graphite network. Accordingly, graphitic corrosion mainly occurs in cast iron pipe buried in soil and having long maintenance cycles. These failures are inconvenient as pipes' failure carrying hazardous materials can irreversibly contaminate the ecosystem or cause a fatal explosion.

Erosion–Corrosion and Fretting

Erosion–corrosion occurs when a surface interacts with a highly corrosive fluid that is moving at a high velocity. The total damage is dependent on factors like the concentration of erosive particles in the fluid and angle of impact on the surface. However, a modest corrosion rate can also be seen when the same fluid is stagnant or slow-flowing. Nevertheless, the corrosive fluid movement can degrade the materials by eroding and removing the protective corrosion film on the surface. The interaction with the eroding particles results in the exposure of the substrate, further accelerating the corrosion.

In addition to the fluids, suspended slurries add to erosion, improving the erosion–corrosion attack. Alloys that rely on corrosion-resistive paint or unique corrosion-resistance treatments, such as low strength alloys, are most susceptible to being corroded. The attack generally depends on the regions of localized flow and the turbulence around surface irregularities.

Cavitation is a particular case of erosion–corrosion. Primarily, mechanical, chemical, and electrochemical characteristics are studied to facilitate the selection of corrosion-resistant alloys and control of solution composition.

Moreover, fretting is another different type of erosion–corrosion that takes place in the vapour phase. The erosion happens from the repeated small motion between the metal gradually corroding. The interaction in the contacts' presence degrades the protective coatings from the metal surface, exposing the vulnerable substrate to the operational conditionals, which can irreversibly damage the material. Moreover, with the presence of the oxide debris, the effect is multiplied, generating additional abrasion.

13.3 PLASMA SPRAYING

Plasma spraying is an adaptive and cheap technique for the fabrication of coating. In 1909, the concept of plasma spraying was conceived and patented, and the initial iterations of the practical set up were seen as a combined effort of two companies, Plasmadyne and Union Carbide, in the year of 1960. The process mainly works when a gas, generally inert gas, is allowed to flow between a cathode and an anode. The anode is a cylindrical nozzle which holds the cathode inside it. An electric arc is generated in the space between the two electrodes with a high-frequency discharge, which is subsequently sustained with dc power. Following this, there is the generation of high-pressure gas plasma as the gas ionizes. With the spray, the powder material is injected in front of the gun nozzle where the plasma plume is present. Here, the powder melts, and is hurled by the gas onto the surface of the substrate. There are two types of plasma-spraying techniques used: atmospheric plasma spraying and vacuum plasma spraying.

Some advantages of plasma spray process are:

- Reduces the energy fluctuation along with the reduction of noise in comparison with other processes.
- Plasma gas velocity is significantly decreased, resulting in increased dwell time of powder particles inside the plasma jet and increased heat transfer. So refractory oxide coatings can be sprayed at increased deposition efficiency.
- With incorporation of robotic arms with 6° of freedom, complex geometry can also be treated conveniently.
- The method has the ability to handle a wide range of materials with applications in fields like corrosion resistance, high-temperature, and ablation resistance coatings along with biocompatible film.
- This type of coating technique can coat materials with very high melting temperatures without causing any significant damage to the substrate due to unnecessary heat transfer.
- This method also generates coatings with higher density when compared to methods like flame spraying.
- Generally, the processing is simple, making it accessible in a wide field of applications.

Along with the above features of the process, proper melting of materials, phase transformation, reduction in porosity, and proper splat formation is attributed to the presence of high temperature at the nozzle and the particle velocity of the melted particles. Therefore, plasma spraying is gaining importance as the process becomes more accessible.

Some applications of plasma spray process are:

- The thermal coating fabricated is found resistant to degradation mechanisms such as wearing, erosion, abrasion, and adhesion, which makes it desirable in many industries.
- This technique finds application in textile industries to handle very abrasive synthetic fibre due to its control on friction during the process.
- As the coatings produced are corrosion-resistant, they can be employed in ship building and bridges.
- Due to the additive nature of the coating, it can be used to restore parts, which are worn.
- Partially stabilized zirconia coatings fabricated with plasma spraying are used in gas turbine combustors, shroud and vanes and on internal combustion cylinders, and valves to increase the effectiveness and reduce metal temperatures or cooling requirements.
- This technique is used for the coating of free-standing components such as rocket nozzles.
- Plasma spraying is widely used in bioactive ceramics such as hydroxyapatite as interfacing osseoconductive layers in metallic surgical implants.
- Plasma spray process offers a wide range of modifications in the powders.

13.4 RELEVANT WORKS ON EROSION AND CORROSION OF PLASMA SPRAY COATING

Various works have been done on erosion and corrosion performance of plasma spray coating. Mishra et al. [1] observed the erosion and corrosion properties of the plasma spray NiCrAlY coatings for boiler application. The coating was kept at a temperature of 540 °C of 10 cycles for 100 hrs duration. It was observed that the presence of Ni avoided the formation of oxygen but still a major amount of oxygen penetrated along the splat boundaries and reacted with aluminium, chromium, and yttrium. Swain et al. [2] developed a NiTi coating on mild steel substrate by atmospheric plasma spray technology for erosion resistance. They noticed that the coatings deposited at higher arc current revealed a higher erosion at 90° erodent impingement angle. Various erosion mechanisms like lip formation, chip formation, crater formation, cutting grooves, and plastic deformation were also observed from the surface micrographs after erosion (Figure 13.5). Wang et al. [3] investigated the cavitation erosion resistance of the plasma-sprayed CoMoCrSi. The coating has dense microstructure with some initial pores and cracks. They have observed that two main mechanisms are responsible for the damage of the coating such as weak adhesion of the splashes formed during plasma spray process and delamination of the fragments resulted from the bubble collapses. Mishra et al. [4] prepared plasma spray

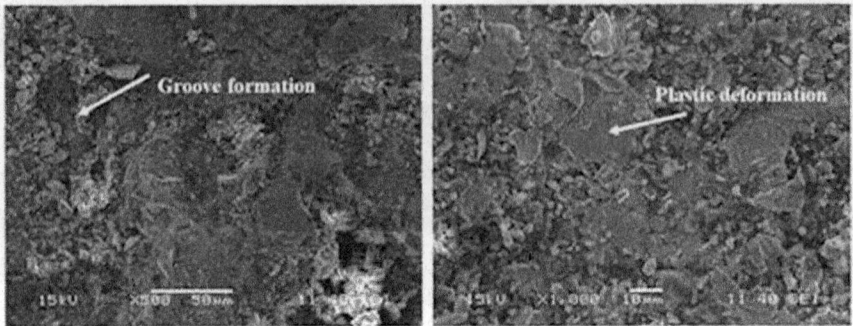

FIGURE 13.5 Solid particle erosion wear of plasma-sprayed NiTi coating.

coating of three different materials such as Ni-22Cr-10Al-1Y, Ni-20Cr, and Ni₃Al on Inconel 718 substrate. They studied the erosion performance of the coating. It was observed that the erosion rate at 30° impact angle on both coated and uncoated samples were higher than the 90° impact angle which indicated its ductile behaviour [5].

13.5 SUMMARY

The chapter revealed that the plasma spray coating is an excellent technique for the enhancement of erosion- and corrosion resistance. The formation of dense splat due to proper melting of coating materials improves the properties of the coating. The erosion- and corrosion-resistance properties of the plasma spray coating widen its application in various applications such as aerospace, biomedical, naval industries, etc.

REFERENCES

1. S.B. Mishra, K. Chandra, S. Prakash, Erosion–corrosion performance of NiCrAlY coating produced by plasma spray process in a coal-fired thermal power plant, *Surf. Coatings Technol.*216 (2013) 23–34. doi:10.1016/j.surfcoat.2012.09.044.
2. B. Swain, P. Mallick, S.K. Bhuyan, S.S. Mohapatra, S.C. Mishra, A. Behera, Mechanical properties of NiTi plasma spray coating, *J. Therm. Spray Technol.*29 (2020) 741–755. doi:10.1007/s11666-020-01017-6.
3. Y. Wang, J. Liu, N. Kang, G. Darut, T. Poirier, J. Stella, H. Liao, M.-P. Planche, Cavitation erosion of plasma-sprayed CoMoCrSi coatings, *Tribol. Int.*102 (2016) 429–435. doi:10.1016/j.triboint.2016.06.014.
4. S.B. Mishra, S. Prakash, K. Chandra, Studies on erosion behaviour of plasma sprayed coatings on a Ni-based superalloy, *Wear*260 (2006) 422–432. doi:10.1016/j.wear.2005.02.098.
5. F. Cai, F. Gao, S. Pant, X. Huang, Q. Yang, Solid particle erosion behaviors of carbon-fiber epoxy composite and pure titanium, *J. Mater. Eng. Perform.*25 (2016) 290–296. doi:10.1007/s11665-015-1848-8.

14 Thermal Spraying Process Kinematics

Importance and Its Control by Advanced Robots

Gaurav Prashar, Hitesh Vasudev, and Lalit Thakur

CONTENTS

14.1 ROLE OF SPRAYING PROCESS KINEMATICS ON FINAL PROPERTIES OF AS-SPRAYED COATINGS

The use of offline-path planning and industrial robots is necessary to ensure the homogeneous thickness of as-sprayed coatings onto the surfaces of complex parts. In the former case, wrong path planning can significantly affect the kinematic and dynamic performance of industrial robots. For instance, it leads to many unwanted

DOI: 10.1201/9781003213185-14

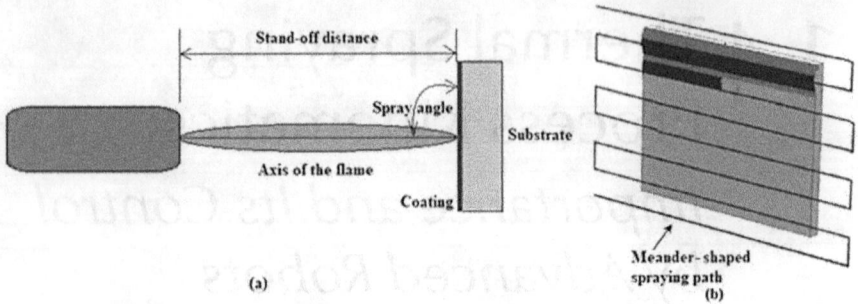

FIGURE 14.1   Schematic showing major handling parameters in thermal spraying.

re-orientations of the robot axis, which changes vital handling parameters and can influence the final coating quality [1]. The main handling parameters that can affect the coatings are schematically shown in Figure 14.1.

14.1.1   SPRAYING ANGLE

The coating properties also depend upon the spraying angle, as the impact angle of molten/semi-molten spraying particles onto the surface of the substrate followed by proper anchorage of these spray particles on the metal surface is controlled by maintaining a proper spraying angle. As shown in Figure 14.1, the spray angle is described as the angle between the axis of the spraying gun and the substrate surface. The angle measures 90° ideally when spray particles impact the component surface and transfer the momentum and thermal energy to the component surface. For coating the complex-shaped geometries, the mobility of a robot system is limited, and it is impossible to maintain a consistent spray angle of 90°, and it needs to be reduced. This change in impact angle can cause undesirable variations in the deposition rate or final coating properties.

Different studies in the literature [2–4] had already discussed that coating porosity increases with reduced spray angles. The feedstock powder particles sprayed at reduced angles indicate a decreased normal velocity component. The impacting particles do not possess the required energy to cover up the entire surface of the substrate effectively, giving rise to more porous coatings [3]. The final deformation of the impacting particles strongly depends on the normal velocity component, and it can be considered that reduced spray angles in thermal spraying will affect the deposition efficiency and microstructure of coating [5]. However, such an effect on microstructure will be entirely different in cold spraying due to the state of the spray particles as they are in a solid state instead of a molten state. Binder et al. discussed the outcome of spraying angle on the microstructure of various cold spray deposits [6]. They observed that the deviations from normal impact conditions could remarkably change the deformation behavior of the particle, resulting in increased porosity levels and reduced tensile and adhesive strength of cold spray deposits (Figure 14.2). Moreover, they also reported that when deviations from normal impacts were less than 20° (Figure 14.2 c and d), spray particles showed deformation behavior same as

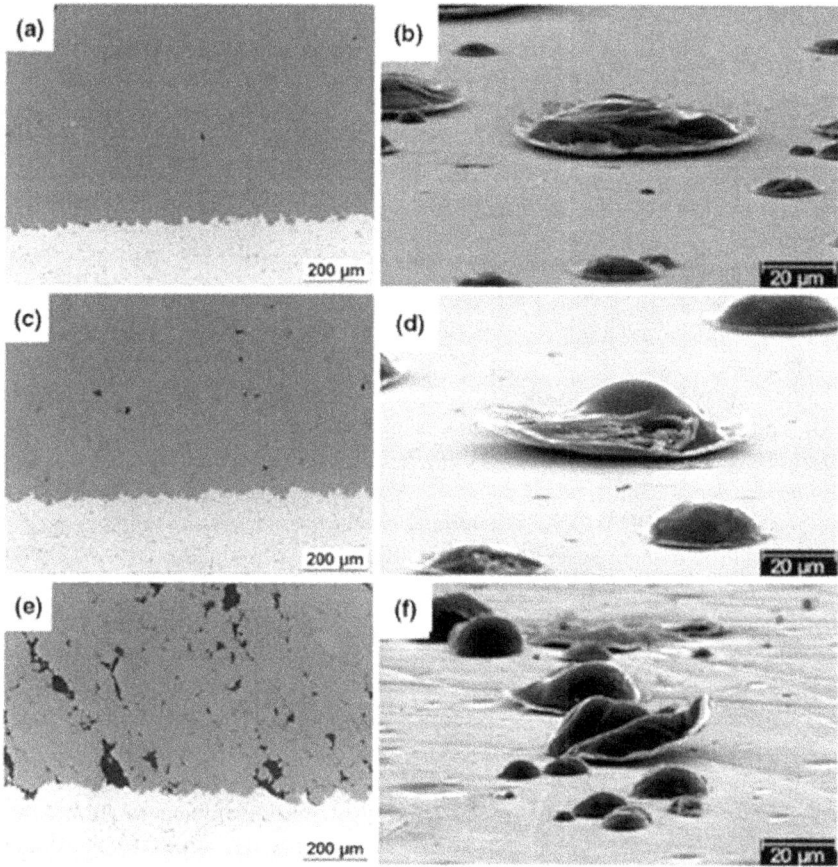

FIGURE 14.2 Microstructures of cold spray Ti on AlMg$_3$ specimens for various spray angles:(a, b) normal angle, (c, d) 70° and, (e, f) 45° [6].

the normal incidence, with no significant alterations in the levels of porosity, and is accountable for most of the applications.

At angular impact (45°; Figure 14.2f), a tangential component of particle momentum at oblique impacts can produce tensile forces at the substrate interface, which is sufficient to remove the particles from the surface of the substrate. Changes in deposition efficiency and coating mechanical properties such as microhardness, surface roughness or porosity levels which can be vital for the final coating quality, occur at spraying angles < 50° as reported by Tillmann et al. [7] and Smith et al. [8]. Furthermore, the type of process and grain size will also have a relation with the spraying angle. [1][9] concluded that HVOF spraying of the fine WC-12Co powders (size 2–10µm) is less prone to changes in the spraying angles in contrast with other methods like plasma or arc spraying. The reduced spraying angles result in a reduction in deposition rates, while no significant degradation in the coating properties is reported up to 30°. However, with further reduction in spraying angle<30°, the development of pores and cracks negatively influences the strength of the coating.

From the above discussion, it can be concluded that the spraying angle should be controlled normally to the surface of the substrate during the operation for the production of quality coatings. However, in the case of complicated geometries, it is a difficult task. For such conditions, there may be some points in the robot's trajectory where a slight reduction in spraying angle can be provided for obtaining a smoother trajectory that is easily reachable during spraying operation by the robot. It improves the entire spraying operation and final quality of the coating.

14.1.2 STAND-OFF DISTANCE

One way to improve the quality of as-sprayed coating during spraying is through the optimization of stand-off distance. The stand-off distance is described as how far away the spray gun is from the substrate to be coated, as exhibited in Figure 14.1 [10]. The stand-off distance has a significant outcome on the performance and properties of the as-sprayed coatings [11]. Shorter stand-off distances lead to higher temperatures, which lead to the production of hard and dense coatings [12]. Models of the as-sprayed HVOF WC-12Co coating were fabricated and simulated using computational fluid dynamics, and outcomes exhibited that the stand-off distance played a significant role in the deposition process [13].

High stand-off distances during the spraying of WC-Co powder result in lower velocities which in turn leads to high porosity levels. As the stand-off distance increases and in-flight particles arrive hotter at target, critical velocity needed for particle deformation and coating build-up is progressively low. This decreases particle compaction and contact among splats, resulting in high levels of porosity. However, on the other hand, larger stand-off distances also result in the re-solidification of partially molten particles in the path and cause low deposition efficiencies. Low deposition efficiencies result in higher porosity levels [10]. These relationships indicate that optimized stand-off distance during the deposition process is a critical parameter.

Studies were conducted on WC-Co-and Ni-based alloys to understand the optimum stand-off distances that can be utilized to enhance the performance of coatings. The stand-off distance for the spraying of Alloy-625 as recommended by the equipment manufacturers (Sulzermetco) is between 9 and 12 inches. If the stand-off distance is significantly less, there is a probability that substrate overheating is more, and if the stand-off distance is too large, the temperature of the in-flight particles will decline before it strikes the substrate surface. This decline in the particle temperature will influence the bond strength of the coating [14]. Experimental studies reported by Stokes have exhibited that stand-off distance has a significant effect on the deposition temperature, and therefore it influences the quality of the deposited coating [15]. WC-Co powder was thermally sprayed at the set range of the stand-off distances (125–260 mm). During spraying, the deposition temperature was controlled and measured continuously. Results exhibited that the deposition temperature relies on stand-off distance. If the temperature of the flame was measured, it should be observed that temperature drops when the distance was increased from gun's head. Consequently, a drop in the temperature of particles takes place when they exit the spray nozzle. As discussed by Stokes and Looney, a larger stand-off distance allows the in-fight-sprayed particles to cool down rapidly, which results in the fall of deposition temperature [16]. During the same experimental study, it was suggested by the authors that any change in the stand-off distances also results in the variation of the magnitude of stress distribution in as-sprayed coatings. Similar experimental studies were also conducted by Guessasma et al. [17], and they supported the relationship among porosity content, microhardness, and phase composition of the HVOF WC-12Co-sprayed coatings.

Experimental research was also conducted by Yilbas et al. to study the outcomes of stand-off distance on a Ni-based alloy [18]. The variation of stand-off distance influenced the stress distribution of the coatings, and outcomes were in context with the outcomes recommended by Stokes [15], which exhibited that when the stand-off distance varies from 180 to 200 mm, residual stress decreases. Hence, optimum spraying distance during deposition is recommended for better-quality coatings.

14.1.3 Spray Path

During the coating process, a meander-shaped spraying path is often used, as shown in Figure 14.1b. It consists of parallel horizontal passes of the spray torch with respect to the surface of the substrate. The distance between the two consecutive passes is called as scanning step. Moreover, the porosity and thickness of the coating are also influenced by the scanning step used [19].

The key to developing high-quality coatings lies in the efficient control of residual stresses in the as-sprayed coatings. Residual stresses negatively influence distinct aspects of features of final coatings (like adhesion strength and resistance to thermal shock, fatigue, and wear) and different categories of failures like buckling deformation, splat de-bonding, and even cracking [20–22]. Some research publications have highlighted the role of spray path in relation to the developed residual stresses. The effect of spray path on the distribution of residual stress in electric arc-sprayed coating on the surface of crankshaft was investigated by Haoliang et al. As the geometry of the crankshaft was complex, offline programming was preferred [23]. Two spray paths were selected: a Z-shaped path and a circular path, as illustrated in Figure 14.3.

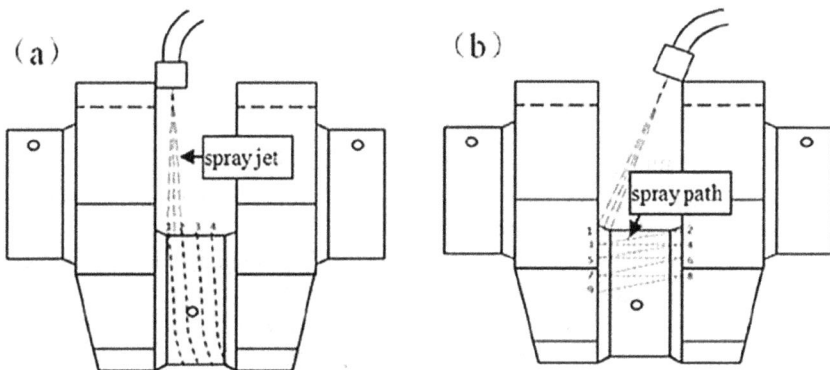

FIGURE 14.3 Diagrammatic set-up of the spray paths: (a) circular and (b) Z-shaped [23].

The authors concluded that the deposition temperature was comparatively less when the spray path was "Z"-shaped during the spraying process, leading to the generation of minor residual stress in the deposited coating. Other studies in the literature have also discussed the effect of spray path employed for the deposition of coating process on both temperature gradient and temperature level in the coating, and relating these with the generation of residual stresses in the coating [24]. These experiments highlight the necessity of appropriate planning of the spray path in thermal spraying and its impact on the coating quality.

14.2   OFFLINE PROGRAMMING FOR THERMAL SPRAY COATING

Advanced thermal spray surface coating techniques for the complicated hot gas structural parts that operate at elevated temperatures like turbine blades have to meet very stringent criteria, for example, uniform coating thickness, as it decides the performance and service life of the component. In specific, thickness in thermal barrier coating governs the temperature gradient throughout the deposited coating. It will provide thermal insulation to the underlying component and affects the thermo-mechanical characteristics of the component. Most of the protective coatings for elevated temperature are commonly sprayed with the techniques of thermal spraying. Even though vast technical literature exists on the studies related to the thermal spray processes, their relationship to the final coating features is not inferred completely. In general, the spraying parameters employed during the spraying process for the development of high-quality coatings are selected by the approach of "trials and errors" involving the numerous process control methods. However, these trials are often conducted in the production booths by using an industrial robot. This approach is very costly and also time-consuming for most cases. Therefore, to simplify the coating production, different models and software tools are used to simulate the entire deposition process and predict specific coating properties required. One such advanced method is offline programming that will provide a complete solution for the thermal spraying process, from the generation of the spray path, the parameters for simulation, and trajectory optimization. The robot's trajectory for the deposition of coating can always be generated with the help of the real geometrical model of the component [25, 26] to guarantee path accuracy and precision. The schematic diagram of this entire process is illustrated in Figure 14.4.

14.2.1   CAD FILE ACQUISITION

Graphical programming needs the CAD geometry of a substrate to develop the trajectories for robots. Therefore, the initial step is to obtain the 3D geometrical model. In case the original CAD model is not available, a simple model must be created in CAD software like Catia, Solid-Works, and Pro-E, etc. If the component is too complex to be designed in CAD software, then another approach known as reverse engineering will be adopted. The required geometrical information regarding the component can be acquired either from a coordinate measuring machine (CMM) or using the laser scanning system [27]. The 3D geometrical model can be developed using these measured points and is a more effective method for complex components.

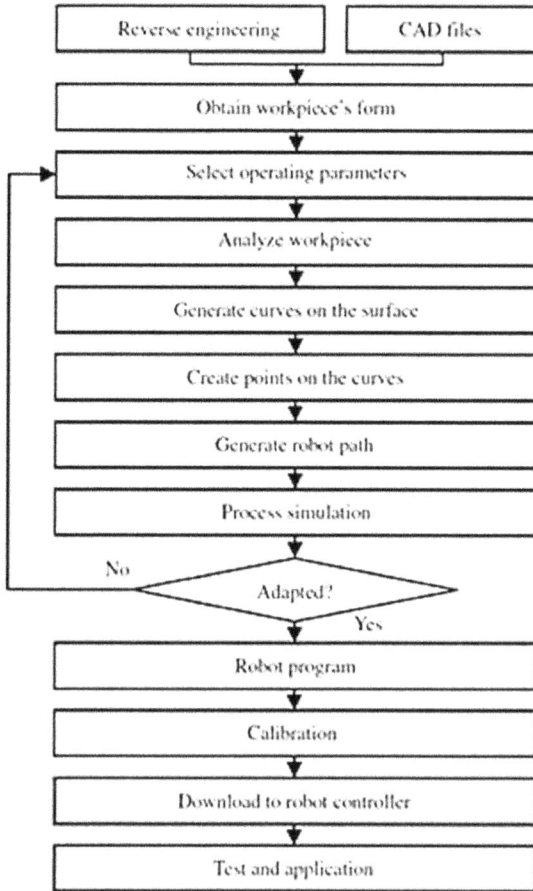

FIGURE 14.4   Method for the generation of an offline trajectory [1].

14.2.2   Selection of Operating Parameters for Thermal Spraying

Thermal spraying process parameters are divided into various groups as claimed in the literature: energy parameters, injection parameters for feedstock powder, and different kinematic parameters. The parameters mentioned above can be controlled (a) directly, such as torch speed, projection distance, and scanning step, or (b) indirectly, such as speed and temperature of in-flight particles, etc. All these handling parameters can affect the process performance and the coating properties. Kout et al. [28] recommended an optimization technique for calculating and approximating the required coating thicknesses with the relative operating parameters. The authors investigated planning-path-oriented spray-coating processes. In another investigation, Trifa et al. [29] studied an interaction among the operating parameters and deposit characteristics, which allows them to select the appropriate settings. In the work presented by Guessasma et al. [30], the authors generated an intelligent system based on fuzzy logic to help choose operating parameters relying upon the required features and the desired deposition.

14.2.3 Robot Trajectory Generation

After defining the kinematic parameters next step is the generation of robot trajectory. The objective is to locate the point coordinates, which define the orientation of every point and the scanning curve. Two methods can be used are as follows.

In the first method, mesh generation of CAD model is done. Thereafter, the robot trajectory is calculated from these mesh points and vectors normal to the substrate surface. The various file formats applicable for this process are STL, IGES, ASCII, and STEP. For instance, the STL file defines a component with the triangular mesh and each triangle is described by the three points and a vector normal to the substrate surface in a 3-D coordinate system. The STL format is ideal for trajectory generation, and robot trajectory can be developed with this process by using software like Matlab. After this, the robot program is composed for implementation in the spraying cell to test the final trajectory. The positive aspect of this process is that robot trajectories on complex components can also be generated easily. The Institutfür Fertigungs technologie Keramische Bauteile (IFKB) has already used this method to generate robot trajectories in thermal spraying on a sample of propeller [31], as shown in Figure 14.5.

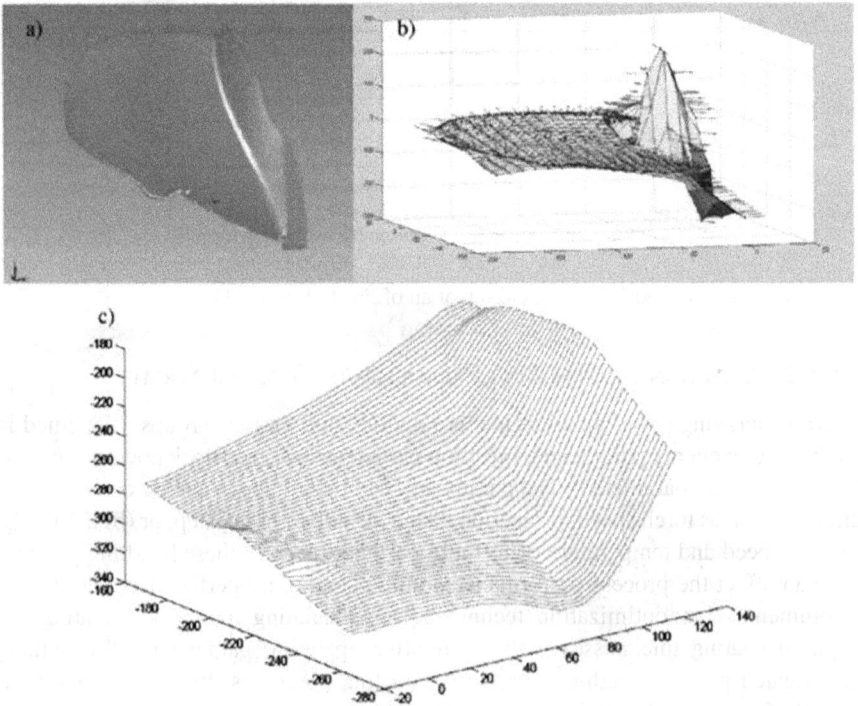

FIGURE 14.5 Trajectory generation in Matlab:(a) reverse engineering results, (b) mesh and normal vectors, and (c) final trajectory [1].

FIGURE 14.6 Offline robot trajectory generation using thermal spray tool kit to deposit coatings for real components [courtesy of LERMPS].

The second method uses the orthogonal planes to cut the substrate surface, and after that, a series of scanning curves are generated. It is simpler than the mesh-generating method, but a satisfactory trajectory is hard to obtain if the component geometry is very asymmetrical and irregular. This method is used by the Laboratoired' Etudes et de Recherchessur les Matériaux, les Procédés et les Surfaces (LERMPS) to generate the trajectories in the process of thermal spraying with commonly used offline programming software known as RobotStudio™ [32, 33]. The LERMPS has designed a designated Thermal Spray Toolkit (TST) based on this software that allows quick and automatic generation of robot trajectory according to the shape of the component and the operating parameters [34, 35], as shown in Figure 14.6.

14.2.4 IMPLEMENTATION OF ROBOT TRAJECTORY IN VIRTUAL ENVIRONMENT OR SIMULATION

The simulation involved two parts: (a) robot kinematics simulation and (b) heat transfer simulation.

As discussed in the previous section, the design of robot trajectories plays a key role during the thermal spray process, and once the trajectory has been generated offline, its implementation is conducted in a real environment. Firstly, a virtual cell is created by taking the measurements of a real cell. This virtual cell simulates the working conditions where the robots are installed. It provides flexibility, and by optimizing the movement of robots, it avoids the collisions of robots with elements in the working cell. Figure 14.7 shows the virtual models of the actual spray booths designed with simulation software's, ROBOGUIDE (Fanuc M710iC, installed in IFKB) and RobotStudio (ABB Ltd.).

However, when some axes reach the limited value of the robotic arm joint, then the robot will not execute any action along the trajectory. In that case, the orientations of specific points or the position of the substrate requires readjustment so that that robot can reach all points [36–38].

Heat transfer simulation in thermal spraying deals with examining the temperature distribution across the surface of the component and the various stresses to which the component is subjected during the coating process. Various studies related

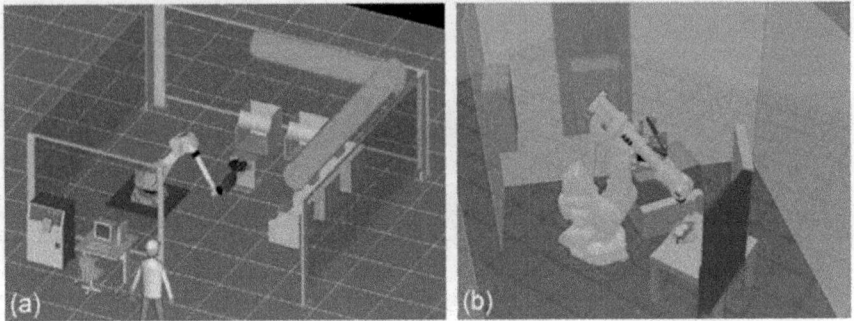

FIGURE 14.7 Virtual cell set-up for simulation with (a) robot Fanuc M710iC and (b) RobotStudio™ (ABB Ltd.) [courtesy of LERMPS].

to the development of simulation in thermal spraying consist of studying the formation of splats, heat transfer, and stress analysis across the interface for thermal spray. For instance, due to the flame impact in plasma spraying at elevated temperatures and due to the projection of the hot particles, the component undergoes a heat load during the spraying process. It results in structural changes, component distortion, and uneven distribution of the residual stresses in the component. Therefore, it is essential to simulate and control the rate of heat transfer during the coating process. Various thermal simulation software in thermal spraying are commercially used, like Ansys (Ansys, Inc.), Abaqus (Abaqus, Inc.), etc. The various studies were already published in literature related to this field [39–43].

14.2.5 CALIBRATION

If the outcomes of simulations are adequate, then the robot program is loaded at the spraying site. Without offline calibration, the program is not executed into the real robot. It is an important step before testing and application.

14.2.6 TESTING AND APPLICATION

After TCP, calibration, and the component position, the testing of the robot program at slow speed is initiated, and finally, it is ready for the actual spraying process. However, if the simulation results are not good, it must go back to the preceding steps to modify and verify the robot program.

14.3 IMPORTANCE OF ROBOT OFFLINE PROGRAMMING IN THERMAL SPRAYING PROCESS

This section discusses the studies related to the importance of offline programming in thermal spraying technology. The need for robot trajectory optimization to achieve uniform coating thickness and the subsequent study is related to trajectory generation in the irregular rotating components using offline programming method.

14.3.1 ROBOT TRAJECTORY OPTIMIZATION

In recent years, the use of high-accuracy robots in thermal spraying applications has increased widely. When a robot system for thermal spraying is used, the trajectory generation of the robot represents a key part of the process. The critical operating parameters involved are the torch speed, spraying distance, spraying angle, etc. Among these, the speed of the torch is the most important operating parameter of the thermal spraying process. For high-quality coatings, the torch movement should be consistent, and the direction of the spraying should be near to normal with respect to the surface of the coating. However, on a curved component, when the spraying torch follows a trajectory, there is a large amount of change in torch orientation, resulting in the apparent decrease in the torch speed, as shown in Figure 14.9. Therefore, to get the uniform thickness of the coating, optimization of the robot movement is essential [44].

As exhibited in Figure 14.8b, there are several spraying points in the robot trajectory, which was prepared in RobotStudio™, and every spray point has a different orientation, blue, green, and red colors represent the z-, y-, and x-axis, respectively.

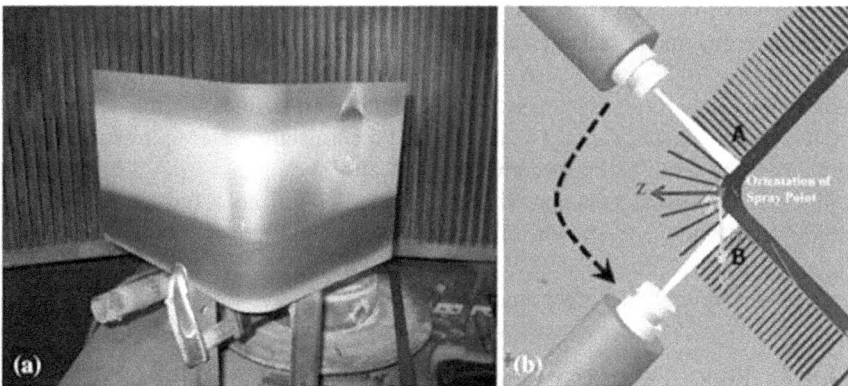

FIGURE 14.8 (a) Real component with greater angle, (b) trajectory with greater angle [44].

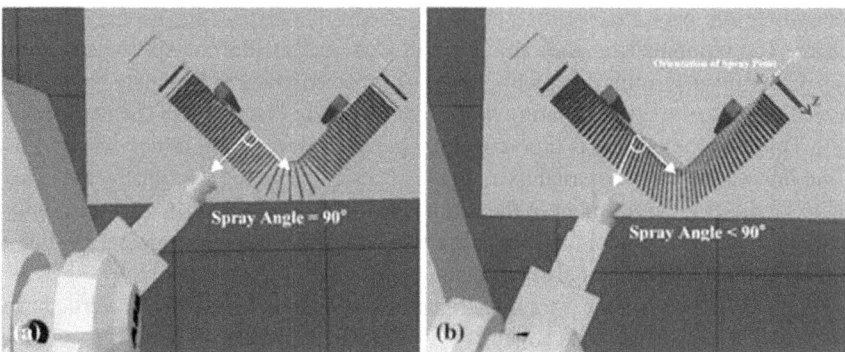

FIGURE 14.9 (a) Trajectory before optimization, (b) after optimization [44].

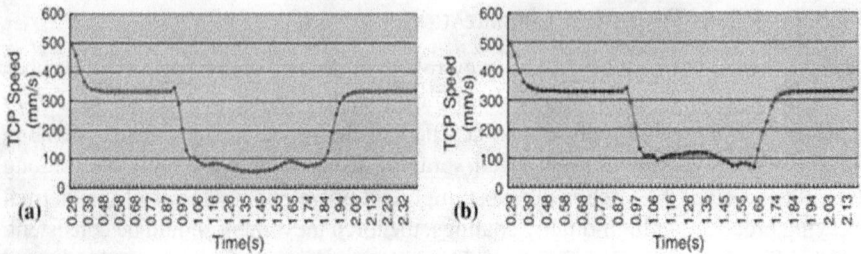

FIGURE 14.10 Simulated TCP speeds (a) before optimization, (b) after optimization [44].

From point A to B, the length is roughly 55 mm, and torch speed work out is 500 mm/s; so to cross these two points, the robot takes 0.11s. This signifies that it will rotate to 90° in just 0.11 s when all six axes are combined. Due to this, TCP speed decreases significantly. One strategy proposed to solve this problem is to optimize the spray trajectory by modifying the angle between the two neighboring points.

For the same, an add-in program in RobotStudio™ was created to optimize spraying trajectory [45]. This particular technique focused on improving the robot trajectory, which encompasses larger angles. The spraying angle will be <90° with this technique, but TCP speed is also maintained near the constant, which is essential to achieve uniform coating thickness. Figure 14.9a signifies the robot trajectory before optimization, whereas Figure 14.9b exhibited the trajectory after optimization, indicating that the torch action becomes smooth after the optimization. Moreover, the simulated TCP speeds with and without the point optimization for the two different coating processes are shown in Figure 14.10, respectively. It is clear from Figure 14.10b that simulated TCP speed was more homogeneous, mainly when the torch passes through the larger angle area.

14.3.2 ROBOT TRAJECTORY GENERATION FOR COMPLEX-SHAPED PARTS

During the coating of some parts having complex-shaped geometries, the application of coordinated external axis of the translation or rotational can be contemplated. Due to this, the working area that a robot can hold out will be extended and is most suitable where the back face of the part is required to be coated, for instance, turbine blades. The combined six axes of the robot will not perform the spray path in one step. Consequently, after completing the coating on one face of the blade, it has to be moved again and then the coating will be done on the back face in the second step [46]. This entire operation is time-consuming. One universal solution to this problem is the integration of an external axis, as it can easily rotate the part during the spraying to facilitate the movement of the robot. To cover the whole surface of the component, as shown in Figure 14.11, there are two different spraying strategies available: (a) vertical and (b) horizontal spray paths.

Figure 14.12 shows the spraying trajectories generated by TST for vertical and horizontal spray paths. This trajectory is simulated in RobotStudio™ to observe whether the robot will be able to reach all the points. The results exhibited that the vertical trajectory gives more constant speed of scanning on the part resulting in the

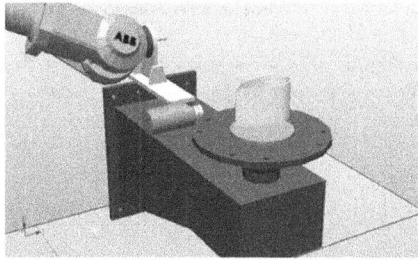

FIGURE 14.11   Spraying with the application of external axis [46].

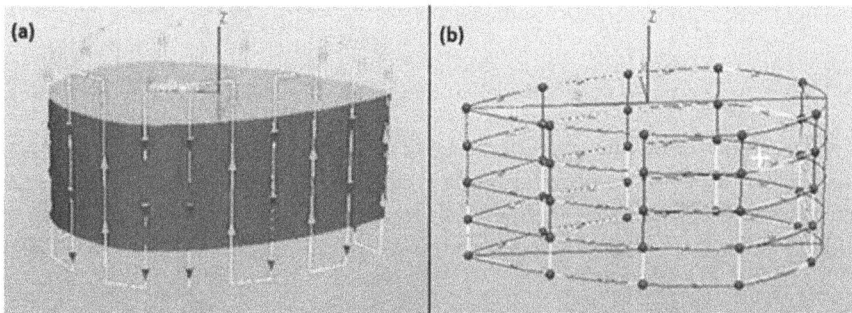

FIGURE 14.12   (a) Vertical spray path, and (b) horizontal spray path [46].

uniform thickness of coating. However, the wastage of powder should be taken into account while spraying onto a larger part. Horizontal trajectory allows for the prevention of excessive powder wastage during the spraying process by permitting a continuous scanning of the part's circumference.

14.4   CONCLUSION

Offline programming of the spraying robots allows for the development of high-quality coatings on complex-shaped components, and it is now used frequently in thermal spraying technology. An extensive range of the operating parameters will affect the final properties of deposited coatings, and for this cause, extensive research activities have been conducted to clarify the relationships among the process handling parameters and final coating properties. Spray angle, distance, and spray path are optimized to fulfill the requirements for coating quality.

Commonly available offline programming software are RobotStudio™ and ROBOGUIDE, combined with modules for generating robot trajectories that permit maintaining the constant kinematic parameters for spraying. Researchers have focused not only on the generation of robot trajectories but also its optimization to assure a uniform coating thickness for complex geometries. The primary objective of offline programming is to predict coating properties before actual spraying by transferring the programmed robot trajectory into the process simulations in a virtual environment.

REFERENCES

1. Deng, S., Cai, Z.H., Fang, D.D., Liao, H.L., Montavon, G.2012a. Application of robot offline programming in thermal spraying. *Surf. Coat. Technol.* 206, 3875–3882.
2. Tillmann, W., Baumann, I., Hollingsworth, P., Laemmerhirt, I.A., 2013. Influence of the spray angle on the properties of HVOF sprayed WC–Co coatings using (210 + 2 μm) fine powders. *J. Therm. Spray Technol.*22(2–3), 272–279.
3. Ilavsky, J., Allen, A.J., Long, G.G., Krueger, S., Berndt, C.C., Herman, H., 1997. Influence ofspray angle on the pore and crack microstructure of plasma-sprayed deposits. *J. Am. Ceram. Soc.*80 (3), 733–742.
4. Leigh, S.H., Berndt, C.C., 1997. Evaluation of off-angle thermal spray. *Surf. Coat. Technol.*89, 213–224.
5. Friis, M., Persson, C., Wigren, J., 2001. Influence of particle in-flight characteristics on themicrostructure of atmospheric plasma sprayed yttria stabilized ZrO2. *Surf. Coat. Technol.* 141, 115–127.
6. Kang, C.W., Ng, H.W., Yu, S.C.M., 2006. Imaging diagnostics study on obliquely impacting plasma-sprayed particles near to the substrate. *J. Therm. Spray Technol.* 15 (1), 118–130.
7. Binder, K., Gottschalk, J., Kollenda, M., Gartner, F., Klassen, T., 2011. Influence of impact angle and gas temperature on mechanical properties of titanium cold spray deposits. *J. Therm. Spray Technol.* 20, 234–242.
8. Tillmann, W., Vogli, E., Krebs, B., 2008. Influence of the spray angle on the characteristics of atmospheric plasma sprayed hard material based coatings. *J. Therm. Spray Technol.* 17 (5–6), 948–955.
9. Smith, M.F., Neiser, R.A., Dykhuizen, R.C., 1994. An investigation on the effects of droplet impact angle in thermal spray deposition. In: Berndt, C.C., Sampath, S. (Eds.), *Proceedingsof the 7th National Thermal Spray Conference*. ASM International, Materials Park, pp. 603–608.
10. Fauchais, P.L., Heberlein, J.V., Boulos, M. I., 2014. *Thermal Spray Fundamentals: From Powder to Part*. Springer Science & Business Media, Berlin/Heidelberg,. Germany.
11. Gil, L., Staia, M., 2002. Influence of HVOF parameters on the corrosion resistance of NiWCrBSicoatings. *Thin Solid Films*. 420, 446–454.
12. Zhao, L., Maurer, M., Fischer, F., 2004. Influence of spray parameters on the particle in-flight properties and the properties of HVOF coating of WC-CoCr. *Wear* 257, 41–46.
13. Li, M., Christofides, P.D., 2006. Computational study of particle in-flight behavior in the HVOF thermal spray process. *Chem. Eng.Sci.* 61, 6540–6552.
14. Oberkampf, W.L., Talpallikar, M., 1996. Analysis of a high-velocity oxygen-fuel thermal spray torch: part 2: Computational results. *J. Thermal Spray Technol.* 5, 62–68.
15. Stokes, J., 2003. Production of Coated and Free-Standing Engineering Components using the HVOF (High Velocity Oxy-Fuel) Process (Doctoral Dissertation, Dublin City University).
16. Stokes, J., Looney, L., 2004. Residual stress in HVOF thermally sprayed thick deposits. *Surf.Coat. Technol.* 177, 18–23.
17. Sahraoui, T., Guessasma, S., Jeridane, M.A., 2010. HVOF sprayed WC–Co coatings: microstructure, mechanical properties and friction moment prediction. *Mater.Design.* 31, 1431–1437.
18. Yilbas, B.S., Al-Zaharnah, I., Sahin, A., 2014. *Flexural Testing of Weld Site and HVOF Coating Characteristics*. Springer, Berlin Heidelberg.

19. Krebs, B., 2011. *KonturgenauBauteilbeschichtungf€ur den Verschleißschutzmittels-AtmospharischenPlasmaspritzens und Lichtbogenspritzens.* VulkanVerlagGmb H, Essen, Germany.

20. Kar, S., Paul, S., Bandyopadhyay, P.P., 2016. Processing and characterisation of plasma sprayed oxides: microstructure, phases and residual stress. *Surf Coat Technol.* 304, 364–374.

21. Wang, Y., Li, K.Y., Scenini, F., Jiao, J., Qu, S.J., Luo, Q., et al. 2016. The effect of residual stress on the electrochemical corrosion behavior of Fe-based amorphous coatings in chloride containing solutions. *Surf Coat Technol.* 302, 27–38.

22. Nazir, M.H., Khan, Z.A., Stokes, K., 2016. Analyzing the coupled effects of compressive and diffusion induced stresses on the nucleation and propagation of circular coating blisters in the presence of micro-cracks. *Eng Fail Anal.* 70, 1–15.

23. Haoliang, T., Changliang, W., Mengqiu, G., Zhihui, T., Hui, T., Xinkun, W., Shicheng, W., Binshi, X., 2019. A residual stresses numerical simulation and the relevant thermal-mechanical mapping relationship of Fe-based coatings. *Res.Phys.* 13, 102195.

24. Killinger, A., 2010. *FunktionskeramischeSchichtendurchthermokinetische-Beschichtungsverfahren.* Shaker VerlagGmb H, Aachen, Germany.

25. Bi, Z.M., Lang, S.Y.T., 2007. A framework for CAD and sensor-based robotic coating automation. *IEEE Trans. Ind. Inform.* 3 (1), 84.

26. Deng, S., 2006. Robot offline programming and real-time monitoring of trajectories: development of an add-in program of robotstudio™ for the thermal spraying, Ph.D. Thesis.Graduate School of "Physics for Engineers and Microtechnology", University of technology of Belfort-Montbéliard, France.

27. Frutos, A., 2009. Numerical analysis of the temperature distribution and offline programming of industrial robot for thermal spraying. Ph.D. Thesis. Institute for Manufacturing Technologies of Ceramics Components and Composites- IMTCCC, University of Stuttgart, Germany.

28. Kout, A., Muller, H., 2009. Parameter optimization for spray coating, *Adv. Eng. Softw.* 40, 1078.

29. Trifa, F.I., Montavon, G.., Coddet, C., 2005. On the relationships between the geometric processing parameters of APS and the Al2O3–TiO2 deposit shapes, *Surf.Coat.Technol.* 195, 54.

30. Guessasma, S., Bounazef, M., Nardin, P., 2006. Neural computation analysis of alumina–titania wear resistance coating, *Int. J. Refract. Met. Hard Mater.* 24, 240.

31. Candel, A., Gadow, R, 2009. Trajectory generation and coupled numerical simulation for thermal spraying applications on complex geometries, *J. Therm. Spray Technol.* 18, 981.

32. Deng, S., Liao, H., Zeng, C., Coddet, C.,2006. *Thermal Spray 2006: building on 100 years ofsuccess, Proceedings of 2006 International Thermal Spray Conference,* ASM International, Materials Park, Ohio, USA, May15–18, 1437.

33. Deng, S., Liao, H., Zeng, C., Charles, P., Coddet, C., 2005a. *Proceedings of Les DeuxièmesRencontresInternationalessur la Projection Thermique,* Lille France, Dec. 1–2, 2005.

34. Fang, D., Deng, S., Liao, H.,Coddet, C., 2010a. In: E. Lugscheider (Ed.), *Proceedings of the 2010 International Thermal Spray Conference, May 3–5, 2010, Singapore,* VerlagfürSchweißen und verwandteVerfahren DVS-Verlag GmbH, Düsseldorf, Germany, 2010.

35. Fang, D., Deng, S., Liao, H.,Coddet, C., 2011. *International Thermal Spray Conference, Proceedings of the 2011 International Thermal Spray Conference,* September 27–29, Hamburg, Germany, 2011.

36. Zha, XF., 2002. Optimal pose trajectory planning for robot manipulators, *Mech. Mach. Theory.* 37, 1063.

37. Hansbo, A., Nylén, P., 1999. Models for the simulation of spray deposition and robot motion optimization in thermal spraying of rotating objects, *Surf.Coat.Technol.* 122, 191.

38. Fang, D., Deng, S., Liao, H.,Coddet, C., 2010b. The effect of robot kinematics on the coating thickness uniformity, *J. Therm. Spray Technol.* 19 (4), 796.

39. Ng, H.W.,Gan, Z.,2005. A finite element analysis technique for predicting as-sprayed residual stresses generated by the plasma spray coating process, *Finite Elem. Anal.Des.* 41, 1235.

40. Zhang, X.C., Xu, B.S., Wang, H.D., Wu, X.Y., 2006. Modeling of the residual stresses in plasma-spraying functionally graded ZrO2/NiCoCrAlY coatings using finite element method, *Mater.Des.* 27, 308.

41. Amara, M., Timchenko, V., El Ganaoui, M., Leonardi, E., de Vahl Davis, G., 2009. A 3D computational model of heat transfer coupled to phase change in multilayer materials with random thermal contact resistance, *Int. J. Therm. Sci.* 48, 421.

42. Khor, K.A., Gu, Y.W., 2000. Effects of residual stress on the performance of plasma sprayed functionally graded ZrO2/NiCoCrAlY coatings, *Mater. Sci. Eng., A* 277, 64.

43. Bolot, R., Planche, M.P., Liao, H., Coddet, C., 2008. A three-dimensional model of the wire-arc spray process and its experimental validation, *J. Mater. Process. Technol.* 200, 94.

44. Fang, D., Deng, S., Liao, H.,Coddet, C., 2009. The effect of robot kinematics on coating thickness uniformity. *J.ThermalSpray Technol.* 19(4), 796–804.

45. Deng, S., Liao, H.,Zeng, C., Charles, P.,Coddet, C., 2005b. *Development of Robotic Trajectories Auto Generation in Thermal Spraying: A New Extended Program of ABB RobotStudio_, Les deuxiemesrencontresInternationalesur la projection thermique*, 2005 (Lille France), 2005, 97–104.

46. Deng, S., Liao, H., Fang, D., Zhenhua, C., 2012b. Application of External Axis in Robot-Assisted Thermal Spraying. *J.Thermal Spray Technol.* 21(6), 1203–1215.

15 A Case Study on the Failure Analysis, Prevention, and Control of Boiler Tubes at Elevated Temperatures

Gaurav Prashar, Hitesh Vasudev, and Lalit Thakur

CONTENTS

15.1 INTRODUCTION

Among all surface-related phenomena, the issue of wear is a serious concern in most industrial components. Various forms of wear exist in these components, such as abrasive, adhesive, corrosive, dry, and wet erosion. The most frequently occurring failure mechanisms in boiler tubes are of two types: (a) external damage of the surface, which includes oxidation, fatigue, corrosion, creep, erosion, degradation of the surface because of overheating [1, 2], and (b) internal damage which includes microstructural damage. Failure of the boiler tubes occurs from time to time. Long-term overheating and fly-ash erosion (also termed as dry erosion) phenomenons are associated with the degradation of boiler tubes. These major factors are responsible

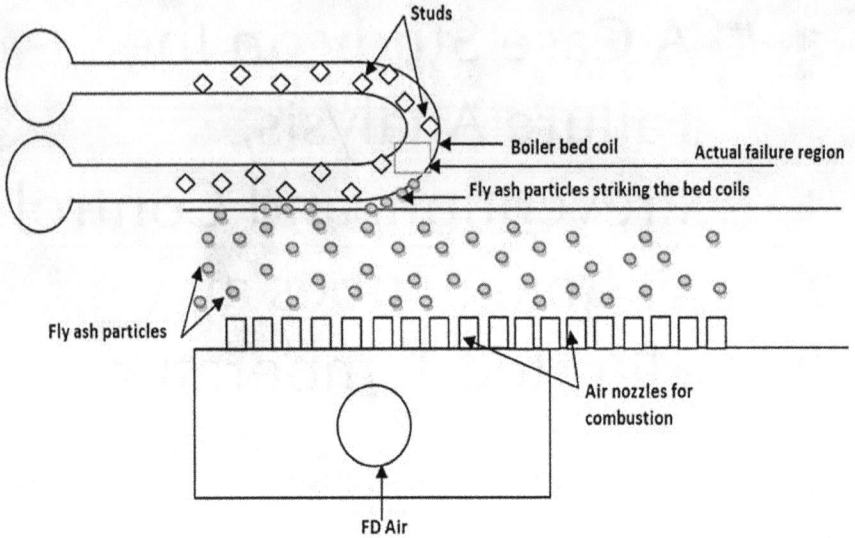

FIGURE 15.1 Schematic outline indicating actual failure region of boiler bed coil.

for the damage of boiler tubes used in the production of energy [3]. The maintenance expenses (damaged tubes) can rise to 54% of the total production cost [4]. Costs associated with erosion alone can rise to around US$150 million/year, as a whole solid particle erosion dramatically reduces the service life and performance of such components. Therefore, to enhance the efficiency and reliability of the boilers, detailed investigations must be done on the failed boiler tubes. Identifying and rectifying the leading cause of tube failures is vital to help to minimize the chances of future problems. A comprehensive assessment is the most effective method of determining the root cause of a failure. A tube failure is usually a symptom of other problems. This article discusses a failure associated with boiler water tubes working at elevated temperatures inside an industry at Hoshiarpur, Punjab. India. A schematic outline indicating the actual failure region is shown in Figure 15.1.

15.2 INVESTIGATION

A detailed investigation was performed to find out the root cause contributing the failure of boiler tubes. The boiler tube specifications and operating conditions are given below.

15.2.1 BOILER TYPE AND SERVICE CONDITIONS

The type of boiler and other service parameters are:

1. Boiler type: Water tube (air-fluidized bed combustion)
2. Make/model: Thermax Ltd./Babcok and Wilcox
3. Boiler bed and furnace temperature: 750–850°C

TABLE 15.1
Details of Fuel Composition

Elements	Design Fuel Analysis in wt.%
Carbon	36.67
Hydrogen	4.57
Sulfur	0.18
Oxygen	32.88
Nitrogen	1.25
Ash	15.01
Water	9.44
Carbon dioxide	
Total	100
Gross calorific value on dry basis	3275Kcal/kg

TABLE 15.2
Details of Chemical Composition of SA210 GrA1

Ni	Cr	Fe	Cu	Mo	Al	Si	C	Mn	S	P	V
0.0216	0.0170	95.09	0.0120	<0.0020	0.0269	0.0943	0.0832	0.2780	0.0061	0.0406	<0.00005

4. Boiler fuels: Rice husk/Indian coal
5. Steam temperature: 485°C±5°C
6. Steam pressure: 65kg/cm²
7. Fly-ash velocity: 20–25m/s
8. Boiler tube material: C.S. seamless tube SA210, GrA1

15.2.2 FUEL COMPOSITION

The tested composition of the fuel samples is summarized in Table 15.1.

15.2.3 CHEMICAL COMPOSITION OF TUBE MATERIAL

The chemical composition in (wt.%) of the tube sample was measured with the help of the optical emission spectrometer (OES) method (Model-1008i, manufactured at Metal Vision, India) and is summed up in Table 15.2.

15.3 FAILURE REPORTED IN BOILER TUBES

15.3.1 CASE-1: OBSERVATION—LONG-TERM OVERHEATING

On a visit to the industry, the boiler's bed coils reported leakage just above the combustion bed. During the detailed investigation, it was concluded that the failure of bed coils occurred due to long-term overheating. The failed tube shown in Figure 15.2a has minimal swelling and a longitudinal split with heavy external scale build-up accompanied by secondary cracking.

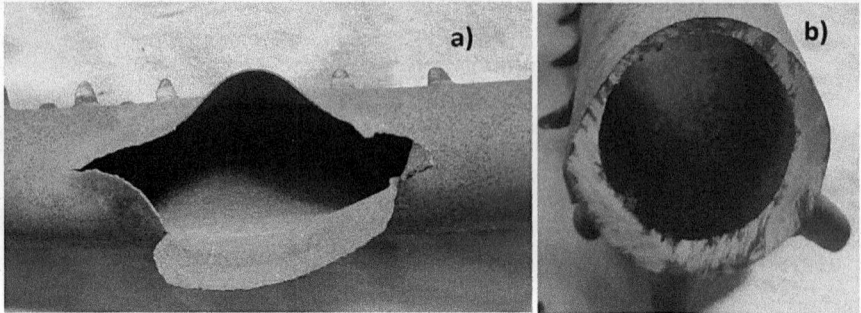

FIGURE 15.2   (a) Bursting of boiler bed coil tube, (b) thinning of wall tube due to erosion.

Causes: Long-term overheating takes place over months or several years and fails due to creep rupture. In boiler water wall tubes, the temperature of the tube increases abnormally, most often from waterside problems like deposits, scale, or restricted flow.

Solutions: Chemically clean tubes to enhance heat transfer rate. Balance furnace/flue gas temperatures with circulation to reduce the temperature of the tube.

15.3.2   CASE-2: OBSERVATION—TUBE EXPERIENCE METAL LOSS FROM THE OUTER SURFACE OF THE TUBE

Damage has occurred on the impact side of the tube. Ultimate failure results from the rupture due to increasing strain as tube material erodes. It was concluded that failure of bed coils occurred due to thinning of bed coil (Figure 15.2b) with time, caused by the abrasive nature of bed material ash which strikes the boiler bed coils with high velocity. As a result, thinning of bed coils occurs from the outer diameter of the tube.

15.4   ROOT CAUSE FINDING AND RECOMMENDATIONS TO PREVENT FAILURE

Fire-side ash erosion is due to the abrasive nature of bed material and boiler design. It is commonly associated with the firing of coal. Firing high ash fuels may lead to more erosion, slagging and fouling problems. Therefore, ash characteristics must be considered during the design of a boiler when establishing its size, geometry, and materials. Combustion gas and temperatures of materials to be used for making bed coils are other important parameters.

15.4.1   RECOMMENDATIONS TO PREVENT THE FAILURE OF BOILER TUBES

It was suggested that the primary reason for failure was fly-ash erosion. The losses associated with it cannot be eliminated but can be minimized by optimum erosion-preventing strategies. Firstly, for fly-ash erosion, distribute flow evenly through the boiler and consider burning a lower ash fuel. Secondly, protective coatings deposited with the appropriate technique effectively enhance the resistance of these structural components with a cost-effective approach. Coal-fired power plants give rise to severe erosive conditions. Plasma and high-velocity oxy-fuel (HVOF)-sprayed thick metallic or metal matrix composite coatings can effectively reduce erosion [5, 6]. A coating may be defined as a layer of a superior material deposited artificially onto the surface of structural components to improve its mechanical or tribological properties [7]. Therefore, structural components operating at high temperatures in these harsh conditions can be coated to protect them from harsh environments.

Nickel-based coatings are the potential candidates, which offer oxidation and erosive wear resistance at high-temperature operating conditions [8–10]. Nickel can be used with Cr, Al, Mo, Mn, Nb, Ta, and Si to increase the mechanical properties (hardness and toughness), hot corrosion, and erosion resistance of the coatings. Among nickel-based alloys, Ni-Cr-based, NiCrAlY-, NiCrBSi-, and Ni-Al-based coatings have gained much attention owing to their better corrosion- and wear-resistance properties. Thus, for efficient use of the coatings for various applications, investigation of its erosion and corrosion performance is essential. Thermal spraying methods are widely used deposition methods for Ni-based coatings in advanced industrial applications.

15.5 RECOMMENDED MATERIALS TO INDUSTRY AS A SOLUTION

Hot corrosion can be prevented and minimized by using suitable materials. Nickel-based Alloy-718 has been recommended due to its superior properties such as high strength, good creep, fatigue resistance, and the ability to resist the wear and corrosion at elevated temperatures for longer duration as compared to other Nickel-based powders [11]. The combination of high-temperature strength and resistance to surface degradation is unmatched by other metallic materials. The Alloy-718 is recommended in the industrial applications where thermomechanical strength is required [12–14]. Further, the addition of Al_2O_3 reinforcement in the Alloy-718 can enhance its hardness and improve its high-temperature erosion performance. An optimized HVOF-sprayed IN718-Al_2O_3 (INAL) composite coating has been recommended to the industry which is developed by varying the alumina content in IN718 alloy feedstock powder material. Al_2O_3 was selected as reinforcement due to its stability at elevated temperatures, high hardness value, and low cost.

15.6 CONCLUSION

1. Long-term overheating and fly-ash erosion are serious problems, more importantly in boiler water wall tubes and should be diagnosed in the early stages to minimize economic losses arising from shutdowns and maintenance expenses.
2. The performance of water wall tubes having low strength can be enhanced by depositing protective thermal spray coatings.
3. At present, different methods have been recognized to minimize the effect of fly-ash erosion; however, research efforts are still needed in this direction.
4. The outcome of the study will help in resolving similar types of failures in boiler industries.

REFERENCES

1. Lobely, G R, Al-Qtaibi, W L "Diagnosing boiler tube failure related to overheating", *Advanced Materials Research*, 41–42, (2008), 175–181.
2. Malik, A U, Ismail, A, Mohammad, M, Fahd, A M, Mohammad, A H "Corrosion of boiler tubes some case studies", *4th SWCC Acquired Experience Symposium held at Jeddah in* 2005, 739–763.
3. Yu, X Q, Fan, M, Sun, Y S "The erosion–corrosion behavior of some Fe3Al-based alloys at high temperatures", *Wear*, 253, (2002), 604–609.
4. Suckling, M, Allen, C "Critical variables in high temperature erosive wear", *Wear*,203–204, (1997), 528–536.
5. Fedrizzi, L, Rossi, S, Cristel, R, Bonora, P L "Corrosion and wear behaviour of HVOF cermet coatings used to replace hard chromium", *Electrochemical Acta*, 49,(2004), 2803–2814.
6. Hidalgo, V H, Varela, J B, Menéndez, A C, Martínez, S P "High temperature erosion wear of flame and plasma-sprayed nickel-chromium coatings under simulated coal-fired boiler atmospheres", *Wear*, 247, (2001), 214–222.
7. Sidhu, B S, Prakash, S "Evaluation of the behavior of shrouded plasma spray coating in the platen superheater of coal fired boilers", *Metallurgical&Materials Transactions*, 37A, (2006), 1927–1936.
8. Knight, R, Smith, R W, "HVOF sprayed 80/20 NiCr coatings-process influence trends" *Thermal Spray: Int. Advances in Coatings Technology*, C.C. Berndt, ed., *ASM International*, Orlando, FL, Materials Park, OH, 159 (1992).
9. Dorfman, M R, DeBarro, J A. *Thermal Spraying: Current and Future Trends*, Akira Ohmori, ed., High Temperature Society of Japan, Kobe, Japan, (1995) 567
10. Edris, H, McCartney, D G, Sturgeon, A J, "Microstructural characterization of high velocity oxy-fuel sprayed coatings of Inconel 625" *Journal of Materials Science*, 32, (1997) 863–872.
11. Prashar, G, Vasudev, H, "Hot corrosion behavior of super alloys," *Materials Today: Proceedings* 4, 2020, 1131–1135.
12. Vasudev, H, Thakur, L, Singh, H,Bansal, A, "A study on processing and hot corrosion behaviour of HVOF sprayed Inconel 718-nano Al_2O_3 coatings," *Materials Today Communications*, 25, no. August, 2020b, 101626.
13. Vasudev, H, Thakur, L, Bansal, A, Singh, H, Zafar, S, "High temperature oxidation and erosion behaviour of HVOF sprayed bi-layerAlloy-718/NiCrAlY coating", *Surface and Coatings Technology*, 362, (2019), 366–380.
14. Vasudev, H, Thakur, L, Bansal, A, Singh, H, "An investigation on oxidation behaviour of high velocity oxy-fuel sprayed IN718- Al_2O_3 composite coatings", *Surface and Coatings Technology*, 393, 2020a, 125770.

For Product Safety Concerns and Information please contact our EU
representative GPSR@taylorandfrancis.com
Taylor & Francis Verlag GmbH, Kaufingerstraße 24, 80331 München, Germany

www.ingramcontent.com/pod-product-compliance
Lightning Source LLC
Chambersburg PA
CBHW060759220326
41598CB00022B/2494